Praise from JMP® Experts

"The social sciences context for this book provides an easy path for any manager, scientist, or engineer in any field to obtain a nearly complete understanding of the entire field of multivariate statistics. This book will become one of my definitive texts, well thumbed and worn. I found the statistics primer at the beginning of the book, with its precise definition of measurement scale types, clear and crisp. The number of realistic examples will build any JMP user's competency and expertise.

"This book is ideal for both beginning and advanced JMP users. For the serious researcher it comes with some weighty topics, such as repeated-measures designs, between-subjects factors, suppressor variables, correlated predictor variables, and Cronbach's alpha. I would recommend this book to anyone who wants to achieve a high level mastery in the field of multivariate statistics in a short period of time."

Wm. Gray McQuarrie
President
Grayrock & Associates LLC

"*JMP for Basic Univariate and Multivariate Statistics: A Step-by-Step Guide* is not only a manual that assists the reader with the use of SAS' JMP software application, but it also leads to a better understanding of the basic research and analyses concepts and procedures. I was particularly impressed with the book's discussion and guidance in the computation of effect size, an analysis that has become quite popular in education and other fields in the last few years."

David Lillie, Ed.D.
Director, Special Education Projects
Center for School Leadership Development
University of North Carolina at Chapel Hill

"There are a number of positive features to the text. I like the inclusion of the first chapter on basic concepts in research and data analysis. Many 'how to' computer guides do not include such material. In the same vein, I like the way they have included a 'case study' at the beginning of the presentation of most statistical procedures. For example, Chapter 7 gives some basic information about *t* tests, provides an example, and reviews/reinforces how to enter data. The 'sample' presentations of results at the ends of the chapters are a nice touch. I look forward to using this book and program more in the future."

William B. Ware, Ph.D.
Professor of Educational Psychology,
Measurement, and Evaluation
School of Social Work
University of North Carolina at Chapel Hill

SAS Press

JMP® for Basic Univariate and Multivariate Statistics
A Step-by-Step Guide

Ann Lehman

Norm O'Rourke

Larry Hatcher

Edward J. Stepanski

The Power to Know.

Contents

Acknowledgments

This book is an adaptation of *A Step-by-Step Approach to Using the SAS System for Univariate and Multivariate Statistics, Second Edition*, by Norm O'Rourke, Larry Hatcher, and Edward J. Stepanski. First and foremost, acknowledgment must be given to these authors for their excellent and comprehensive discussions of basic and advanced statistical concepts.

I would also like to extend special thanks to John Sall for making time within the JMP Division for the writing of this adaptation. Also, invaluable technical review was provided by Lee Creighton, Duane Hayes, Catherine Truxillo, Lori Rothenberg, Paul Marovich, and Annie Zangi.

Further acknowledgment goes to the outstanding support provided by the Books by Users department at SAS Institute. In particular, thanks to Stephenie Joyner for the daily exchange that kept things moving, Ed Huddleston for careful final editing, and Candy Farrell for final composition.

Using This Book

Purpose

This book provides you with what you need to know to manage JMP data and to perform the statistical analyses that are most commonly used in research in the social sciences and other fields. *JMP for Basic Univariate and Multivariate Statistics: A Step-by-Step Guide* shows you how to

- understand the basics of using JMP software
- enter and manage JMP data
- understand the correct statistic for a variety of study situations
- perform an analysis
- interpret the results
- prepare tables, figures, and text that summarize the results according to the guidelines of the *Publication Manual of the American Psychological Association* (the most widely used format in the social science literature).

Audience

This text is designed for students and researchers with a limited background in statistics but can also be useful for more experienced researchers. An introductory chapter reviews basic concepts in statistics and research methods. The chapters on data and statistical analysis assume that the reader has no familiarity with JMP; all statistical concepts are conveyed at an introductory level. The chapters that deal with specific statistics clearly describe the circumstances under which each statistic is used. Each chapter provides at least one detailed example of data, shows how to analyze the data, and demonstrates how to interpret the results for a representative research problem. Even users whose only exposure to data analysis was an elementary statistics course taken years previously should be able to use this guide to perform statistical analyses successfully.

Organization

Although no single text can discuss *every* statistical procedure, this book covers the statistics that are most commonly used in research in psychology, sociology, marketing, organizational behavior, political science, communication, and the other social sciences.

Chapter 1: Basic Concepts in Research and Data Analysis

Fundamental issues in research methodology, statistics, and JMP software need to be reviewed first. This chapter defines and describes the differences between concepts such as *variables* and *values*, *quantitative variables* and *classification variables*, *experimental research* and *nonexperimental research*, and *descriptive analysis* and *inferential analysis*. It also describes the various scales of measurement, called modeling types in JMP (continuous, nominal, and ordinal), and covers the basic issues in hypothesis testing. This chapter gives you the fundamentals and terminology of data analysis needed to learn about using JMP for statistical analysis.

Chapter 2: Getting Started with JMP

Students and researchers who are new to JMP should begin with this chapter. It discusses the general approach to analyzing data using JMP platforms. A step-by-step example takes you through a simple JMP session. You see how to start JMP, open a table, perform an exploratory analysis, and end the session. If you have used JMP previously and feel familiar with it, use this chapter for review and proceed to Chapter 4, which begins with statistical explanations and examples.

Chapter 3: Working with JMP Data

All statistical analyses begin with data. This chapter covers the basics of data input and managing data in JMP. Topics include

- simple input such as keying in data
- how to copy and paste data
- reading simple and complex raw text files
- reading SAS data sets
- reading data from other external files
- creating data values with a formula.

This chapter also introduces shaping JMP tables by stacking and splitting columns, creating subsets, concatenating tables, and joining tables.

Chapter 4: Exploring Data with the Distribution Platform

The first step in analyzing data is to become familiar with the data by looking at descriptive statistical information. This chapter illustrates the JMP Distribution platform, which is used to calculate means, standard deviations, and other descriptive statistics for quantitative variables, and to construct frequency distributions for categorical variables. Features in the Distribution platform can test for normality and produce stem-and-leaf plots.

You see how the Distribution platform can be used to screen data for errors, identify outliers, select subsets of data, and provide other useful preliminary information about a set of data.

Chapter 5. Measures of Bivariate Association

This chapter discusses ways to study the relationship between two variables and determine if the relationship is statistically significant. You see how the JMP Fit Y by X platform chooses the correct statistic based on the level of measurement (data type and modeling type) of the variables. There are examples of using the JMP Fit Y by X platform to prepare bivariate scatterplots and perform the chi-square test of independence, and using the JMP Multivariate platform to compute Pearson correlations and Spearman correlations.

Chapter 6: Assessing Scale Reliability with Coefficient Alpha

This chapter shows how to use the Multivariate platform in JMP to compute the coefficient alpha reliability index (Cronbach's alpha) for a multiple-item scale. You review basic issues regarding the assessment of reliability, and learn about the circumstances under which a measure of internal consistency is likely to be high. Fictitious questionnaire data are analyzed to demonstrate how you can perform an item analysis to improve the reliability of scale responses.

Chapter 7: *t* Tests: Independent Samples and Paired Samples

In this chapter you begin by learning the differences between the independent-samples *t* test and the paired-samples *t* test, and see how to perform both tests. An example of a research design is developed that provides data appropriate for each type of *t* test. With respect to the independent-samples test, this chapter shows how to use JMP to determine whether the equal-variances or unequal-variances *t* test is appropriate, and how to interpret the results. There are analyses of data for paired-samples research designs, with discussion of problems that can occur with paired data. You learn when it is appropriate to perform either the independent-samples or the paired-samples *t* test, and understand what steps to follow in performing both analyses.

Chapter 8: One-Way ANOVA with One Between-Subjects Factor

The one-way analysis of variance is one of the most flexible and widely used procedures in the social sciences and other areas of research. In this chapter you learn how to prepare data using JMP to perform a one-way analysis of variance (ANOVA). It focuses on the *between-subjects* design, in which each participant is exposed to only one condition under the independent variable. This chapter discusses

- the R^2 statistic from the results of an analysis of variance, which represents the percentage of variance in the response that is accounted for or explained by variability in the predictor variable

- how to interpret the graphical results produced by JMP for a one-way ANOVA

- Tukey's HSD multiple comparison test, for comparing group means

- a systematic format to use when summarizing the results of an analysis

- the construction and meaning of the *F* statistic used in the ANOVA.

Chapter 9: Factorial ANOVA with Two Between-Subjects Factors

The factorial design introduced in this chapter has a single dependent response variable and two independent predictor (*between-subject*) variables. This chapter shows how to use the JMP Fit Model platform to perform a two-way analysis of variance (ANOVA). The two predictor variables are manipulated so that treatment conditions include all combinations of levels of the predictor variables. Each subject is exposed to only one condition under each independent variable.

Guidelines are given for interpreting results that do not display a significant interaction, and separate guidelines are given for interpreting results that do display a significant interaction. After completing this chapter, you should be able to determine whether an interaction is significant and to summarize the results involving main effects in the case of a nonsignificant interaction. For significant interactions, you should be able to display the interaction in a figure and perform tests for simple effects (test slices).

Chapter 10: Multivariate Analysis of Variance (MANOVA) with One Between-Subjects Factor

This chapter examines the situation where groups of subjects produce response measurements for two responses. The focus is on the *between-subjects* design, in which each subject is exposed to only one condition (level) of a single nominal (grouping) independent predictor variable.

You see how to use the Fit Model platform in JMP to perform a one-way multivariate analysis of variance (MANOVA). You can think of MANOVA as an extension of ANOVA that allows for the inclusion of multiple response variables in a single test. Examples show how to summarize both significant and nonsignificant MANOVA results.

Chapter 11: One-Way ANOVA with One Repeated-Measures Factor

This chapter focuses on *repeated-measures* designs in which each participant is exposed to every condition (level) of the independent variable. This design is compared to the between-subjects design described in Chapter 8, "One-Way ANOVA with One Between-Subjects Factor." You also learn how problems such as the lack of control group, order effects, and carryover effects can affect the validity of this kind of design.

This chapter also introduces both the univariate approach and the multivariate approach to analysis of repeated-measures designs, and discusses the homogeneity of variance necessary for a valid univariate analysis.

After completing this chapter, you should be familiar with

- necessary conditions for performing a valid repeated-measures ANOVA

- alternative analyses to use when the validity conditions are not met

- strategies for minimizing sequence effects.

Chapter 12: Factorial ANOVA with Repeated-Measures Factors and Between-Subjects Factors

This chapter introduces designs that have both repeated-measures factors and between-subjects factors. This two-way mixed design extends the one-way repeated-measures design presented in the previous chapter by adding one or more groups. Example data include one additional group, used as a control group. Adding a control group lets you test the plausibility of alternative explanations that could account for study results.

There are example analyses for data with significant interaction and with nonsignificant interaction. The analyses illustrate using multivariate fitting (MANOVA) and explain how the MANOVA approach to analysis of repeated-measures data automatically uses the correct error term for statistical tests. A detailed description shows how to perform the analysis and interpret the MANOVA results.

When there is a significant main effect with no interaction, you learn how to test each level of the main effect with a one-way repeated-measures ANOVA.

A univariate approach to analyzing a two-way mixed design is shown as an alternative analysis method.

Chapter 13: Multiple Regression

This chapter discusses the situation in which a response variable is being predicted from continuous predictor variables, all of which display a linear relationship with the response. You learn how to use the Fit Model platform in JMP to perform multiple regression analysis that investigates the relationship between the continuous response variable and multiple continuous predictor variables.

Chapter 13 describes the different components of the multiple regression equation and discusses the meaning of R^2 and other results from a multiple regression analysis. It also shows how bivariate correlations, multiple regression coefficients, and uniqueness indices can be reviewed to assess the relative importance of predictor variables.

After completing the chapter, you should be able to use the Fit Model platform in JMP to conduct the multiple regression analysis, and be able to summarize the results of a multiple regression analysis in tables and in text.

Chapter 14: Principal Component Analysis

This chapter presents principal component analysis as a way to reduce the number of observed variables to a smaller number of uncorrelated variables that account for most of the variance in a set of data. You learn how to use the Multivariate platform in JMP to do a principal component analysis. Several methods are presented to determine the subset of meaningful components to retain and use for further analysis. Example data (fictitious) show that factor rotation can facilitate interpretation of the relationship between the components and possible underlying characteristics in the data.

You also see how to use the Spinning Plot platform to do principal component analysis, and see a biplot that helps you visualize the relationship the components to the predictor variables.

By the end of the chapter, you should be able to perform a principal component analysis, determine the correct number of components to retain, interpret the rotated solution, create component scores, and summarize the results.

Appendix: Choosing the Correct Statistic

Although JMP uses the correct statistics for analyses based on the data type and modeling of the variables you are analyzing, it is useful to see a structured overview of the correct statistical procedure for use in analyzing data. This approach bases the choice of a specific statistic on the number and modeling type of the response (criterion or dependent) variables and the predictor (independent) variables. The appendix groups commonly used statistics into three tables based on the number of criterion and predictor variables in the analysis.

Basic Concepts in Research and Data Analysis

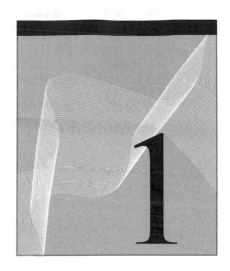

> **Overview.** This chapter reviews basic concepts and terminology from research design and statistics. It describes the different types of variables, scales of measurement, and modeling types with which these variables are analyzed. The chapter reviews the differences between nonexperimental and experimental research and the differences between descriptive and inferential analyses. Finally, it presents basic concepts in hypothesis testing. After completing this chapter, you should be familiar with the fundamental issues and terminology of data analysis, and be prepared to learn about using JMP for data analysis.

Introduction: A Common Language for Researchers

Research in the social sciences is a diverse topic. In part, this is because the social sciences represent a wide variety of disciplines, including (but not limited to) psychology, sociology, political science, anthropology, communication, education, management, and economics. Further, within each discipline, researchers can use a number of different methods to conduct research. These methods can include unobtrusive observation, participant observation, case studies, interviews, focus groups, surveys, ex post facto studies, laboratory experiments, and field experiments.

Despite this diversity in methods used and topics investigated, most social science research still shares a number of common characteristics. Regardless of field, most research involves an investigator gathering data and performing analyses to determine what the data mean. In addition, most social scientists use a common language in conducting and reporting their research: researchers in psychology and management speak of "testing null hypotheses" and "obtaining significant *p* values."

The purpose of this chapter is to review some of the fundamental concepts and terms that are shared across the social sciences. You should familiarize (or refamiliarize) yourself with this material before proceeding to the subsequent chapters, as most of the terms introduced here will be referred to again and again throughout the text. If you are currently taking your first course in statistics, this chapter provides an elementary introduction. If you have already completed a course in statistics, it provides a quick review.

Steps to Follow When Conducting Research

The specific steps to follow when conducting research depend, in part, on the topic of investigation, where the researchers are in their overall program of research, and other factors. Nonetheless, it is accurate to say that much research in the social sciences follows a systematic course of action that begins with the statement of a research question and ends with the researcher drawing conclusions about a null hypothesis. This section describes the research process as a planned sequence that consists of the following six steps:

1. Developing a statement of the research question
2. Developing a statement of the research hypothesis
3. Defining the instrument (questionnaire, unobtrusive measures)
4. Gathering the data
5. Analyzing the data
6. Drawing conclusions regarding the hypothesis.

The preceding steps reference a fictitious research problem. Imagine that you have been hired by a large insurance company to find ways of improving the productivity of its insurance agents. Specifically, the company would like you to find ways to increase the dollar amount of insurance policies sold by the average agent. You begin a program of research to identify the determinants of agent productivity.

The Research Question

The process of research often begins with an attempt to arrive at a clear statement of the research question (or questions). The research question is a statement of what you hope to have learned by the time you complete the program of research. It is good practice to revise and refine the research question several times to ensure that you are very clear about what it is you really want to know.

For example, in the present case, you might begin with the question

> "What is the difference between agents who sell more insurance and agents who sell less insurance?"

An alternative question might be

> "What variables have a causal effect on the amount of insurance sold by agents?"

Upon reflection, you realize that the insurance company really only wants to know what things management can do to cause the agents to sell more insurance. This realization eliminates from consideration certain personality traits or demographic variables that are not under management's control, and substantially narrows the focus of the research program. This narrowing, in turn, leads to a more specific statement of the research question, such as

> "What variables under the control of management have a causal effect on the amount of insurance sold by agents?"

Once you have defined the research question more clearly, you are in a better position to develop a good hypothesis that provides an answer to the question.

The Hypothesis

A *hypothesis* is a statement about the predicted relationships among events or variables. A good hypothesis in the present case might identify which specific variable has a causal effect on the amount of insurance sold by agents. For example, the hypothesis might predict that the agents' level of training has a positive effect on the amount of insurance sold. Or, it might predict that the agents' level of motivation positively affects sales.

In developing the hypothesis, you can be influenced by any of a number of sources, such as an existing theory, related research, or even personal experience. Let's assume that you are influenced by *goal-setting theory*. This theory states, among other things, that higher levels of work performance are achieved when difficult work-related goals are set for employees. Drawing on goal-setting theory, you now state the following hypothesis:

> "The difficulty of the goals that agents set for themselves is positively related to the amount of insurance they sell."

Notice how this statement satisfies the definition for a hypothesis: it is a statement about the relationship between two variables. The first variable could be labeled Goal Difficulty, and the second, Amount of Insurance Sold. Figure 1.1 illustrates this relationship.

Figure 1.1 Hypothesized Relationship between Goal Difficulty and Amount of Insurance Sold

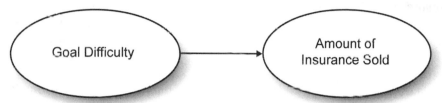

The same hypothesis can also be stated in a number of other ways. For example, the following hypothesis makes the same basic prediction:

> "Agents who set difficult goals for themselves sell greater amounts of insurance than agents who do not set difficult goals."

Notice that these hypotheses have been stated in the present tense. It is also acceptable to state hypotheses in the past tense. For example, the preceding could have been stated,

> "Agents who set difficult goals for themselves sold greater amounts of insurance than agents who did not set difficult goals."

You should also note that these two hypotheses are quite broad in nature. In many research situations, it is helpful to state hypotheses that are more specific in the predictions they make. A more specific hypothesis for the present study might be,

> "Agents who score above 60 on the Smith Goal Difficulty Scale sell greater amounts of insurance than agents who score below 40 on the Smith Goal Difficulty Scale."

Defining the Instrument, Gathering Data, Analyzing Data, and Drawing Conclusions

With the hypothesis stated, you can now test it by conducting a study in which you gather and analyze some relevant data. *Data* can be defined as a collection of scores obtained when a subject's characteristics and/or performance are assessed. For example, you could choose to test your hypothesis by conducting a simple correlational study.

Suppose you identify a group of 100 agents and determine

- the difficulty of the goals set for each agent
- the amount of insurance sold by each agent.

Different types of instruments result in different types of data. For example, a questionnaire can assess goal difficulty, but company records measure amount of insurance sold. Once the data are gathered, each agent has one score that indicates difficulty of the goals, and a second score that indicates the amount of insurance the agent sold.

With the data gathered, an analysis helps tell if the agents with the more difficult goals did, in fact, sell more insurance. If yes, the study lends some support to your hypothesis; if no, it fails to provide support. In either case, you can draw conclusions regarding the tenability of the hypotheses, and you have made some progress toward answering your research question. The information learned in the current study might then stimulate new questions or new hypotheses for subsequent studies, and the cycle repeats. For example, if you obtained support for your hypothesis with the current correlational study, you could follow it up with a study using a different method, perhaps an experimental study. The difference between correlational and experimental studies is described later. Over time, a body of research evidence accumulates, and researchers can review this body to draw general conclusions about the determinants of insurance sales.

Variables, Values, and Observations

When discussing data, you often hear the terms *variables*, *values*, and *observations*. It is important to have these terms clearly defined.

Variables

For the type of research discussed here, a *variable* refers to some specific characteristic of a subject that assumes one or more different values. For the subjects in the study just described, amount of insurance sold is an example of a variable—some subjects sold a lot of insurance and others sold less. A different variable was goal difficulty—some subjects had more difficult goals, while others had less difficult goals. Age was a third variable, and gender (male or female) was yet another.

Values

A *value* refers to either a subject's relative standing on a quantitative variable, or a subject's classification within a classification variable. For example, Amount of Insurance Sold is a quantitative variable that can assume many values. One agent might sell $2,000,000 worth of insurance in one year, another sell $100,000 worth of policies, and another sell nothing ($0). Age is another quantitative variable that assumes a wide variety of values. In the sample shown in Table 1.1, these values ranged from a low of 22 years to a high of 56 years.

Quantitative Variables versus Classification Variables

You can see that, in both amount of insurance sold and age, a given value is a type of score that indicates where the subject stands on the variable of interest. The word "score" is an appropriate substitute for the word "value" in these cases because both are *quantitative variables*. They are variables in which numbers serve as values.

A different type of variable is a *classification variable*, also called a *qualitative variable* or *categorical variable*. With classification variables, different values represent different groups to which the subject belongs. Gender is a good example of a classification variable, as it assumes only one of two values—a subject is classified as either male or female. Race is another example of a classification variable, but it can assume a larger number of values—a subject can be classified as Caucasian American, African American, or Asian American, or as belonging to another group. These variables are classification variables and not quantitative variables because values only represent group membership; they do not represent a characteristic that some subjects possess in greater quantity than others.

Observations

In discussing data, researchers often make references to observational units (or *observations*), which can be defined as the individual subjects (or other objects) that serve as the source of the data. Within the social sciences, a person is usually the observational unit under study (although it is also possible to use some other entity, such as an individual school or organization, as the observational unit). In this text, the person is the observational unit in all examples. Researchers often refer to the number of observations (or cases) included in their data, which simply refers to the number of subjects who were studied. For a more concrete illustration of the concepts discussed so far, consider the data in Table 1.1.

Table 1.1 Insurance Sales Data

Observation	Name	Gender	Age	Goal Difficulty Score	Rank	Sales
1	Bob	M	34	97	2	$598,243
2	Walt	M	56	80	1	$367,342
3	Jane	F	36	67	4	$254,998
4	Susan	F	24	40	3	$80,344
5	Jim	M	22	37	5	$40,172
6	Mack	M	44	24	6	$0

This table reports information about six research subjects: Bob, Walt, Jane, Susan, Jim, and Mack—the data table includes six observations. Information about a given observation (subject) appears as a *row* running from left to right across the table. The first *column* of the data set (running vertically) indicates the observation number, and the second column reports the name of the subject who constitutes or identifies that observation. The remaining five columns report information on the five research variables under study.

- The Gender column reports subject gender, which assumes either "M" for male or "F" for female.

- The Age column reports the subject's age in years.

- The Goal Difficulty Score column reports the subject's score on a fictitious goal difficulty scale. Assume that each participant completed a 20-item questionnaire that assessed the difficulty of the work goals. Depending on how they respond to the questionnaire, subjects receive a score that can range from a low of 0 (meaning that the subject's work goals are quite easy to achieve) to a high of 100 (meaning that they are quite difficult to achieve).

- The Rank column shows how the supervisor ranked the subjects according to their overall effectiveness as agents. A rank of 1 represents the most effective agent, and a rank of 6 represents the least effective.

- The Sales column lists the amount of insurance sold by each agent (in dollars) during the most recent year.

The preceding example illustrates a very small data table with six observations and five research variables (Gender, Age, Goal Difficulty, Rank, and Sales). Gender is a classification variable and the others are quantitative variables. The numbers or letters that appear within a column represent some of the values that these variables can have.

Scales of Measurement and JMP Modeling Types

One of the most important schemes for classifying a variable involves its *scale of measurement*. Researchers generally discuss four scales of measurement: nominal, ordinal, interval, and ratio. In JMP, scales of measurement are designated using three *modeling types*. Modeling types are discussed later, in the section "Modeling Types in JMP."

Before analyzing a data set, it is important to determine each variable's scale of measurement (modeling type) because certain types of statistical procedures require certain scales of measurement. For example, one-way analysis of variance generally requires that the independent variable be a nominal-level variable and the dependent variable be an interval or ratio (continuous) variable. In this text, each chapter that deals with a specific statistical procedure indicates what scale of measurement is required by the variables under study. Then, you must decide whether your variables meet these requirements.

Nominal Scales

A nominal scale is a classification system that places people, objects, or other entities into mutually exclusive categories. A variable measured using a nominal scale is a classification variable that indicates the group to which each subject belongs. The examples of classification variables provided earlier (Gender and Race) also serve as examples of nominal variables. They tell us to which group a subject belongs, but they do not provide any quantitative information about the subjects. That is, the Gender variable might tell us that some subjects are males and other are females, but it does not tell us that some subjects possess more of a specific characteristic relative to others. However, the remaining three scales of measurement provide some quantitative information.

Ordinal Scales

Values on an ordinal scale represent the rank order of the subjects with respect to the variable being assessed. For example, the preceding table includes one variable called Rank that represents the rank ordering of subjects according to their overall effectiveness as agents. The values on this ordinal scale represent a hierarchy of levels with respect to the construct of *effectiveness*. That is, we know that the agent ranked "1" was perceived as being more effective than the agent ranked "2," that the agent ranked "2" was more effective than the one ranked "3," and so forth.

Caution: An ordinal scale has a limitation, in that equal differences in scale values do not necessarily have equal quantitative meaning. For example, look at the following rankings:

Rank	Name
1	Walt
2	Bob
3	Susan
4	Jane
5	Jim
6	Mack

Notice that Walt is ranked "1" while Bob is ranked "2." The rank difference between these two rankings is 1 (2 – 1 = 1), so there is one unit of rank difference between Walt and Bob. Now notice that Jim is ranked "5" while Mack is ranked "6." The rank difference between them is also 1 (6 – 5 = 1), so there is also 1 unit of difference between Jim and Mack. Putting the two together, the rank difference between Walt and Bob is equal to the rank difference between Jim and Mack. However, that does not necessarily mean that the difference in overall effectiveness between Walt and Bob is equal to the difference in overall effectiveness between Jim and Mack. It is possible that Walt is just barely superior to Bob in effectiveness, while Jim is substantially superior to Mack. These rankings reveal very little about the quantitative differences between the subjects with regard to the underlying construct (effectiveness, in this case). An ordinal scale simply provides a rank order of the subjects.

Interval Scales

With an interval scale, equal differences between scale values do have equal quantitative meaning. For this reason, an interval scale provides more quantitative information than the ordinal scale. A good example of an interval scale is the Fahrenheit degree scale used to measure temperature. With the Fahrenheit scale, the difference between 70 degrees and 75 degrees is equal to the difference between 80 degrees and 85 degrees: The units of measurement are equal throughout the full range of the scale.

However, the interval scale also has an important limitation: it does not have a true zero point. A *true zero point* means that a value of zero on the scale represents zero quantity of the construct being assessed. The Fahrenheit scale does not have a true zero point. When a Fahrenheit thermometer reads 0 degrees, that does not mean there is absolutely no heat present in the environment.

Researchers in the social sciences often assume that many of their man-made variables are measured on an interval scale. For example, in the preceding study involving insurance agents, you probably assume that scores from the goal difficulty questionnaire constitute an interval level scale. That is, you assume that the difference between a score of 50 and 60 is approximately equal to the difference between a score of 70 and 80. Many researchers also assume that scores from an instrument such as an intelligence test are measured at the interval level of measurement.

On the other hand, some researchers are skeptical that instruments such as these have true equal-interval properties, and prefer to call them *quasi-interval* scales. Disagreement about the level of measurement achieved with such instruments continues to be a controversial topic within the social sciences.

However, it is clear that neither of the preceding instruments has a true zero. A score of 0 on the goal difficulty scale does not indicate the complete absence of goal difficulty, and a score of 0 on an intelligence test does not indicate the complete absence of intelligence. A true zero point is found only with variables measured on a ratio scale.

Ratio Scales

Ratio scales are similar to interval scales in that equal differences between scale values have equal quantitative meaning. However, ratio scales also have a true zero point, which gives them an additional property. With ratio scales, it is possible to make meaningful statements about the ratios between scale values. For example, the system of inches used with a common ruler is an example of a ratio scale. There is a true zero point because zero inches does in fact indicate a complete absence of length. With this scale, it is possible to make meaningful statements about ratios. It is appropriate to say that an object four inches long is twice as long as an object two inches long. Age, as measured in years, is also on a ratio scale—a 10-year-old house is twice as old as a 5-year-old house. Notice that it is not possible to make these statements about ratios with the interval-level variables discussed above. You would not say that a person with an IQ of 160 is twice as intelligent as a person with an IQ of 80 because there is no true zero point on the IQ scale.

Although ratio-level scales might be easiest to find in the physical properties of objects, such as height and weight, they are also common in the type of research discussed in this manual. For example, the study discussed previously included the variables for age and amount of insurance sold (in dollars). Both of these have true zero points, and are measured as ratio scales.

Modeling Types in JMP

In JMP, each variable has a modeling type that designates its scale of measurement. In a JMP analysis, the modeling types of the variables convey their scale of measurement. The JMP modeling types are called *nominal*, *ordinal*, and *continuous*. Nominal and ordinal modeling types have the same characteristics as those described above for the same scales of measurement. The continuous modeling type in JMP encompasses the characteristics of both the ratio and interval scales of measurement. Modeling types are used by JMP analysis platforms to help determine the correct analysis that needs to be done.

The discussions that follow refer to JMP modeling types, which are discussed in detail in Chapter 3, "Working with JMP Data."

Basic Approaches to Research

Nonexperimental Research

Much research can be described as being either nonexperimental or experimental in nature. In *nonexperimental* research (also called *nonmanipulative, correlational,* or *observational* research), the researcher studies the naturally occurring relationship between two or more naturally occurring variables. A *naturally occurring variable* is a variable that is not manipulated or controlled by the researcher. It is measured as it normally exists.

The insurance study described previously is a good example of nonexperimental research in that you measured two naturally occurring variables (goal difficulty and amount of insurance sold) to determine whether they were related. Another example of nonexperimental research would be an investigation of the relationship between IQ and college grade point average (GPA).

With nonexperimental designs, researchers sometimes refer to response variables and predictor variables.

- A *response variable* is an outcome variable or criterion variable, whose values you want to predict from one or more predictor variables. The response variable is often the main focus of a study because it is mentioned in the statement of the research problem. In the previous example, the response variable is Amount of Insurance Sold. In some experimental research, the response variable is also called the dependent variable.

- A *predictor variable* is the variable used to predict values of the response. In some studies, you might even believe that the predictor variable has a causal effect on the response. In the insurance study, for example, the predictor variable was Goal Difficulty. Because you believed that Goal Difficulty could positively affect insurance sales, you conducted a study in which Goal Difficulty was the predictor and Sales was the response. You do not necessarily have to believe that there is a causal relationship between two variables to conduct a study such as this—you might only be interested in determining whether it is possible to predict one variable from the other. In experimental research, the predictor variable is also known as the independent variable.

Notice that nonexperimental research, which investigates the relationship between just two variables, does not provide evidence concerning cause-and-effect relationships. The reason for this can be seen by reviewing the insurance sales study. If a psychologist conducts this study and finds that the agents with the more difficult goals also tend to sell more insurance, it is not necessarily true that having difficult goals causes them to sell more insurance. Perhaps selling a lot of insurance increases the agents' self-confidence, and this causes them to set higher work goals for themselves. Under this second scenario, it is the insurance sales that had a causal effect on goal difficulty.

As this example shows, with nonexperimental research it is often possible to obtain a single result that is consistent with a number of contradictory causal explanations. Hence, a strong inference that variable A had a causal effect on variable B is rarely if ever valid when you conduct simple correlational research with just two variables. To obtain stronger evidence of cause and effect, researchers either analyze the relationships between a larger number of variables using sophisticated statistical procedures that are beyond the scope of this text, or drop the nonexperimental approach entirely and use experimental research methods instead. The nature of experimental research is discussed in the following section.

Experimental Research

Most experimental research can be identified by three important characteristics:

- Subjects are randomly assigned to experimental conditions.
- The researcher manipulates an independent predictor variable.
- Subjects in different experimental conditions are treated similarly with regard to all variables except the independent variable.

To illustrate these concepts, assume that you conduct an experiment to test the hypothesis that goal difficulty positively affects insurance sales. Assume that you identify a group of 100 agents to serve as subjects. You randomly assign 50 agents to a "difficult goal" condition. Subjects in this group are told by their superiors to make at least 25 cold calls (unexpected sales calls) to potential policyholders per week. The other 50 agents have been randomly assigned to the "easy goal" condition. They have been told to make just 5 cold calls to potential policyholders per week. The design and results of this experiment are illustrated in Table 1.2.

After one year, you determine how much new insurance each agent sold that year. Assume that the average agent in the difficult goal condition sold $156,000 of new policies, while the average agent in the easy goal condition sold just $121,000 worth.

Table 1.2 Design of the Experiment Used to Assess the Effects of Goal Difficulty

Group	Treatment Conditions under the Independent Variable (Goal Difficulty)	Results Obtained with the Dependent Variable (Amount of Insurance Sold)
Group 1 (N = 50)	Difficult Goal Condition	$156,000 in Sales
Group 2 (N = 50)	Easy Goal Condition	$121,000 in Sales

It is possible to use some of the terminology associated with nonexperimental research when discussing this experiment. For example, it is appropriate to continue to refer to Amount of Insurance Sold as being a response variable, because this is the outcome variable of interest. You can also refer to Goal Difficulty as the predictor variable because you believe that this variable, to some extent, predicts the amount of insurance sold.

Notice, however, that Goal Difficulty is now a different kind of variable. In the nonexperimental study, Goal Difficulty was a naturally occurring variable that could take on a wide variety of values (whatever score the subject received on the goal difficulty questionnaire). In the present experiment Goal Difficulty is a *manipulated variable*, which means that you (as the researcher) determine what value of the variable is to be assigned to each subject. In this experiment, Goal Difficulty assumes only one of two values—subjects are in either the difficult goal group or the easy goal group. Therefore, Goal Difficulty is now a classification variable, with a nominal modeling type.

Although it is acceptable to speak of predictor and response variables within the context of experimental research, it is more common to speak in terms of independent variables and dependent variables.

- An *independent variable* is that variable whose values (or levels) the experimenter selects to determine what effect this independent variable has on the dependent variable. The independent variable is the experimental counterpart to a predictor variable.

- A *dependent variable* is some aspect of the subject's behavior assessed to reflect the effects of the independent variable. The dependent variable is the experimental counterpart to a response variable.

In the example shown in Table 1.2, Goal Difficulty is the independent variable and Sales is the dependent variable.

Remember that the terms predictor variable and response variable can be used with almost any type of research, but that the terms independent and dependent variable should be used only with experimental research.

Researchers often refer to the different *levels* of the independent variable. These levels are also referred to as *experimental conditions* or *treatment conditions* and correspond to the different groups to which a subject can be assigned. The present example includes two experimental conditions, a "difficult goal condition" and an "easy goal condition."

With respect to the independent variable, you can speak in terms of the experimental group versus the control group. Generally speaking, the *experimental group* receives the experimental treatment of interest, while the *control group* is an equivalent group of subjects that does not receive this treatment. The simplest type of experiment consists of one experimental group and one control group. For example, the present study could have been redesigned so that it consisted of an experimental group that was assigned the goal of making 25 cold calls (the difficult goal condition) and a control group in which no goals were assigned (the no-goal condition).

You can expand the study by creating more than one experimental group. You could do this in the present case by assigning one experimental group the difficult goal of 25 cold calls and the second experimental group the easy goal of just 5 cold calls, and include a third group as the control group assigned no goals.

Descriptive versus Inferential Statistical Analysis

To understand the difference between descriptive and inferential statistics, you must first understand the difference between populations and samples.

- A *population* is the entire collection of a carefully defined set of people, objects, or events. For example, if the insurance company in question employed 10,000 insurance agents in the U.S., then those 10,000 agents would constitute the population of agents hired by that company.

- A *sample* is a subset of the people, objects, or events selected from that population. For example, the 100 agents used in the experiment described earlier constitute a sample.

Descriptive Analyses: What Is a Parameter?

A *parameter* is a descriptive characteristic of a population. For example, if you found the average amount of insurance sold by all 10,000 agents in this company (the population of agents in this company), the resulting average (also called the *mean*) would be a population parameter. To obtain this average, you first need to tabulate the amount of insurance sold by each and every agent. When calculating this mean, you are engaging in descriptive statistical analysis. *Descriptive statistical analysis* focuses on the exhaustive measurement of population characteristics. You define a population, assess each member of that population, and compute a summary value (such as a mean or standard deviation) based on those values.

Most people think of populations as being very large groups, such as all of the people in the U.S. However, a group does not have to be large to be a population; it only has to be the entire collection of the people or things being studied. For example, a teacher might define as a population all 23 students taking an English course, and then calculate the average score of these students on a measure of class satisfaction. The resulting average is a parameter.

Inferential Analyses: What Is a Statistic?

A *statistic* is a numerical value that is computed from a sample, describes some characteristic of that sample such as the mean, and can be used to make inferences about the population from which the sample is drawn. For example, if you were to compute the average amount of insurance sold by your sample of 100 agents, that average would be a statistic because it summarizes a specific characteristic of the sample. Remember that the word "statistic" is generally associated with samples, while "parameter" is generally associated with populations.

In contrast to descriptive statistics, *inferential* statistical analysis involves using information from a sample to make inferences, or estimates, about the population. For example, assume that you need to know how much insurance is sold by the average agent in the company. Suppose it is impossible (or very difficult) to obtain the necessary information from all 10,000 agents and then calculate a mean. An alternative is to draw a random (and ideally representative) sample of 100 agents and determine the average amount sold by this subset. If this group of 100 sold an average of $179,322 worth of policies last year, then your best guess of the amount of insurance sold by all 10,000 agents would be $179,322. You have used characteristics of the sample to make inferences about characteristics of the population. Using some simple statistical procedures, you can even compute confidence intervals around the estimate, which allows you to make statements such as

> "There is a 95% chance that the actual population mean lies somewhere between $172,994 and $185,650."

This is the real value of inferential statistical procedures—they allow you to review information obtained from a relatively small sample and then to make inferences about a population.

Hypothesis Testing

Most of the procedures described in this manual are inferential procedures that let you test specific hypotheses about the characteristics of populations. As an illustration, consider the simple experiment, described earlier, in which 50 agents are assigned to a difficult goal condition and 50 other agents to an easy goal condition. Assume that, after one year, the difficult-goal agents sold an average of $156,000 worth of insurance, while the easy goal agents sold only $121,000 worth. On the surface, this seems to support your hypothesis that agents sell more insurance when they have difficult goals. But, even if goal setting had no effect at all, you don't really expect the two groups of 50 agents to sell exactly the same amount of insurance. You expect one group to sell somewhat more than the other due to chance alone. The difficult-goal group did sell more insurance, but did it sell *enough* more to make you confident that the difference was due to placing the agents in different goal groups?

What's more, you can argue that you don't even care about the amount of insurance sold by these two small samples. What really matters is the amount of insurance sold by the larger populations they represent. Define the first population as "the population of agents assigned difficult goals" and the second as "the population of agents assigned easy goals." Your real research question is whether the first population sells more than the second. To address this question, you need hypothesis testing.

Types of Inferential Tests

Generally speaking, there are two types of tests conducted when using inferential procedures:

- tests of group differences
- tests of association.

With a *test of group differences*, you want to know whether two populations differ with respect to their mean scores on some response variable. The present experiment leads to a test of group differences because you want to know whether the average amount of insurance sold in the population of difficult-goal agents is different from the average amount sold in the population of easy-goal agents. A different example of testing group differences might involve a study in which the researcher wants to know whether Caucasian Americans, African Americans, and Asian Americans differ with respect to their mean scores on a locus of control scale. (Locus of control refers to the extent to which people believe that their own actions determine the rewards they obtain.) Notice that in both cases, two or more distinct populations are being compared with respect to their mean scores on a single response variable.

With a *test of association*, there is a single population of individuals and you want to know whether there is a relationship between two or more variables within this population. Perhaps the best-known test of association involves testing the significance of a correlation coefficient. Assume that you conduct a simple correlational study in which you ask 100 agents to complete the 20-item goal difficulty questionnaire. Remember that, with this questionnaire, subjects can receive a score that ranges from a low of 0 to a high of 100. You can then correlate these goal difficulty scores with the amount of insurance sold by the agents that year. Here, the goal difficulty scores constitute the predictor variable and the amount of insurance sold serves as the response. Obtaining a strong positive correlation between these two variables means that the more difficult the agents' goals, the more insurance they tend to sell. This is called a test of association because you determine whether there is an association, or *relationship*, between the predictor and response variables. Notice also that there is only one population being studied—there is no experimental manipulation that creates a difficult-goal population versus an easy-goal population.

For the sake of completeness, it is worth mentioning that there are some relatively sophisticated procedures that also let you test whether the association between two variables is the same across two or more populations. Analysis of covariance (ANCOVA) is one procedure that allows such a test.

For example, you could form a hypothesis that the association between self-reported goal difficulty and insurance sales is stronger in the population of agents assigned difficult goals than it is in the population assigned easy goals. To test this hypothesis, you randomly assign a group of insurance agents to either an easy-goal condition or a difficult-goal condition (as described earlier). Each agent completes the 20-item self-report goal difficulty scale and is then given the group assignment (treatment) to make more or fewer cold calls. Subsequently, you could record each agent's sales. Analysis of covariance allows you to determine whether the relationship between questionnaire scores and sales is stronger in the difficult-goal population than it is in the easy-goal population.

ANCOVA also allows you to test a number of additional hypotheses and is beyond the scope of this text. For more information about ANCOVA in JMP, see the *JMP Statistics and Graphics Guide* (2003).

Types of Hypotheses

Two different types of hypotheses are relevant to most statistical tests. The first is called the *null hypothesis*, which is often abbreviated as H_0. The null hypothesis is a statement that, in the population(s) being studied, there are either (a) no differences between the group means, or (b) no relationships between the measured variables. For a given statistical test, either (a) or (b) applies, depending on whether the test is to detect group differences or is a test of association.

With a test of group differences, the null hypothesis states that, in the population, there are no differences between any of the groups being studied with respect to their mean scores on the response variable. In the experiment in which a difficult-goal condition is being compared to an easy-goal condition, the following null hypothesis might be used:

H_0: In the population, the amount of insurance sold by individuals assigned difficult goals does not differ from the amount of insurance sold by individuals assigned easy goals.

This null hypothesis can also be expressed with symbols in the following way:

H_0: $M_1 = M_2$

where

H_0 represents the null hypothesis
M_1 represents mean sales for the difficult-goal population
M_2 represents mean sales for the easy-goal population.

In contrast to the null hypothesis, there is also an *alternative hypothesis* (H_1) that states the opposite of the null. The alternative hypothesis is a statement that there is a difference between the means, or that there is a relationship between the variables, in the population(s) being studied.

Perhaps the most common alternative hypothesis is a *nondirectional alternative hypothesis*. With a test of group differences, a nondirectional alternative hypothesis predicts that the means for the various populations differ, but makes no specific prediction as to which mean will be relatively high and which will be relatively low. In the preceding experiment, the following nondirectional null hypothesis might be used:

H_1: In the population, individuals assigned difficult goals differ from individuals assigned easy goals with respect to the mean amount of insurance sold.

This alternative hypothesis can also be expressed with symbols in the following way:

H_1: $M_1 \neq M_2$

In contrast, a *directional alternative hypothesis* makes a more specific prediction regarding the expected outcome of the analysis. With a test of group differences, a directional alternative hypothesis not only predicts that the population means differ, but also predicts which population means will be relatively high and which will be relatively low.

Here is a directional alternative hypothesis for the preceding experiment.

H_1: The average amount of insurance sold is higher in the population of individuals assigned difficult goals than in the population of individuals assigned easy goals.

This hypothesis can be symbolically represented as follows:

H_1: $M_1 > M_2$

If you believe that the easy-goal population sells more insurance, you replace the "greater than" symbol (>) with the "less than" symbol (<) in the alternative hypothesis, as follows:

$H_1: M_1 < M_2$

Null and alternative hypotheses are also used with tests of association. For the study in which you correlated goal-difficulty questionnaire scores with the amount of insurance sold, you might use the following null hypothesis:

H_0: In the population, the correlation between goal difficulty scores and the amount of insurance sold is zero.

You could state a nondirectional alternative hypothesis that corresponds to this null hypothesis as follows:

H_1: In the population, the correlation between goal difficulty scores and the amount of insurance sold is not equal to zero.

Notice that the preceding is an example of a nondirectional alternative hypothesis because it does not specifically predict whether the correlation is positive or negative, only that it is not zero. A directional alternative hypothesis, on the other hand, might predict a positive correlation between the two variables. You could state such a prediction as follows:

H_1. In the population, the correlation between goal difficulty scores and the amount of insurance sold is greater than zero.

There is an important advantage associated with the use of directional alternative hypotheses compared to nondirectional hypotheses. Directional hypotheses allow researchers to perform *one-sided* statistical tests (also called one-tailed tests), which are relatively powerful. Here, "powerful" means the ability of a test to detect significant differences between group means when differences really do exist. In contrast, non-directional hypotheses allow only *two-sided* statistical tests (also called two-tailed tests), which are less powerful.

Because they lead to more powerful tests, directional hypotheses are generally preferred over nondirectional hypotheses. However, directional hypotheses should be stated only when they can be justified on the basis of theory, prior research, or some other grounds. For example, you should state the directional hypothesis that

"The average amount of insurance sold is higher in the population of individuals assigned difficult goals than in the population of individuals assigned easy goals," only if there are theoretical or empirical reasons to believe that the difficult-goal group will indeed score higher on insurance sales. The same should be true when you specifically predict a positive correlation rather than a negative correlation (or vice versa).

The *p* Value

Hypothesis testing is a process to determine whether you can reject a null hypothesis with an acceptable level of confidence. When analyzing data with JMP, you look at the results for two pieces of information that are critical for this purpose:

1. the obtained (calculated) statistic
2. the probability (*p*) value associated with that statistic.

Consider the experiment in which you compared the difficult-goal group to the easy-goal group. One way to test the null hypothesis associated with this study is to perform an independent samples *t* test. When the data analysis for this study is complete, you compute a *t* statistic and its corresponding *p* value. The *p* value indicates the probability that you would obtain the present results if the null hypothesis were true. If the *p* value is very small, you reject the null hypothesis. Recall that the null hypothesis states that there is no difference between groups.

For example, assume that you obtain a *t* statistic of 0.14 and a corresponding *p* value of 0.90. This *p* value means that there are 90 chances in 100 that you would obtain a *t* statistic of 0.14 (or larger) if the null hypothesis were true. Because this probability is so high, you report that there is very little evidence to refute the null hypothesis. In other words, you fail to reject the null hypothesis, and, instead, conclude that there is not sufficient evidence to support a statistically significant difference between the two groups.

On the other hand, assume that the research project instead produces a *t* value of 8.45 and a corresponding *p* value of 0.001. The *p* value of 0.001 means that there is only one chance in 1000 that you would obtain a *t* value of 8.45 (or larger) if the null hypothesis were true. This is so unlikely that you are fairly confident that the null hypothesis is *not* true. You therefore reject the null hypothesis and conclude that there is a difference in mean sales between the two populations. In rejecting the null hypothesis, you have tentatively accepted the alternative hypothesis.

Technically, the *p* value does not really provide the probability that the null hypothesis is true. Instead, it provides the probability that you would obtain the present results (the present *t* statistic, in this case) if the null hypothesis were true. This might seem like a trivial difference, but it is important to know the meaning of the *p* value.

Notice that you are able to reject the null hypothesis only when the *p* value is a small number (0.001, in the above example). But how small must a *p* value be before you can reject the null hypothesis? A *p* value of 0.05 is one of the most commonly accepted cutoff values. Typically, when researchers obtain a *p* value *larger* than 0.05 (such as 0.13 or 0.37), they fail to reject the null hypothesis, and instead conclude that the differences or relationships being studied are not statistically significant (there is no significant difference between groups). When researchers obtain a *p* value *smaller* than 0.05, they reject the null and conclude that differences or relationships being studied are statistically significant (there is a significant difference between groups). The 0.05 level of significance is not an absolute rule that must be followed in all cases, but it is serviceable for most types of investigations likely to be conducted in the social sciences and other areas.

Fixed Effects versus Random Effects

Experimental designs can be represented by mathematical models described as fixed-effects models, random effects models, or mixed-effects models. The use of these terms refers to the way that the levels of the independent variable (or predictor variable) were selected.

When the researcher arbitrarily selects the levels of the independent variable, the independent variable is called a *fixed-effects factor,* and the resulting model is a *fixed-effects model.* For example, assume that in the current study you arbitrarily decided that the subjects in your easy-goal condition would be told to make just 5 cold calls per week, and that the subjects in the difficult-goal condition would be told to make 25 cold calls per week. In this case, you have *fixed* (arbitrarily selected) the levels of the independent variable. Your experiment therefore represents a fixed-effects model.

In contrast, when the researcher randomly selects levels of the independent variable from a population of possible levels, the independent variable is called a *random-effects factor,* and the model is a *random-effects model.* For example, assume you know that the number of cold calls an insurance agent could possibly place in one week ranges from 0 to 45. This range represents the population of cold calls that you could possibly research. Assume you use some random procedure to select two values from this population of possible calls, and that those two randomly selected values are 12 and 32. In conducting your study, one group of subjects is assigned to make at least 12 cold calls per week, while the second is assigned to make 32 calls. In this case, your study represents a random-effects model because the levels of the independent variable were randomly selected from all possible levels.

As an illustration of a fixed-effects model, assume that you want to conduct research on the effectiveness of hypnosis in reducing anxiety among subjects who suffer from phobias. Specifically, you want to perform an experiment that compares the effectiveness of 10 sessions of relaxation training versus 10 sessions of relaxation training plus hypnosis. In this study, the independent variable might be labeled "Type of Therapy." Notice that you did not randomly select these two treatment conditions from the population of all possible treatment conditions because you know which treatments you want to compare, and design the study accordingly. This is experimental research and your study represents a fixed-effects model.

To provide a nonexperimental example, assume that you want to conduct a study to determine whether Caucasian Americans score significantly higher than African Americans on internal locus of control. The predictor variable in your study is race, and the response variable is scores on some index of locus of control. Most likely, you chose "Caucasian American" versus "African American" as predictor variable groups because you are particularly interested in these two races. You did not randomly select these groups from all possible races. Therefore, the study is another example of a fixed-effects model.

Random-effects factors do sometimes appear in research. For example, in a repeated-measures investigation more than one measure of the response variable is taken from each subject. Subjects are viewed as a random-effects factor, assuming they were randomly selected. Some studies include both fixed-effects factors and random-effects factors. Those models are called *mixed-effects models*.

This distinction between fixed and random effects has important implications for the types of inferences that can be drawn from statistical tests. When analyzing a fixed-effects model, you can generalize the results of the analysis only to the specific levels of the independent variable manipulated in that study. This means that if you arbitrarily selected 5 cold calls versus 25 cold calls for your two treatment conditions, once the data are analyzed you can draw conclusions only about the population of agents assigned 5 cold calls versus the population assigned 25 cold calls.

On the other hand, if you randomly selected two values for your treatment conditions (say, 12 versus 32 cold calls) from the population of possible numbers of calls, the model is a random-effects model. This means that you can draw conclusions about the entire population of possible values that the independent variable can assume. Inferences are not restricted to just the two treatment conditions investigated in the study. In other words, you can draw inferences about the relationship between the population of the possible number of cold calls that agents could be assigned and the response variable (insurance sales).

Summary

Regardless of discipline, researchers need a common language to use when discussing their work with others. This chapter has reviewed the basic concepts and terminology of research that will be referred to throughout this text. Now that you can speak the language, you are ready to move on to Chapter 2, "Getting Started with JMP," where you learn how to do simple JMP analyses.

References

SAS Institute Inc. 2003. *JMP Statistics and Graphics Guide.* Cary, NC: SAS Institute Inc.

Getting Started with JMP

Overview. This chapter is for those who are new to JMP software. It discusses the general approach to analyzing data using JMP platforms. A step-by-step example takes you through a simple JMP session. You see how to start JMP, open a table, perform an exploratory analysis, and end the session.

If you have used JMP previously and feel familiar with it, use this chapter as a review and proceed to Chapter 4, which begins with statistical explanations and examples.

Starting the JMP Application

There are several ways to start the JMP application:

- Double-click on the JMP application icon.
- Click on the JMP application icon to highlight it, and select **Open** from the **File** menu.
- Double-click on an existing JMP table or script.

If you start from the JMP application, the first thing you see is the JMP Starter window. If you open a table to start, then that data table is also displayed. Figure 2.1 shows the first tab of the JMP Starter window. Buttons on the tabs access commands on the main menu bar. You can close the JMP Starter window and use menu bar commands, but the starter tabs provide brief descriptions of many commands and are a good way to start learning how to use JMP.

Figure 2.1 The JMP Starter Window, Main Menu Bar, and File Menu

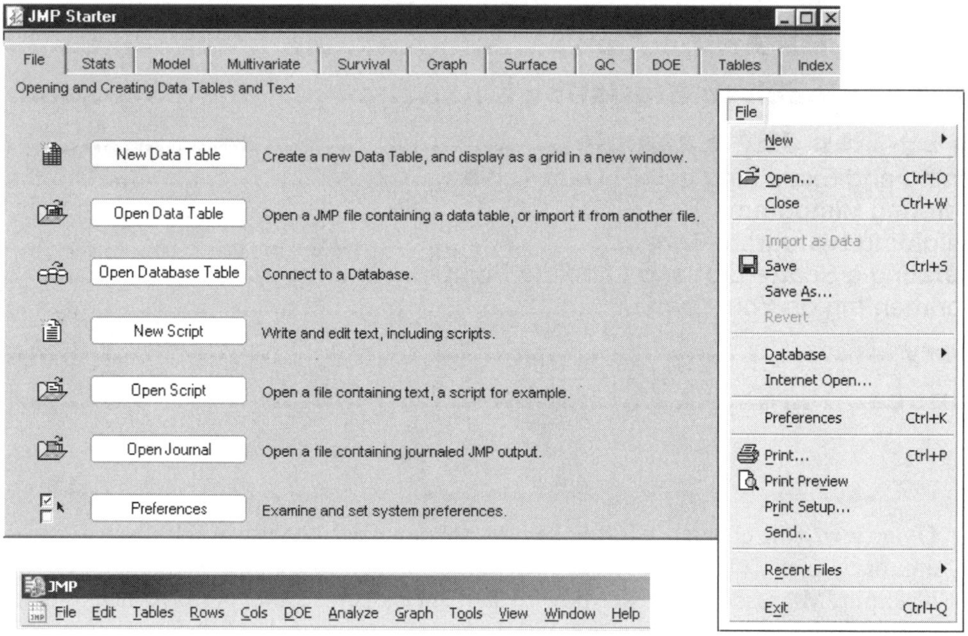

Figure 2.1 also shows the main menu bar and the **File** menu. All buttons on the **File** tab can be found on the **File** menu.

Note: The **Index** tab in the JMP Starter window is a scrolling alphabetic list of JMP features with explanations and scripts to run examples.

The JMP Approach to Statistics

Before any analysis work can be done, data must be in the form of a JMP data table. The next chapter, "Working with JMP Data," talks about creating new tables; this chapter assumes you already have a JMP table. To open an existing table, double-click on the JMP table icon, or use the **Open** command on the **File** menu (**File → Open**) and navigate to the table you want to open.

After data are in the form of a JMP table, data exploration and analyses are available that include simple histograms, a broad selection of plots and graphs, univariate and multivariate modeling, specialized techniques needed for time series analysis, survival analysis, quality control, multivariate methods that include correlations, clustering, and exploratory modeling tools.

When you select **Analysis** or **Graph** menu commands, a launch dialog prompts you to specify variable roles such as Y (response, dependent, or criterion variable) and X (factor, independent, or predictor variable). Most platforms also let you choose a grouping variable (or BY variable), which results in a separate analysis for each level of that variable. The type of analysis is a function of the modeling type (continuous, nominal, or ordinal) of each variable specified for analysis in the launch dialog. Modeling types are discussed further in Chapter 3, "Working with JMP Data."

Most JMP platforms begin with plots or graphs and a preliminary data analysis when appropriate. The analysis platforms have options and commands to request additional information pertinent to the analysis. These options let you see as much or as little detail as you want. The results are presented in an outline form that lets you open or close whole sections of an analysis.

The analysis results are highly interactive. When you click on data points in plots, the corresponding rows are highlighted in the data table and in all other plots and graphs. You have control of the axes specification in plots, color and marker representation of data points, what information to include in statistical tables, how many decimal places to display, and so forth.

There is also a scripting language in JMP (JSL) that lets you program an analysis and rerun it at any time. You can add JSL commands to a completed analysis to tailor the graphical results. Also, you can use platform commands to save the JSL commands for a completed analysis and store the JSL script with the data table.

Other main menu commands let you save results in a journal or create an editable layout of results, suitable for organizing information into a presentation report.

The step-by-step example in the next section shows how to do simple data exploration in JMP. You can follow each mouse step ( ) to familiarize yourself with the JMP approach. This short example illustrates the basic features that make exploration and analysis of data in JMP fast, flexible, informative, and intuitive.

A Step-by-Step JMP Example

The purpose of the following example is to show you how to get started in JMP and how to use basic JMP features available in most analyses, simple or complex. This example uses a sample data table called colleges.jmp, found on the companion Web site for this book (support.sas.com/companionsites). The data were compiled by a prospective college student, and no guarantee is made that the data are accurate or current. They are for example purposes only.

The data are for 39 colleges and have 7 columns (variables) with values available at the time the data were collected:

- **college** names the college and is used as a label column.

- **size** is the size of the college, measured by number of students.

- **tuition** is the yearly tuition in dollars.

- **accept rate** is the proportion of applications accepted into the college.

- **mean SAT** is the mean SAT score of students enrolled at the college.

- **grad program** is a categorical variable with values "yes" and "no" that classifies the school according to whether or not it has a graduate program.

- **> 500 freshmen** is a categorical variable with values "yes" and "no" that classifies the school according to the number of freshmen students enrolled (greater than 500, or less than or equal to 500).

You can use interactive histograms to identify relationships between these variables and identify schools with certain characteristics. For example, suppose you are looking for a large school, with lowest tuition possible, and available graduate courses. The Size, Tuition, and Grad Program variables can identify such schools. On the other hand, you might just be concerned about *getting in* somewhere and are looking for the highest acceptance rate and lowest SAT scores, so you look at the Accept Rate and Mean SAT variables.

Mouse steps (🖰) lead you through the example, and explanations follow each mouse step, as needed.

> **Note:** You can download any JMP data table or other data used in examples from the companion Web site for this book (support.sas.com/companionsites).

Opening and Examining a JMP Data Table

🖰 Use **File → Open** to open the colleges.jmp table. Figure 2.2 shows a partial listing of the Colleges data table. Alternatively, double-click on the data table file icon to open it. There is information for 39 colleges, so there are 39 rows in the table.

You can see in the Columns panel that the College (college name) variable is designated as a label column—the label icon shows to the right of the variable name. When you generate a plot and pause the cursor over a data point, the label value for that point appears until you move the cursor. Or, if you highlight data points and select **Rows → Label/Unlabel**, the label values persist on the plot.

Figure 2.2 Partial Listing of the Colleges Data Table

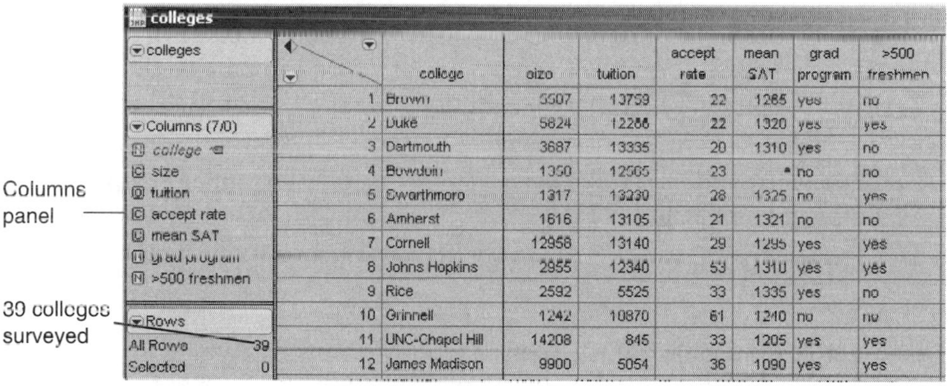

	college	size	tuition	accept rate	mean SAT	grad program	>500 freshmen
1	Brown	5507	13759	22	1285	yes	no
2	Duke	5824	12288	22	1320	yes	yes
3	Dartmouth	3687	13335	20	1310	yes	no
4	Bowdoin	1350	12565	23	•	no	no
5	Swarthmore	1317	13230	28	1325	no	yes
6	Amherst	1616	13105	21	1321	no	no
7	Cornell	12958	13140	29	1295	yes	yes
8	Johns Hopkins	2955	12340	53	1310	yes	yes
9	Rice	2592	5525	33	1335	yes	no
10	Grinnell	1242	10870	61	1240	no	no
11	UNC-Chapel Hill	14208	845	33	1205	yes	yes
12	James Madison	9900	5054	36	1090	yes	yes

Generating Histograms

🖰 On the main menu bar, select **Distribution** from the **Analyze** menu (Figure 2.3).

🖰 When the Distribution launch dialog appears, select all columns in the column selector list on the left of the dialog (except **college**) and click **Y, Columns**. You should see the completed dialog shown on the right in Figure 2.2.

Figure 2.3 Distribution Platform Dialog

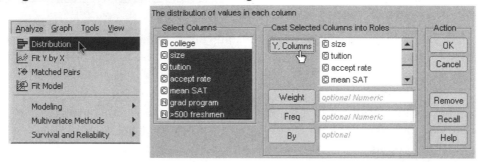

🖰 Click **OK** in the launch dialog to see the initial results of a distribution analysis for all the columns assigned as Y variables in the launch dialog.

The usual default settings for the Distribution platform include plots and report tables in addition to histograms. The type of results depends on the modeling type of the analysis variables. The icon to the left of the variable name in the launch dialog and in the data table tells you each variable's modeling type.

- For continuous variables [☉], analysis results include an outlier box plot and tables of quantiles and moments.

- A frequency table is given for nominal [☒] or ordinal [☉] variables.

This example only looks at histograms. Options accessed by the red triangle icon on the histogram title bar, as shown to the right, can display or suppress any part of the distribution analysis. All results other than the histograms are suppressed in the following graphics. Later chapters discuss the Distribution platform in more detail.

Highlighting Histogram Bars

🖱 When the histograms appear, use shift-click to highlight the bars for the three highest values of Size, as shown next in Figure 2.4.

When you highlight a histogram bar, the corresponding portions of bars in all other histograms are highlighted, and the observations (rows) corresponding to these bars are selected in the data table. Highlighting shows relationships between large schools and the other factors of interest:

- As expected, the larger schools have greater numbers (> 500) of freshmen and have graduate programs.

- However, the acceptance rate in larger schools seems to be lower.

- There appears to be little relationship between larger schools and tuition or mean SAT score.

Figure 2.4 Histograms with Larger Colleges Selected

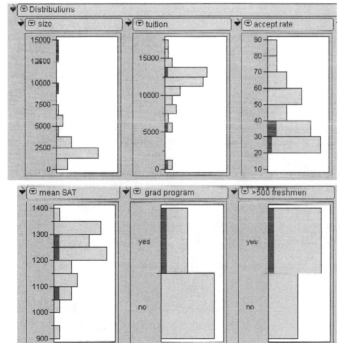

🖰 Next, click on the highest **accept rate** bar and shift-click on the second highest bar to see the results shown next in Figure 2.5. Schools with the highest acceptance rate also have large numbers of freshmen but no graduate programs (in this sample of schools).

- As expected from the previous histograms, these schools are smaller.
- You can quickly hypothesize that the schools in this sample with higher acceptance rates have the lowest mean SAT, have lower tuition, and are generally smaller in size.

Figure 2.5 Histograms with Highest Acceptance Rate Selected

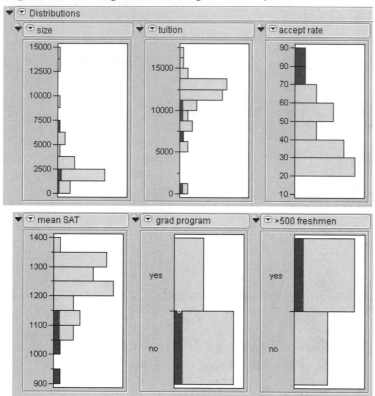

🖰 Finally, click on the "no" bar for **> 500 freshmen** (shown next in Figure 2.6) to see several clear relationships. Schools with fewer freshmen have

- smaller numbers of students (**size**) in general
- mostly higher **tuition**

- medium to low acceptance rates (**accept rate**)
- higher **mean SAT** scores.

Figure 2.6 Histograms with Small Number of Freshmen Highlighted

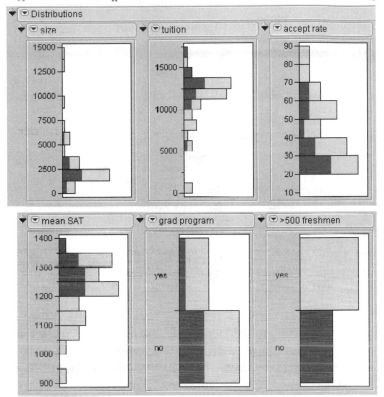

Generating a Scatterplot and Labeling Points

Given the information in the previous histograms, which schools best fit the school you are looking for? If just *getting in* is the most important, then Figure 2.5 clearly identifies the schools you want to know more about. There are several ways to find out which schools are selected in Figure 2.5, and how they relate to other schools.

🖑 Re-create the highlighted bars in Figure 2.5. That is, shift-click the two highest bars for **accept rate**.

🖑 Choose **Fit Y by X** from the **Analyze** menu.

🖰 When the launch dialog appears, select **accept rate** in the column selector list on the left of the dialog and click **Y, Response**. Select **mean SAT** and click **X, Factor**. Figure 2.7 shows the completed Fit Y by X launch dialog.

🖰 Click **OK** to see the scatterplot results at the top in Figure 2.8.

Figure 2.7 Completed Launch Dialog to Produce Scatterplot

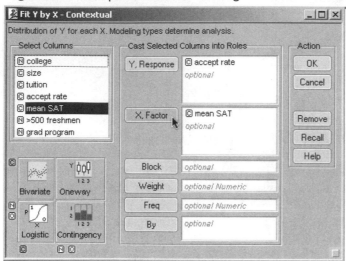

Notice that the points corresponding to the highlighted histogram bars are also highlighted in the scatterplot. You can see a distinct trend in the data—as the mean SAT score increases, the acceptance rate decreases. Let's identify the highlighted points. Recall that the college variable is a Label variable (see Figure 2.2).

🖰 Choose the **Label/Unlabel** command from the **Rows** menu. This command displays the value of one or more label columns to highlighted points.

🖰 It is often interesting to compare extreme points. To look at points with the highest mean SAT scores and lowest acceptance rate, shift-click any points at the lower right of the plot to highlight them, and again choose **Rows → Label/Unlabel**.

The (resized) scatterplot at the bottom in Figure 2.8 shows the names of colleges at the extremes of the distribution of acceptance rate and mean SAT scores.

Figure 2.8 Scatterplot with Labels

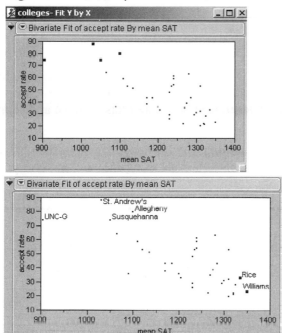

Experimenting on Your Own

Using dynamically linked data, histograms, and scatterplots, you can easily identify colleges that fall into any category of interest. At this point there are a variety of simple things to do within JMP to experiment on your own:

- Continue to investigate the data by clicking on any histogram bar. Shift-click to extend the selection over multiple bars or points.

- Use the **Journal** or **Layout** command on the **Edit** menu to save any windows and edit them for presentation purposes.

- Use **Save → Save Script to Data Table**, available from the menu on the title bar, to save the JSL commands with the data table. The script is then displayed in the Table panel at the left of the data grid, and you can rerun the analysis at a later time.

- Access JMP Help from the **Help** main menu to investigate any aspect of the JMP product.

When you are finished, close the analysis windows and the data table, and quit JMP.

Summary

The purpose of this chapter is to give a jump start to the new JMP user with a brief description of JMP software:

- The general approach to analyzing data uses modeling types assigned to variables. Variables are cast into Y (response) and X (factor) roles. JMP statistical platforms use these modeling types and roles to determine appropriate plots, graphs, and supporting tables.

- The step-by-step example of a simple exploratory data analysis helps the new user learn the "look and feel" of interactive data analysis using JMP.

References

SAS Institute Inc. 2003. *JMP Introductory Guide.* Cary, NC: SAS Institute Inc.
SAS Institute Inc. 2003. *JMP User's Guide.* Cary, NC: SAS Institute Inc.

Working with JMP Data

Overview. This chapter covers the basics of data input in JMP. Topics include simple input such as keying in data, copying and pasting data, reading simple and complex raw text files, reading SAS data sets, and reading data from other external files. Also, the chapter includes an overview of characteristics of JMP data, including data created by formulas.

If you are familiar with JMP, use this chapter as a review and move on to the subsequent data analysis chapters.

Structure of a JMP Table

Before any analysis work can be done, data must be in the form of a JMP data table. The next section talks about creating new tables. This section assumes there is an existing JMP table. To open an existing table, double-click on the JMP table icon, or use the **Open** command on the **File** menu and navigate to the table you want to open.

The Open File dialog box in JMP is similar to those in other applications. Notice that in the Open file dialog box JMP can open many different kinds of file types, including SAS files, Microsoft Excel files, text files, and others.

An open JMP table called **social survey** is shown in Figure 3.1, with annotations of the components you need to know about. These features are discussed below.

Figure 3.1 Data Table Elements

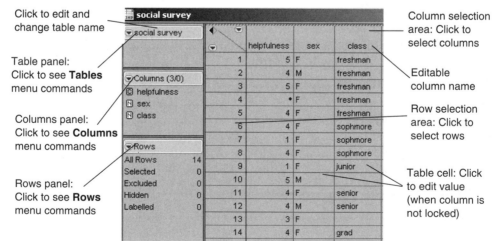

The data table is divided into two parts: the left side of the table consists of three panels, and the right side is the data grid.

- The Table panel, at the top left of the table, shows the data table name; the icon on its title bar accesses the **Tables** main menu and has several additional specialized commands.

- The Columns panel lists the columns in the table; the icon in its title bar accesses the **Cols** main menu.

- The Rows panel shows the total number of rows and summarizes characteristics assigned to rows; the icon on its title bar accesses the **Rows** main menu.

- The data grid holds the data values in a spreadsheet fashion.

Using the Data Table Panels

The panels on the left side of the data grid show overall information about the data table and give quick access to the **Tables**, **Rows**, and **Columns** main menu commands. Drag the divider that separates the panels from the data grid to make the panels wider or narrower.

The Table Panel

The Table panel is at the top of the data table's panels. The icon on the Table panel title bar accesses commands for manipulating the table and creating and storing table information. The first command (**Tables**) is the same as the **Tables** command on the main menu. Table panel commands let you document the table with special table variables, lock the table, attach scripts (JMP programs) to the table, edit and run attached scripts, and determine when formulas should be evaluated.

The Columns Panel

The Columns panel lists the columns (also called variables) in the data table in the order in which they appear from left to right in the data grid. You can rearrange the columns by grabbing and dragging the column names up or down in the Columns panel. The Columns panel title bar shows the total number of columns and the number of columns that are selected (highlighted). The section "Selecting (Highlighting) Rows and Columns" later in this chapter gives details about selecting rows and columns. The icon on the Columns panel title bar accesses the **Columns** main menu commands.

To the left of each column name is an icon that indicates the modeling type of each variable. The modeling type of a variable gives information to the JMP analysis platforms about how to analyze the variable. Details about modeling types are discussed later in the chapter, in the section "Data Types and Modeling Types."

The Rows Panel

The Rows panel tells how many rows are in the table and gives information about certain row states. Rows can be selected, excluded, hidden, and labeled. The **Exclude/Unexclude**, **Hide/Unhide**, and **Label/Unlabel** commands assign row states and are found on the **Rows** main menu or the **Rows** menu on the Rows panel title bar. These commands only affect selected (highlighted) rows.

To select (highlight) rows, click in the row number area to the left of the data in the data grid. The row number area is then highlighted, and the number of selected rows is displayed in the Rows panel. To select multiple rows, drag down the rows (or shift-click on the first and last row you want to select). To make a noncontiguous selection, control-click (⌘-click on the Macintosh) on the rows to be selected.

Using the Data Grid

The data grid is the set of rows and columns that contain the data you want to analyze. A data cell is the intersection of a row and a column. If the column does not have a formula that computes its values, and isn't locked, you can enter or edit data in a cell.

Both rows and columns can be rearranged by dragging them in the data grid. To move one or more rows or columns, first select them, then click and drag to the position in the table you want the selected items to precede. Alternatively, the **Move Rows** command on the **Rows** menu and the **Reorder Columns** command on the **Columns** menu operate on selected rows and columns.

The upper-left corner of the data table offers several useful table actions:

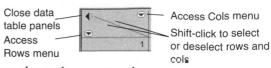

Close data table panels
Access Rows menu
Access Cols menu
Shift-click to select or deselect rows and cols

- To select all rows or all columns at the same time, shift-click in the triangular area above the row numbers or next to the column names. Likewise, to deselect all rows or columns, click in the appropriate triangular area in the upper-left corner of the data grid.

- The red icons access the **Rows** and **Columns** main menus. You can also see these menus by using a right click (control-click on the Macintosh) anywhere in the triangular rows and columns areas.

- The disclosure icon in the upper-left corner of the data grid hides the data table panels section, leaving only the data grid showing.

The following sections discuss creating a data table and entering different kinds of data.

JMP Tables, Rows, and Columns

The previous section, "Structure of a JMP Table," describes the components of a JMP data table and briefly talks about the rows and columns in a table. The current section expands on creating and using rows and columns, describes types and properties of data, and shows how to get data into JMP.

Data for the examples are in the form of a JMP table. You can access the example tables from the companion Web site for this book (support.sas.com/companionsites).

To open an example JMP table, double-click on its icon or use the **Open** command on the **File** menu. Then proceed with the examples. However, to process your own data, they must be in a JMP data table.

Creating a New Table

Follow these steps to create and name a new
data table, and enter data.

New Untitled Table

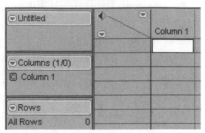

- ⍟ Choose the **New** command from the
  File menu. A new JMP table appears
  with one column, called Column 1,
  showing.

- ⍟ Give the new table a name. The default name, Untitled, is displayed in the upper-
  left corner of the table. When you double-click on the table name it becomes
  editable, and you can enter a new table name. Alternatively, use the **Save As**
  command from the **File** menu and name the table in the Save As dialog.

- ⍟ Click any cell in the data grid to see a cursor for typing data. Enter a value and
  press the Enter key to move the cursor to the next row.

- ⍟ To edit or delete cell values, drag over them in the cell and retype the data, or press
  the Delete key to erase them.

- ⍟ Type data into cells as you would in any spreadsheet.

Double-click in the empty areas of a data grid to create new rows and columns. Or, use
the **Add Rows** command on the **Rows** menu to create a specified number of new rows.
Use the **New Column** or **Add Multiple Columns** command on the **Columns** menu to
add new columns. To name a new column, click on the column name at the top of the
column and begin typing. You can also modify a column name and other column
characteristics with the **Column Info** command on the **Cols** menu.

> **Note:** You can drag the lines between columns to change the column width in the
> table, and drag to make the rows area wider or narrower.

Selecting (Highlighting) Rows and Columns

Many commands on the **Rows** menu and **Cols** menu only apply to selected rows and
columns. Editing values takes place in selected cells. Figure 3.2 shows examples of
highlighted rows, columns, and cells.

- To select a row, click on the space to the left of the row number.

- To select a column, click on the space at the top of the column or on the column name in the Columns panel.

- Click in a cell to select it for data entry or data editing.

Figure 3.2 Rows and Columns

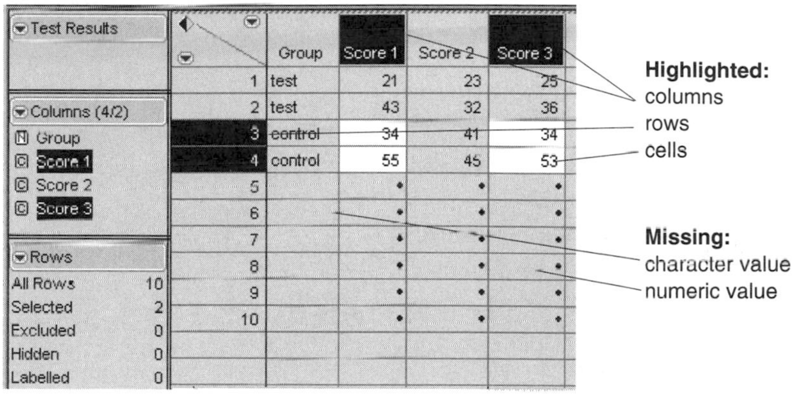

Select multiple rows by dragging up or down the row space to the left of the row numbers. Likewise, select multiple columns by dragging across the space at the top of the columns or on the column names in the Columns panel. You can select rows or columns that are not next to each other (noncontiguous) by using control-click (command-click on the Macintosh).

To deselect (unhighlight) all selected rows, click on the triangular area in the upper left of the data grid, above the row numbers. Likewise, to deselect all selected columns, click on the triangular area in the upper left of the data grid next to the column names. To deselect a single row or column, control-click on a row or column space.

You can also select multiple rows and columns by dragging across the cells that form the intersection of those rows and columns. Use control-click to drag across noncontiguous cells.

You can also use the **Row Selection** command on the **Rows** menu (see Figure 3.3) to select specific subsets of rows. **Row Selection** has a list of subcommands that do almost any kind of selection you need. Several of the most commonly used selection commands are as follows:

Go to Row deselects all currently selected rows, then moves the cursor to the row you specify and selects it.

Invert Row Selection reverses the selection status of all rows.

Select All Rows selects all rows.

Select Randomly randomly selects the proportion of rows or sample size you specify.

Figure 3.3 Row Selection Dialog

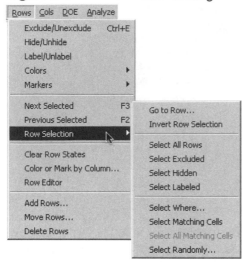

Note: The data table and all analyses on that table are linked. For example, when you click on a histogram bar, the corresponding rows in the data table are then highlighted. Selecting points in a plot also highlights them in the data table.

Creating and Deleting Columns

You can add columns as needed to a JMP table in the following ways:

- Double-click anywhere in the empty grid area to the right of the current columns.

- Use the **New Column** command from the **Cols** menu. This command displays the dialog shown in Figure 3.4, for naming the column and assigning column properties (described later).

- Use the **Add Multiple Columns** command from the **Cols** menu. This method of adding columns lets you add any number of sequentially numbered columns with a naming prefix you specify, and lets you position them anywhere in the table. All columns created this way have the same data type (character or numeric).

- Duplicate an existing column. You can drag a column to change its position by highlighting and dragging the column name at the top of the data grid or in the Columns panel. To duplicate a column, highlight the column in the data grid, press the Control key (Option key on the Macintosh), and drag to a new position.

To change the column name from the default, Column 1 (or to change any column's name), highlight the column, then type a name that describes the column values. Or highlight the column and select **Column Info** from the **Cols** menu. Enter a new name in the **Name** text box in the dialog. Other properties of data are discussed later in this chapter.

To delete one or more columns, highlight them and choose **Delete Columns** from the **Cols** menu.

Note: By default, new columns created by double-clicking in the data grid are numeric and accept only numeric data. If you enter alphabetic or special characters, a prompt appears asking if you want to change the column to a character column or retype the data.

Creating and Deleting Rows

Whenever you double-click in a cell beneath the current row and then press the Enter key, new empty rows form down to the position of the cursor. Click in any cell to begin entering data.

You can also create rows with the **Add Rows** command on the **Rows** menu. **Add Rows** displays the dialog shown to the right, which lets you enter the number of rows you specify at the beginning of the table or the end of the table. When you select **After row**, a text box appears in which you enter the row number where you want the new rows to begin.

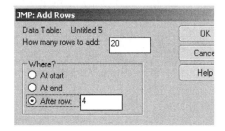

Use the **Delete Rows** command from the **Rows** menu to delete one or more selected rows.

Data Types and Modeling Types

Data types, modeling types, and other properties are characteristics of columns in the data grid. Columns of data are also called *variables*.

The Column Info dialog gives all the characteristics and properties of a column. To see the Column Info dialog for a specific column, highlight the column in the data table (click on the space at the top of the column) and choose **Column Info** from the **Cols** menu (see Figure 3.4).

The columns in a JMP data table hold different kinds of data, to be used in different ways. The kind of data in a column is called its *data type*. Data can be any numeric value, alphabetic characters, special keyboard characters, and row state values that JMP creates when you request them. Most of the examples in this book use numeric and character data.

Every column also has a *modeling type*. The modeling type provides information to the JMP analysis platforms that helps define the analyses. In JMP, it is important to understand modeling types, as well as the relationship between data types and modeling types of variables.

   ◎ *Continuous* means the data values are numeric only and are analyzed as numeric values. The continuous modeling type includes both the interval and ratio measurement scales.

   ◎ *Ordinal* means the data may be either numeric or character, but are to be analyzed as categorical values that have an implied order, an ordinal measurement scale.

   ◎ *Nominal* means the data can have either numeric or character values, have no order, and are analyzed as character or categorical data, a nominal measurement scale.

A variable with the continuous data type can be assigned any one of the three modeling types. However, a variable with the character data type cannot have a continuous modeling type—it must be assigned as either nominal or ordinal. It is not possible to perform statistical calculations on character values, but it is possible to use numbers as categories. Table 3.1 summarizes the relationship between data types and modeling types.

Table 3.1 Summary of Relationship between Data Types and Modeling Types

		Modeling Type		
		Continuous	Ordinal	Nominal
Data	Numeric	yes	yes	yes
Type	Character	no	yes	yes

Assigning Properties to Columns

By default, when you create a new column, it is a numeric column with a continuous modeling type. You can change properties of a column or add new properties with the Column Info dialog. Figure 3.4 shows the Column Info dialog for a variable named Score 1 in a data table named Test Results.

Data Formats

The default format for numeric variables is Best, which displays the data in the best way that the numeric information fits the default field width of 10. However, you can change the field width of both numeric and character variables to be as large as you need.

- Numbers larger than the specified field width are displayed in scientific notation, showing as many decimal places as possible.
- Negative numbers are displayed with a leading minus sign that uses one position in the field width.
- Missing numeric values are displayed as a dot, or period, in the data table.
- Missing character values are displayed as an empty cell in the data table.

There are useful numeric formats (Probability, Currency, Date, Time, and others) that are not needed in this book. For a description of these formats, see the *JMP User's Guide* or the JMP online help facility.

Figure 3.4 Column Options in Column Info Dialog

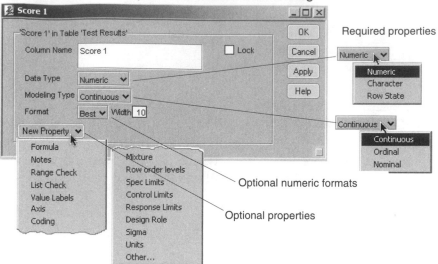

Column Formulas

Optionally, columns can have many other properties, but only a few are used in this book. The **New Property** menu gives a list of other column properties you can assign to a column. The most important additional property used by columns is **Formula**. The section "Computing Column Values with a Formula" later in this chapter shows how to create a column formula that computes values for that column.

Other Column Properties

The following are other optional column properties you might find useful when storing data in a JMP table.

Notes displays a dialog that lets you enter information about the column. For example, you might want to document a column named Temperature with the phrase "temperature in degrees Celsius taken at 12:00 am." Whenever you look at the Column Info dialog, the notes act as column documentation.

Range Check lets you enter a range of acceptable values for a numeric column, which limits the values you can enter into the column. If you try to enter a value outside this range, a dialog prompts you to change the value to one within the acceptable range or enter a missing value. If you select **Range Check** for a column that has values, the

Range Check dialog automatically contains the value range using the low and high values found in the column.

List Check lets you set up a list of acceptable values for a column. As with **Range Check**, you can only enter values into the column from the specified list. **List Check** can be used with either numeric or character columns. If you select **List Check** for a column that has values, the List Check dialog automatically contains the list of values found in the column.

Value Labels let you display a descriptive label instead of the original value wherever the value appears. For example, the value labels appear in the data table but the original value is not lost. You can double-click in a cell to see the original values. All analyses use the original values.

The remaining properties are specific to types of analyses not covered in this book. For more details, see the *JMP User's Guide* or the JMP online help facility.

Getting Data into JMP

You can enter data into a new JMP table just as you do in any spreadsheet program. When you click on a cell, the cursor becomes an I-beam, and the cell is ready for you to type in data. To edit or delete cell values, drag over them in the cell and retype the data, or press the Delete key to erase them. Two other common ways to fill a JMP table with values is by pasting data from text files, other JMP files, or other applications, and reading data from external files such as SAS and Excel.

Copying and Pasting Data

The **Copy** and **Paste** commands on the **Edit** menu perform standard operations. Values you copy to the clipboard from other applications can be pasted into JMP, and values copied from JMP can be pasted into other applications.

The **Copy** command acts on JMP data as follows:

- The entire table is copied to the clipboard when no rows or columns are selected. The same is true if all rows but no columns are selected, or if no rows and all columns are selected.

- **Copy** copies all selected rows if no columns are selected, and all selected columns if no rows are selected.

- When both rows and columns are selected, the subset of values that is the intersection of those rows and columns is copied to the clipboard; select a single row and a single column to copy a single value.

The **Paste** command pastes the values from the clipboard into the current JMP table. If no rows or columns are selected, the entire contents of the clipboard are pasted into the table. If the table has values, the contents of the clipboard are appended to the bottom of the existing data. To paste values into selected rows and columns, you must select the same number of rows and columns as the data on the clipboard have. If the selected rows and columns contain data, the data from the clipboard replace the selected data.

> **Note:** If you use Shift-**Copy** (Option-**Copy** on the Macintosh), the column headers as well as the data values are copied to the clipboard. The column headers are displayed when you paste the data into another application. If you use Shift-**Paste** to paste the clipboard data into another JMP table, the column headers are preserved in the new table.

Reading Data into JMP from Other Files

One common situation is the need to read text data or data from other applications into JMP. Often, the form of the data can be identified by the three-character suffix on the filename, or by the way they are labeled in the Open File dialog file list.

Here are a few of the most commonly encountered file types that JMP can read:

- Text Document (txt)
- Text or Data files on Windows (dat)
- Text with comma-delimited values (csv, or cvs)
- SAS Versions 5 to 9 (sd2, sd5, sd7, sas7bdat) on Windows—open with **Open** command or double-click on file icon to open without dialog
- SAS Version 6 (sas7, bdat, ssci, ssd01, sadeb$data) on Macintosh and Linux—open with the **Open** command or double-click on file icon to open without dialog
- SAS Transport files (xpt, stx)
- Microsoft Excel (xls)—use **Open** command.

If a JMP file, Excel file, or SAS file is visible, you can double-click on the file icon to open it directly, instead of using the Open File dialog. If a text or data file has spaces or tab-delimited fields, the Open File dialog often opens them without further information.

However, a common situation occurs when you have text data with other field delimiters, and variable or column names in the first line of the text file. If the Import/Export preferences are not set to interpret this kind of file, you must provide further information about the file to open it. The file called Animals.txt is an example of raw text data. To see this data in raw form, open it with your system application called Notepad on Windows, TextEdit on the Macintosh, or Gedit/Kedit on Linux. The figure to the right is an illustration of how simple text data appears. You can access this text file from the companion Web site for this book, which you can access at support.sas.com/companionsites.

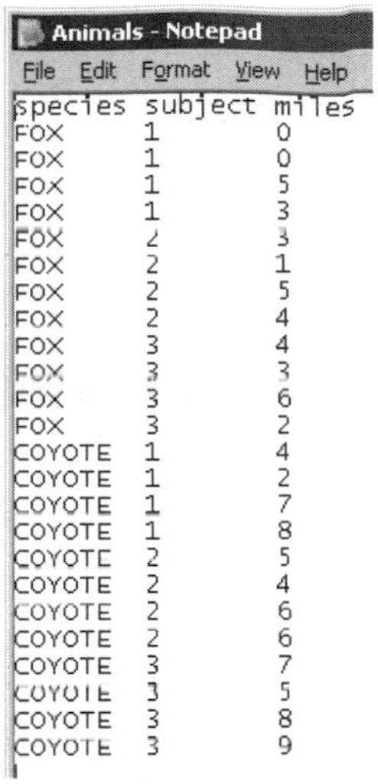

Animals - Notepad

File Edit Format View Help

species	subject	miles
FOX	1	0
FOX	1	0
FOX	1	5
FOX	1	3
FOX	2	3
FOX	2	1
FOX	2	5
FOX	2	4
FOX	3	4
FOX	3	3
FOX	3	6
FOX	3	2
COYOTE	1	4
COYOTE	1	2
COYOTE	1	7
COYOTE	1	8
COYOTE	2	5
COYOTE	2	4
COYOTE	2	6
COYOTE	2	6
COYOTE	3	7
COYOTE	3	5
COYOTE	3	8
COYOTE	3	9

To give information to JMP about how to open this text file, do the following:

- Select **File → Open**. Navigate to the Animals (or Animals.txt) file found on the companion Web site for this book (see above) to see the dialog in Figure 3.5.

- Be sure to select **Text Import Preview (*.TXT, *.CSV, *.DAT)** from the **Files of Type** menu, which lets you see and change (preview) information about how to read the incoming text file.

Figure 3.5 Open File Dialog Showing Text File Only

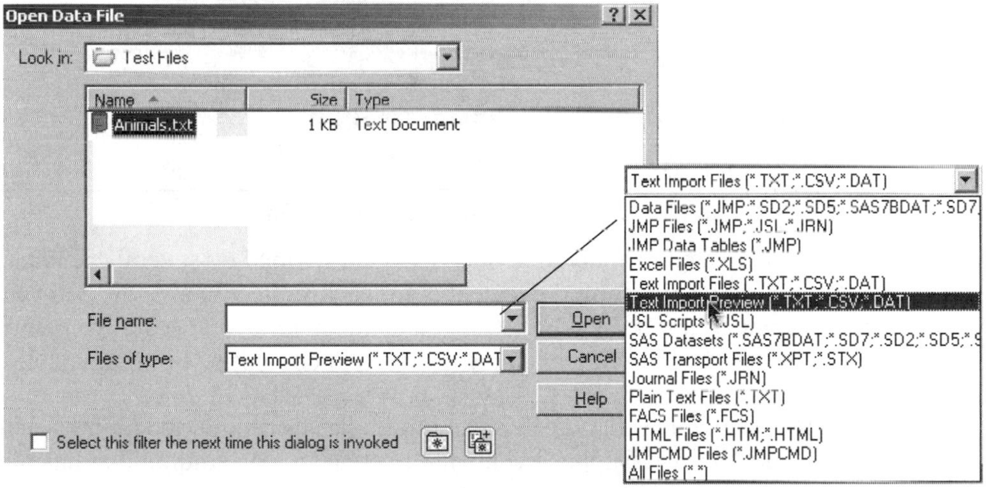

🖱 Select the **Animals** (or **Animals.txt**) table and click **Open** in the Open File dialog. Because you selected the Text Import Preview file type, a prompt appears asking if the incoming file is delimited or fixed width. When you click **Delimited**, the Text Import Preview dialog in Figure 3.6 appears.

This dialog reflects the way JMP sees the incoming data, showing the field delimiter(s), column name, data type, and first two rows of data. If the dialog does not match the incoming data, you change the specifications and click **Apply Settings**. If you change the number of columns, a dialog prompts you to select only the ones you want.

In this example, both **Tab** and **Space** are checked as possible end-of-field delimiters. Fields delimited by either of these characters will be read correctly.

Figure 3.6 Text Import Preview Dialog

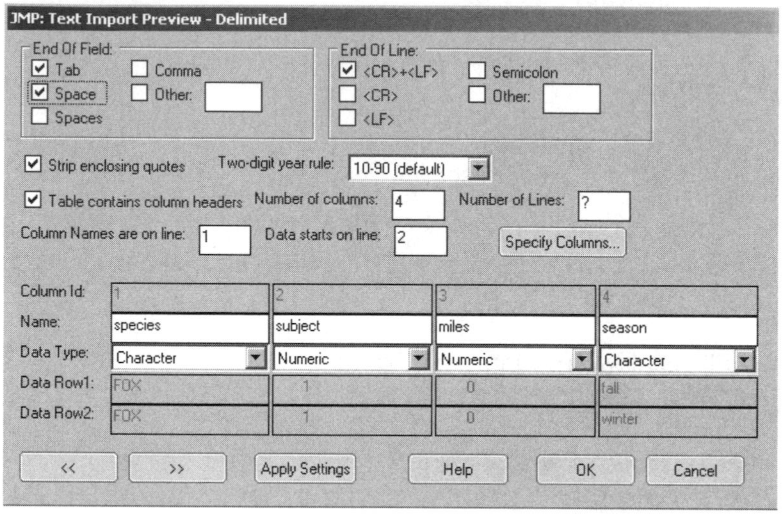

Computing Column Values with a Formula

Often, you are interested in analyzing values that are defined by several columns. For instance, you might have recorded two scores for each subject in a study, and want to look at improvement in scores. To do this, you could subtract one score from another. The following steps take you through a simple example of computing column values.

🖱 Open the data table called Difference.jmp. It has two numeric columns, score1 and score2, and 10 rows. Suppose you want to create a new column that has the difference between the two columns as its values.

⁀ Create a new column, and call it score2-score1 (or any name you like). See the previous section, "Creating and Deleting Columns," for details about creating and naming new columns.

⁀ Click at the top of the new column to highlight it, and choose **Column Info** from the **Cols** menu to see its Column Info dialog (see Figure 3.4).

⁀ Select **Formula** from the **New Property** menu in the Column Info dialog, as shown in Figure 3.7.

The Formula Editor creates a formula that is permanently stored with a specific column. Once a formula is assigned to a column, that column becomes *locked,* which means you can no longer manually enter values. The column is also *linked* to (dependent on) all columns used in the formula to compute its values. The column's values change only when its formula reevaluates.

Formulas consist of column names, operators (+, −, etc.) and functions that can manipulate character or numeric values, trigonometric functions, functions that compare values, conditional functions, probability and statistical functions, and much more. Complete documentation of the Formula Editor is not needed in this book, and can be found in the *JMP User's Guide* (2003). A brief description of the components of the Formula Editor is included here, with a simple example that computes the difference between two columns.

The left side of the Formula Editor lists all the columns in the active data table. In this example, the columns are score1, score2, and the new column to be computed, score2-score1. The keypad in the middle of the Formula Editor lets you insert standard operators into a formula. The Function Browser lists the function categories.

To build a formula, select from the list of columns, the operators on the keypad, or a function from the Function Browser. Continue inserting operators, columns, and functions until you have the desired formula. To see the results of the formula construction, stop at any time and click **Apply** on the right of the Formula Editor. The computed results appear in the Formula column.

Figure 3.7 Column Info Dialog and Formula Editor

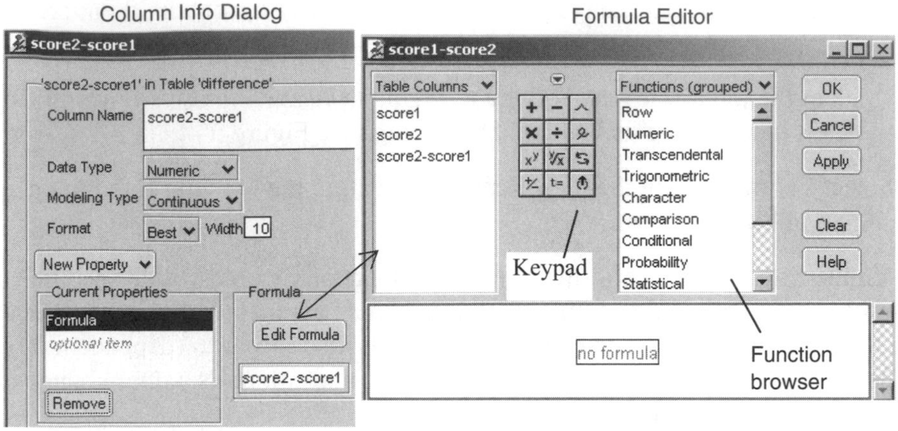

Continue with these steps to build a simple formula that computes the difference between two columns:

🖱 Click on the empty box in the formula to highlight it, and then select **score2** in the column list; score2 appears in the formula. `score2`

🖱 Click the minus sign in the keypad to insert the minus operator into the formula. The minus sign appears and a second empty box is highlighted to the right of the minus sign. `score2-☐`

🖱 Click **score1** in the list of columns to complete the formula. `score2-score1`

🖱 Click **Apply** or close the Formula Editor to see the results in the data table as shown in Figure 3.8.

Figure 3.8 Computing the Difference between Two Columns

	score1	score2	score 2-score 1
1	9	13	•
2	9	22	•
3	10	17	•
4	11	21	•
5	12	10	•
6	12	15	•
7	14	22	•
8	15	28	•
9	17	29	•
10	19	31	•

	score1	score2	score 2-score 1
1	9	13	4
2	9	22	13
3	10	17	7
4	11	21	10
5	12	10	-2
6	12	15	3
7	14	22	8
8	15	28	13
9	17	29	12
10	19	31	12

Go one step further and express the change in scores as a percentage of the base score (score1). To do this:

- Display the formula for the computed column, score2-score1. To quickly show the Formula Editor, right-click on the column header and select **Formula** from the menu that appears.

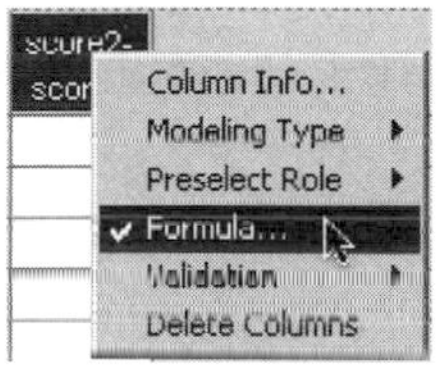

- The entire formula in the Formula Editor is selected. It appears enclosed in a red box. Click the divide operator (÷).

- Click **score1** in the variable list to insert it into the denominator. Then click inside the largest box to highlight the entire formula. Alternatively, type a right parenthesis to highlight an entire formula.

- Click the multiply operator (**✕**) on the keypad to add a term that multiplies the entire formula. The new term appears highlighted, ready to accept a value.

- On your keyboard, type "100" to enter it as the final term. Click **Apply** and **OK** to compute the difference as a percentage of score1.

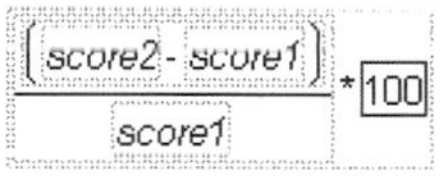

🖰 Finally, clean up the table. Right-click in the header area of the new computed column and select the **Column Info** command.

🖰 In the Column Info dialog, change the column name to **percent change**. Also, select **Fixed Dec** from the **Format** menu to display the percentages with two decimal places.

Figure 3.9 shows these changes entered into the Column Info dialog and the data table.

Figure 3.9 Changing Variable Name and Showing Difference as a Percentage

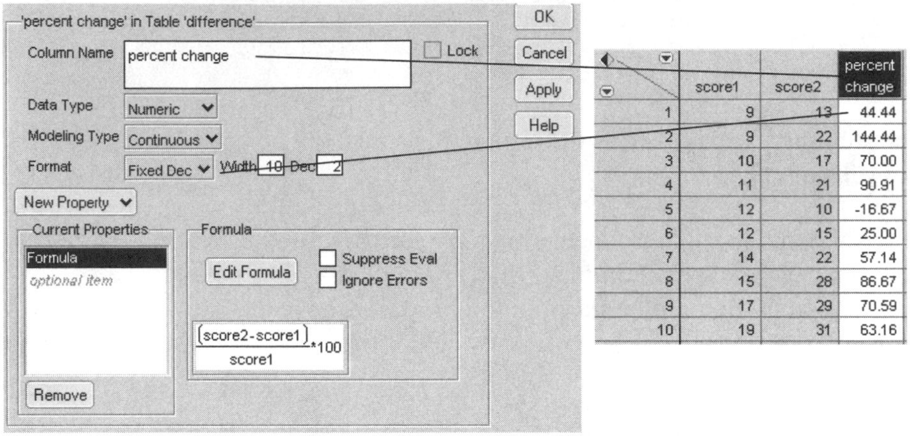

Data Table Management

The JMP data grid holds a rectangular array of data. Rows of data usually constitute information for one subject. The columns (or variables) list information for each row. However, there are different ways to arrange the data grid, and a particular type of analysis might first require manipulating the data.

The following kinds of data rearrangement are often needed to prepare data for analysis:

- sorting tables
- stacking or splitting columns within a single table
- subsetting a table
- joining multiple tables side by side to create a single new table
- concatenating (appending) multiple tables end to end to create a single new table

The next sections explain rearranging data and include simple examples.

Sorting a Table

The **Sort** command in the **Tables** menu gives standard sorting options. There is rarely a need to sort JMP tables in preparation for a statistical analysis, but you might want to view data sorted by one or more variables.

Figure 3.10 illustrates a completed Sort dialog to sort a table by a variable called **Name**. To sort, select one or more variables in the variable selection list and click **By** to see them in the sort list. By default, the variables sort in ascending order. To sort by a variable in descending order, select it in the sort list and click the **a...Z/Z...a** button. Note that you can sort a table in place by checking the **Replace Table** box. If you want to preserve the original order, enter a name in the Output Table edit box, as shown here.

Figure 3.10 Sort Dialog

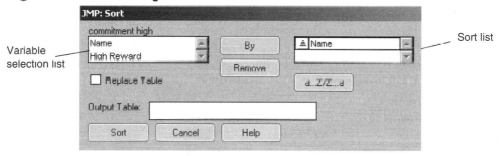

Stacking or Splitting Columns

Sometimes an analysis platform looks for a specific arrangement of data, or can handle data in more than one type of arrangement. Consider the example JMP tables in Figure 3.11. These tables contain the same information arranged different ways. Both tables have values on a commitment scale for 10 people under both low-reward and high-reward conditions.

- The table on the left *splits* the commitment response values into two columns called **Low Reward** and **High Reward**. The subject's name identifies each observation.

- The table on the right *stacks* the commitment response values into a single column called **Commitment Value**. Each observation is identified by both the subject's name and the reward condition that corresponds to the recorded commitment value.

In the chapters that follow, data tables are in the arrangement needed for examples, but you might receive data with multiple columns that you want rearranged into a single column. Or, there could be values listed in a single column that you want split into multiple columns. The **Tables** menu has **Stack** and **Split** commands that rearrange data into a new data table.

Figure 3.11 Split Columns (left) and Stacked Columns (right)

	Name	Low Reward	High Reward
1	John	9	20
2	Mary	9	22
3	Tim	10	23
4	Susan	11	23
5	Maria	12	24
6	Fred	12	25
7	Frank	14	26
8	Edie	15	28
9	Jack	17	29
10	Shirley	19	31

Response columns are split

Response columns are stacked

	Name	Reward	Commitment Value
1	John	Low Reward	11
2	Mary	Low Reward	9
3	Tim	Low Reward	10
4	Susan	Low Reward	9
5	Maria	Low Reward	12
6	Fred	Low Reward	12
7	Frank	Low Reward	17
8	Edie	Low Reward	15
9	Jack	Low Reward	14
10	Shirley	Low Reward	19
11	John	High Reward	19
12	Mary	High Reward	22
13	Tim	High Reward	23
14	Susan	High Reward	18
15	Maria	High Reward	21
16	Fred	High Reward	25
17	Frank	High Reward	22
18	Edie	High Reward	25
19	Jack	High Reward	24
20	Shirley	High Reward	31

The following steps show how to stack and split columns.

- ↺ Open the **commitment paired.jmp** table, shown on the left in Figure 3.11.
- ↺ Choose **Stack** from the **Tables** menu.
- ↺ Complete the Stack dialog as shown in Figure 3.12. The name of the new table is **stacked commitment values**. If you don't give it a name, the new table is called **Untitled 1**.
- ↺ Click **Stack** in the Stack dialog to see the data table on the right in Figure 3.11.

The entries in the Stack dialog cause a single column to be formed by stacking the columns you add to the Stack Columns list (**Low Reward** and **High Reward**). The new column, called **Commitment Value** as specified in the Stack dialog, lists all the commitment scores. A new column, called **Reward**, as seen in Figure 3.11, lists the original column name that corresponds to the response value in each row. You can stack as many columns as you want.

Figure 3.12 Stack Columns Dialog

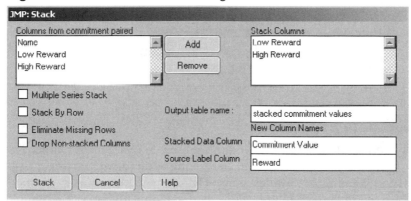

Next, use the **Split** command to convert the stacked table (on the right in Figure 3.11) to the original table shown on the left.

- ✒ With the newly created data table (**stacked commitment values**) active, choose **Split** from the **Tables** menu.

- ✒ Complete the Split dialog as shown in Figure 3.13. Enter **Commitment Value** into the **Split** columns list because you want the single column of commitment scores to be split into a high-reward column and a low-reward column.

- ✒ Enter **Reward** as the **Split Label Col** variable. The values in the **Reward** column become the new column names. There are as many new (split) columns as there are values of the **Split Label Col** variable.

- ✒ Click the **Split** button. Splitting the **Commitment Value** column according to the values of **Reward** gives the original table, shown on the left in Figure 3.11.

Figure 3.13 Split Columns Dialog

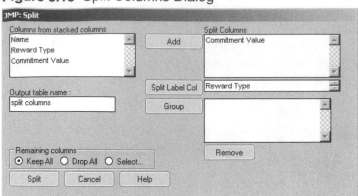

Creating Subsets of Data

It is not unusual to want to analyze one or more subsets of a larger data table. The most common way to create a subset from an existing table (or *source table*) is to use the **Subset** command on the **Tables** menu. The source table in this example is **difference.jmp**, used in the previous section, "Computing Column Values with a Formula."

Using the Subset Command to Create a Subset

The **Subset** command creates a new table that consists of the highlighted rows and columns in the source table. The Subset dialog, shown in Figure 3.14, offers these subsetting options.

- The dialog tells you to select the columns in the table to include in the new table. If no columns are selected, the new table will include all columns from the source table.

- The new table's name is displayed as **Subset of *Source Table***. You can change the name of the new table in the **Subset Name** text box.

- If you check **All Rows**, the result is a duplicate of the entire data table.

- To create a subset of just those rows you select in the source table, click **Selected Rows**. This is the default if any rows are selected.

- To create a random subset, click **Random Sample** and enter either a decimal sample rate or the actual sample size you want.

- By default, the **Linked to original data table** box is checked. The subset is linked to its source table. This means that actions taken in the source table, such as changing data values or highlighting rows, also change the data in the subset. Clear this box, as shown in Figure 3.14, if you want an independent subset.

Figure 3.14 Subset Dialog

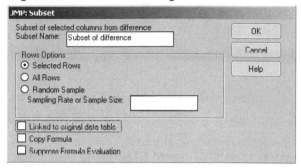

If a column in the source table has a formula associated with it, the Subset dialog shows the **Copy Formula** and **Suppress Formula Evaluation** chcck boxes. If **Copy Formula** is not chccked, the column formula is not included in the subset, but the computed column still appcars locked, as shown in Figure 3.15. To unlock the column, use the Column Info dialog for the column.

Follow these steps to create a subset with the **Subset** command.

- Open the **difference.jmp** table used in the section "Computing Column Values with a Formula," and compute the **percent change** column as dcscribed in that scction.

- Highlight the **gender** column and the **percent change** columns. (Click in the column name area to highlight these columns.)

- IIighlight the rows whcrc **gcndcr** is "male." Use control-click (option-click on thc Macintosh) to highlight noncontiguous rows.

The table with highlighted rows and columns appears on the left in Figure 3.15.

- Select **Tables → Subset** and complete the dialog as shown in Figure 3.14. Click **OK** to see the subset table on the right in Figure 3.15.

Figure 3.15 Subset Example

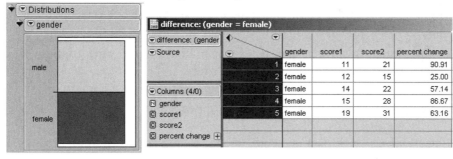

	gender	score1	score2	percent change
1	male	9	13	44.44
2	male	9	22	144.44
3	male	10	17	70.00
4	female	11	21	90.91
5	male	12	10	-16.67
6	female	12	15	25.00
7	female	14	22	57.14
8	female	15	28	86.67
9	male	17	29	70.59
10	female	19	31	63.16

Subset of difference
Source
Columns (2/0)
gender
percent change

	gender	percent change
1	male	44.44
2	male	144.44
3	male	70.00
4	male	-16.67
5	male	70.59

The **Subset** command subsets highlighted rows and columns (or all rows if none are selected, and all columns if none are selected). There are many ways to highlight rows other than manually clicking on row numbers, which is often tedious. The next section presents one simple way to create subsets using the Distribution platform.

Using a Histogram to Create a Subset

If you create a histogram and double-click on one of its bars, JMP creates a subset of the rows represented by that bar. This action highlights the corresponding rows in the analysis table, and immediately creates a subset of those rows and all selected columns (or all columns if none are selected). This "no frills" subset has all formulas in the source table, as indicated by the formula icon next to **percent change** in Figure 3.16.

Figure 3.16 Double-Clicking a Histogram Bar to Create a Subset

Distributions
gender

male

female

difference: (gender = female)
difference: (gender
Source
Columns (4/0)
gender
score1
score2
percent change

	gender	score1	score2	percent change
1	female	11	21	90.91
2	female	12	15	25.00
3	female	14	22	57.14
4	female	15	28	86.67
5	female	19	31	63.16

Note: To analyze subsets of data without creating subsets, use the **By** option found on most analysis platforms. This option automatically analyzes data for each level of a variable (*BY variable*) you specify. An analysis report labeled for each BY-group level is displayed in a single window.

Concatenating Tables

Research in any field often involves results from more than one group or location. The data must then be combined into a single table for analysis. In the simplest case, different tables have the same variables and you want to *concatenate* them—attach or append them to each other end to end.

Suppose scores for males and females are in different tables, as illustrated previously in Figures 3.15 and 3.16. You can experiment with the sample files called **difference male.jmp** and **difference female.jmp**, to see what happens when you concatenate tables.

- Open the **difference male.jmp** and the **difference female.jmp** data tables. Note that they have the same variables with the same names.
- Choose **Concatenate** from the **Tables** menu and complete the dialog as shown in Figure 3.17.
- Click **Concatenate** in the Concatenate dialog to see the single combined table.
- Now change the variable name from **gender** to **males** in the males source table. The tables should look like the ones at the bottom in Figure 3.17.
- Again choose **Tables** → **Concatentate** to combine the tables.

Note that the concatenate action preserves all variable names. It has no way of knowing that columns with different names represent the same variable. You can correct this problem by changing the column names in the original tables to be the same and concatenating again. Or, copy the "male" values from the males column in the combined table and paste them into the empty cells of the gender column. Then use **Cols** → **Delete Columns** to eliminate the spurious males column.

Figure 3.17 Concatenating Tables with Same or Different Variable Names

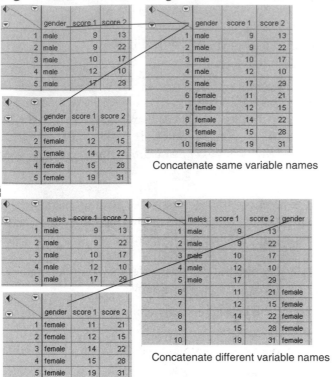

Concatenate same variable names

Concatenate different variable names

Joining Tables

It is also not unusual to combine tables side by side, or *join* tables. Joining tables is necessary when different information for the same subjects is in separate tables and you want to analyze the data together. Suppose the commitment data in Figure 3.11 originally existed in the two tables shown in Figure 3.18. The tables contain most of the same subjects, although the tables have different numbers of subjects. These sample data tables are called **commitment low.jmp** and **commitment high.jmp**. Note that the names are not sorted, and not in the same order in the two tables. The JMP **Join** command does not require that the names be sorted in order to identify matching names in the two tables.

The **Join** command on the **Tables** menu can combine these tables to produce the **commitment paired.jmp** table shown previously in Figure 3.11. Using options in the Join dialog, the combined table eliminates subjects that are not in both tables.

Joining tables can be a multi-step process:

1. Identify the two tables to be joined.
2. Identify the variable(s) whose values must match for observations to be joined.
3. Select the variables to include in the new joined table.

Figure 3.18 Tables with High-Reward and Low-Reward Data for Same Subjects

	Name	Low Reward			Name	High Reward
1	Mary	9		1	John	10
2	Frank	17		2	Mary	22
3	Tim	10		3	Sam	22
4	Jenny	9		4	Tim	23
5	John	11		5	Susan	18
6	Susan	9		6	Maria	21
7	Maria	12		7	Fred	25
8	Shirley	19		8	Frank	22
9	Fred	12		9	Edie	25
10	Edie	15		10	Jack	24
11	Jack	14		11	Sandra	30
				12	Shirley	31

You can see how the **Join** command works by doing this.

- Open the two tables called **commitment high.jmp** and **commitment low.jmp** to see the tables shown in Figure 3.18.

- With the **commitment high.jmp** table active, choose **Join** from the **Tables** menu.

- Click **commitment low** in the selection list on the left of the dialog. Note in Figure 3.19 that **commitment high** shows as the **join** table in the initial dialog, and **commitment low** shows as the **with** table in this initial dialog.

The Matching Specification radio buttons let you choose from three kinds of join.

- **By Row Number** combines the first row in the **join** table with the first row in the **with** table, the second rows in each table, and so on. No consideration is given to the values of any variables.

- **Cartesian Join** combines each row in the **join** table with each row in the **with** table. This can produce a large table if there are many observations in the tables to be joined. The total number of observations in the Cartesian joined table is the number of rows in the **join** table multiplied by the number of rows in the **with** table.

- **By Matching Columns** combines rows from each table only if the values of the columns you specify are the same.

Because the data tables in this example contain information for the same subjects, identified by name, you want to match the observations by name. That is, observations should only be joined when the names match.

- Click **By Matching Columns** in the Join dialog. When you choose this method a second dialog, shown on the right in Figure 3.19, lets you select the variable names from each table whose values must match.
- Select **Name** in each table and click **Match** in the dialog. The variables to be matched appear in the lower portion of the Match Columns dialog.
- Click **Done** in the Match Columns dialog to return to the Join dialog.

Note that the check boxes at the bottom of the Match Columns dialog are not checked. In particular, by leaving **Include Non Matches** unchecked, only those observations that occur in both tables are included in the new joined table.

Figure 3.19 Intitial Join Dialog and Match Columns Dialog

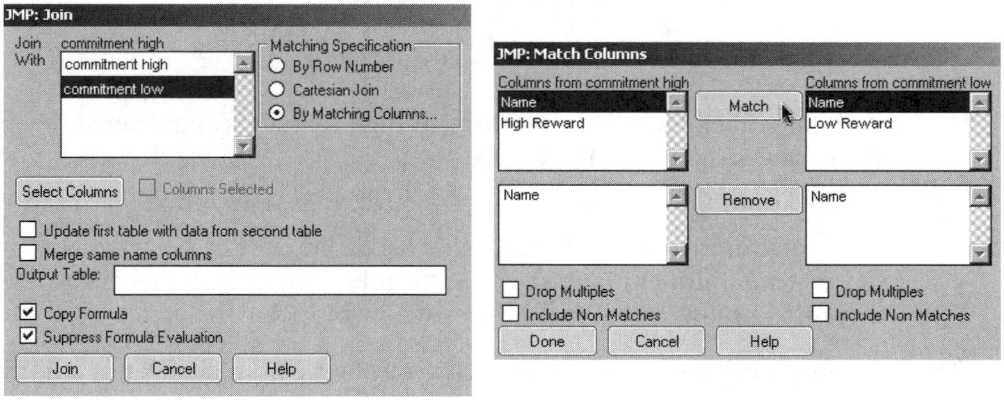

If you finish the join now, all variables from both tables appear in the new joined table, but it isn't necessary to include the matching variable (Name) from both tables.

- Click the **Select Columns** button in the Join dialog. Complete the **Select Columns** dialog as shown in Figure 3.20.
- Click **Done** in the Select Columns dialog to again return to the Join dialog.

Now you are ready to complete the join. Optionally, enter a name for the new joined table in the **Output Table** text box.

🖱 Click the **Join** button in the Join dialog to see the table in Figure 3.20.

Figure 3.20 Select Columns Dialog

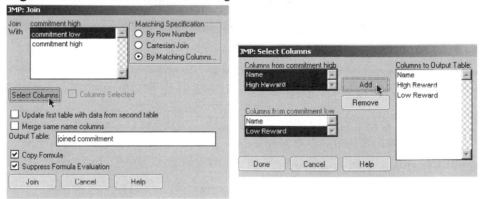

The final joined table has the 10 observations that had matching values. The observations are now sorted by the matching variable.

Figure 3.21 Table Created by Joining Two Tables

	Name	High Reward	Low Reward
1	Edie	25	15
2	Frank	22	17
3	Fred	25	12
4	Jack	24	14
5	John	19	11
6	Maria	21	12
7	Mary	22	9
8	Shirley	31	19
9	Susan	18	9
10	Tim	23	10

Joined Commitment
 Source

Columns (3/0)
 N Name
 ◎ High Reward
 ◎ Low Reward

Rows
All Rows   10
Selected   0
Excluded   0

Summary

All computer programs that do statistical analyses require that data be in a form recognized by the program. Although most tables used as examples can be found on the companion Web site for this book (support.sas.com/companionsites), your own data must be entered or read into a JMP data table. This chapter presented an overview of JMP data tables, including

- the structure and form of a JMP data table
- how to create and delete rows and columns
- how to get data into a JMP table
- how to compute column values
- how to stack and split columns in a data table
- how to create subsets of data
- how to combine tables side by side (join) and end to end (concatenate).

The information presented is designed to get you started. Details and extensive examples can be found in the *JMP User's Guide* (2003). The *JMP User's Guide* is available as a written document or accessed online in JMP Help.

References

SAS Institute Inc. 2003. *JMP Statistics and Graphics Guide*, *Version 5.1*. Cary, NC: SAS Institute Inc.

Exploring Data with the Distribution Platform

Overview. This chapter illustrates the JMP Distribution platform, which can be used to calculate means, standard deviations, and other descriptive statistics for quantitative variables, and to construct frequency distributions for categorical variables. The Distribution platform can also be used to test for normality and produce stem-and-leaf plots. Once data are entered into a JMP data table, features of the Distribution platform can be used to screen for errors, test statistical assumptions, and obtain simple descriptive statistics.

Why Perform Simple Descriptive Analyses?

This chapter focuses on the Distribution platform in JMP. Looking at distributions is useful for (at least) three important purposes:

- The first purpose involves the concept of *data screening*. Data screening is the process of carefully reviewing the data to ensure that they were entered correctly and are being read by the computer as you intended. Before conducting any of the more sophisticated analyses to be described in this book, you should carefully screen your data to make sure that you are not analyzing *garbage*—numbers that were accidentally entered incorrectly, impossible values on variables such as negative ages, and so forth. The process of data screening does not guarantee that your data are correct, but it does increase the likelihood.

- Second, a distribution analysis is useful because it lets you explore the shape of your data. Among other things, understanding the shape of data helps you choose the appropriate measure of central tendency—the mean versus the median, for example. In addition, some statistical procedures require that sample data be drawn from a normally distributed population, or at least that the sample data not display a marked departure from normality. You can use the procedures discussed in this chapter to produce plots of the data, as well as test the null hypothesis that the data come from a normal population.

- The nature of an investigator's research question itself might require the use of the Distribution platform in JMP to obtain a desired statistic. For example, assume your research question is

  > "What is the average age of marriage for women living in the United States in 2004?"

  You can obtain data from a representative sample of women living in the United States who married in that year, analyze their ages with the Distribution platform, and review the results to determine the mean age.

Similarly, in almost any research article it is desirable to report demographic information about the sample. For example, if a study is performed on a sample that includes subjects from a variety of demographic groups, it is desirable to report the percent of subjects of each gender, the percent of subjects by race, the mean age, and so forth. You can use the Distribution platform to obtain this information.

Example: The Helpfulness Social Survey

To help illustrate these procedures, assume that you conduct a social survey on helpfulness. You construct a questionnaire, like the one in Figure 4.1, that asks just one question related to helping behavior. The questionnaire also contains an item that identifies the subject's gender, and another that determines the subject's class in college (freshman, sophomore, etc.).

Figure 4.1 Social Survey Questionnaire

```
Please indicate the extent to which you agree or disagree
with the following statement:
1. "I feel a personal responsibility to help needy people
in my community." (please check your response below)
    (5)_____Agree Strongly
    (4)_____Agree Somewhat
    (3)_____Neither Agree nor Disagree
    (2)_____Disagree Somewhat
    (1)_____Disagree Strongly
2. Your gender (please check one):
    (F)_____Female
    (M)_____Male
3. Your classification as a college student:
    (1)_____Freshman
    (2)_____Sophomore
    (3)_____Junior
    (4)_____Senior
    (5)_____Graduate
```

Notice that this instrument is designed so that entering the data is relatively simple. For each variable, the value to be entered appears to the left of the corresponding subject response. For example, with question 1 the value "5" appears to the left of "Strongly Agree." This means that the number "5" is to be entered for any subject checking that response. For subjects checking "Strongly Disagree," a "1" will be entered. Similarly, notice that, for question 2, the letter "F" appears to the left of "Female," so an "F" is to be entered for subjects checking this response. Suppose you administer the questionnaire to 14 students and enter the results into a JMP table. To do this:

- Start JMP and select **File → New** to see a new Untitled table, like the one shown at the top left in Figure 4.2.

- Double-click in the Table panel title bar to change the table name. Change it to **Social Survey**, as shown.

- Double-click anywhere in the empty area above the data grid to form new columns. Alternatively, you can use the **Add Columns** command on the **Cols** menu to add new columns to a table.

- Click in the column name area to highlight it, and key in the column names, **helpfulness**, **gender**, and **class**. You should see the table on the right in Figure 4.2.

- Be sure to assign the **gender** column and the **class** column a data type of character instead of the default numeric. For each column, choose **Cols → Column Info** and select **Character** from the **Data Type** menu in the Column Info dialog.

- Click in a data grid cell to begin entering data. Use the mouse to click in any data cell, or tab to automatically move from cell to cell. Enter the data as shown in the completed table at the bottom in Figure 4.2.

Figure 4.2 Entering the Social Survey Data into a JMP Table

Use **File → New** to create a new empty table. Double-click on the title bar in the Table panel and type a new table name, Social Survey.

Use **Cols → New Column** to create two new columns. Click in the column header area to highlight it, and enter column names.

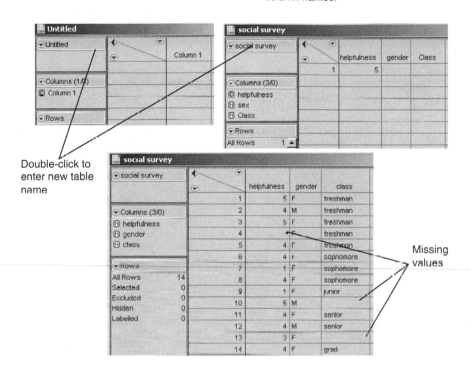

Double-click to enter new table name

Missing values

The data obtained from the first subject have a value of "F" for the Gender variable (a female), show a value of "5" for the Helpfulness variable (indicating that she checked "Agree Strongly"), and show a value of "freshman" for the Class variable.

Notice that there are some missing data in this data table. In row 4 you can see that this subject indicated she was a female freshman, but did not answer question 1. That is why the corresponding cell in the **helpfulness** column is left blank. Subject 10 gave no answers for either the **helpfulness** or the **class** column, and subject 13 gave no answer for the **class** column. Questionnaire data often have missing data.

The **Social Survey.jmp** table can be found on the companion Web site for this book (support.sas.com/companionsites).

Computing Summary Statistics

You can use the Distribution platform to analyze both quantitative (numeric) variables and qualitative (character) variables.

When the Distribution platform analyzes the numeric variable, helpfulness, it gives a table of selected quantiles and the following information:

- mean
- standard deviation (Std Dev)
- standard error of the mean (Std Err Mean)
- upper 95% confidence level (upper 95% mean)
- lower 95% confidence level (lower 95% mean)
- the number of useful cases on which calculations were performed (N).

Optionally, more moments can be displayed, and are discussed later in this chapter.

There is a frequency table for each character variable (such as gender) that shows the number of rows that have each value (level) of the character variable, and the total number of levels.

The Distribution Platform

To use the Distribution platform:

    ⌐ Select **Analyze** → **Distribution** as illustrated in Figure 4.3. The Distribution command displays the launch dialog that has a variable selection list of all the variables in the active table, as shown on the right in Figure 4.3.

    ⌐ Select **helpfulness** and **gender** from the selection list and click **Y, Columns** to see them in the analysis variables list on the right in the dialog. Note that the icons to the left of the variable names tell you the modeling type, continuous (◎) or nominal (◎).

Figure 4.3 Launching the Distribution Platform and Selecting Analysis Variables

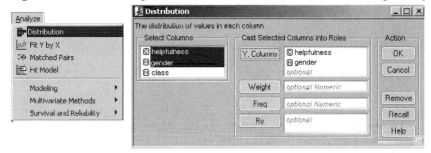

    ⌐ When you click **OK**, the results of the Distribution analysis appear in a new window, as shown in Figure 4.4.

The distribution of each analysis variable you selected in the launch dialog is initially shown as a histogram. Tables appropriate for the data type of each analysis variable follow the histograms. Figure 4.4 shows the analysis for the questionnaire variables, helpfulness and gender. The next section describes these results.

Figure 4.4 Results of the Distribution Analysis

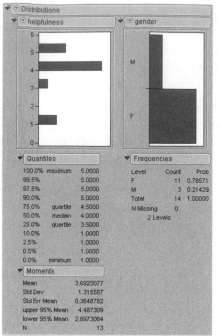

Reviewing the Distribution Results for Continuous Numeric Variables

Figure 4.4 contains the results created by the Distribution platform. Before doing any more sophisticated analyses, it is usually helpful to look at distributions of the variables and carefully review the results to ensure that everything looks reasonable.

The analysis window has a report for each variable being analyzed. The histograms give you a quick feel for the shape of the distributions. The statistics for each variable appear beneath its histogram. Quantitative (numeric) variables have a table of selected quantiles and a Moments table. Character variables have a Frequency table.

First, note the item called **N** at the bottom of the Moments table. **N** is the number of valid or nonmissing cases on which calculations are performed. In this instance, calculations are performed on only 13 cases for helpfulness. This might come as a surprise, because the data set contains 14 cases. However, recall that one subject did not respond to the helpfulness question, so **N** is equal to 13 instead of 14 for helpfulness in these analyses.

Next, you should review the mean for the variable, to verify that it is a reasonable number. The mean is the sum of the responses for a variable divided by the number of nonmissing responses for that variable. Remember that the values for helpfulness range from 1 (disagree strongly) to 5 (agree strongly). Therefore, the mean response should be somewhere between 1.00 and 5.00 for the helpfulness variable. If the mean is outside this range, you know there is some type of error. In the present case the mean for helpfulness is 3.692 (between 1.00 and 5.00)—everything looks correct so far.

Use the same reasoning and check the Quantiles table. The value next to the item called Minimum is the lowest value of helpfulness that appeared in the data table. If this is less than 1, you know there is some type of error, because 1 is the lowest value that could have been assigned to a subject. In the analysis report, the minimum value is 1, which indicates no problems. The largest value observed for that variable is shown next to the item called Maximum. The maximum should not exceed 5, because that is the largest helpfulness score a subject can have. The reported maximum value is 5, so again there appears to be no obvious errors in the data.

Reviewing the Distribution Results for Character Variables

The analysis results on the right in Figure 4.4 show the histogram for gender, followed by a Frequency table. In this report, the column called Level lists the possible values ("F" and "M") for the gender variable. When reviewing a frequency distribution, it is useful to think of these different values as representing categories to which a subject can belong.

The column called Count shows that 11 subjects were female and 3 were male. The column called Prob lists the proportion (probability) of the count in each level of the categorical variable. The table shows that 0.78571 (78.6%) of the subjects are female and 0.21429 (21.4%) are male.

Changing the Modeling Type

Recall that all JMP variables have both a data type (numeric or character) and a modeling type (continuous, nominal, or ordinal). A character variable, such as gender and class in this example, whose values are alphabetic characters must have a data type of character and a modeling type of either nominal or ordinal. But the variable helpfulness has numeric values. It was assigned a numeric data type when it was created. However, numeric variables can be analyzed using any modeling type.

The previous example treats the helpfulness variable as continuous with a modeling type of numeric and gives moments (mean, standard deviation, and so forth) for the responses. However, you can use the Distribution platform to determine the proportion of subjects who "agreed strongly" with a statement on a questionnaire, the proportion who "agreed somewhat," and so on. In other words, you want to analyze the helpfulness variable using a nominal modeling type.

To change the modeling type of helpfulness, click on the modeling type icon next to its name in the Columns panel to the left of the data grid. Then select **Nominal** from the menu that appears, as shown here. For analysis purposes, this variable is now treated as categorical. You can change modeling types in the data table at any time.

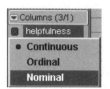

Again select **Analyze → Distribution** and complete the launch dialog by selecting **helpfulness** and **class** as the analysis variables. Then click **OK** to see the results in Figure 4.5.

The results for helpfulness, on the left in Figure 4.5, treat the values as categories instead of continuous numeric values. If no subject appears in a given category, the value representing that category does not appear in the frequency distribution. For example, you see only the values "1," "3," "4," and "5." The value "2" does not appear because none of the subjects checked "Disagree Somewhat" for question 1.

The results on the right in Figure 4.5 are for the class variable. Notice there are only 12 total responses even though there are 14 observations in the data table. This is because two respondents did not provide a value for class. The existence of two missing values means that there are only 12 valid cases for the class variable.

The menu accessed by the icon on the response variable title bar (shown on the right in Figure 4.5) gives options to modify the appearance of results. In this example, the options selected for the helpfulness histogram show the standard error bars, probability axis, percents, and counts. The class variable analysis only uses the default options.

Figure 4.5 Distribution Results for Categorical Variables

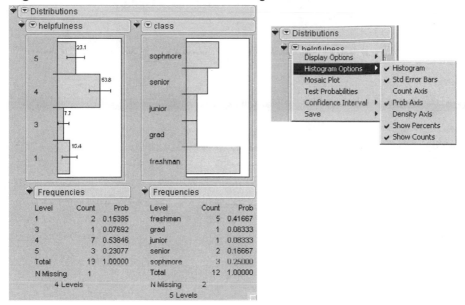

Ordering Histogram Bars

By default, the order of histograms bars for categorical values begins at the bottom of the chart with the lowest sort-order (alphabetic) value, and each bar continues in sort order. Often, this is not the arrangement you want. The histogram for class in Figure 4.5 is an example of a histogram presented in sorted order, but you would like to see the bars in order of class level—that is, "freshman," "sophomore," "junior," "senior," and "grad." Another common example occurs when month of the year is a table column and a histogram displays bars in alphabetic order.

One way to tell the Distribution platform how you want bars ordered is to use the Column Info dialog and assign a special property called List Check to the column. To do this:

- Right-click in the **class** column name area and select **Column Info** from the popup menu that appears. Or, click in the **class** column name area to highlight it and select **Column Info** from the **Cols** dialog on the main menu.

- When the Column Info dialog appears, select **List Check** from the New Property menu, as shown in Figure 4.6. Notice that the list of values is in the default (alphabetic) order.

Figure 4.6 Column Info Dialog and Using List Check Dialog

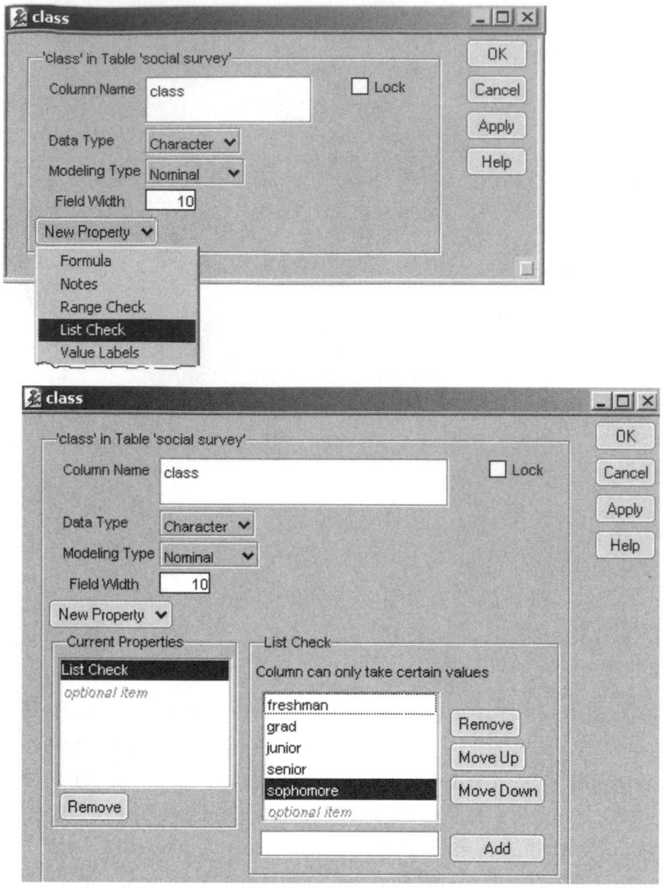

🖰 Highlight values in the List Check list and use the **Move Up** and **Move Down** buttons to rearrange the values in the list.

🖰 Click **Apply**, then **OK**.

Note: The List Check column property prevents you from entering any value into the column other than those listed. However, you can add or remove values at any time.

Figure 4.7 shows the list check values ordered the way you want them, and the new histogram that results from a distribution analysis of the class variable.

Figure 4.7 Values Ordered as List Check Items and Resulting Histogram

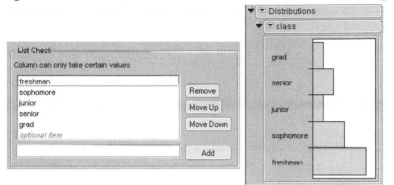

Missing Data

When a respondent doesn't provide information for a variable, the result is a missing value. For a categorical variable or a numeric variable analyzed as a categorical variable (such as Helpfulness in Figure 4.5), the last line in the Frequency table shows the number of missing values for that distribution. You can see which respondents did not give values by checking the data table (see Figure 4.5).

Testing for Normality

The normal distribution is a symmetrical, bell-shaped distribution of values, as illustrated in Figure 4.8.

To understand the distribution, assume that you are interested in conducting research on people who live in retirement communities. Imagine that you know the age of every person in this population. To summarize this distribution, you prepare a figure similar to Figure 4.8. The variable, Age, is plotted on the horizontal axis, and the frequency of persons at each age is plotted on the vertical axis. Figure 4.8 shows that many of the subjects are around 71 years of age, because the distribution of ages peaks near the age of 71. This suggests that the mean of this distribution will likely be close to 71. Notice also that most of the subjects' ages are between 67 (near the lower end of the distribution) and 75 (near the upper end of the distribution). This is the approximate range of ages that you expect for persons living in a retirement community.

Figure 4.8 The Normal Distribution

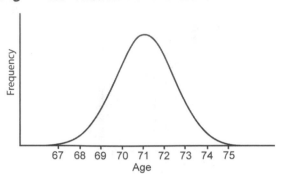

Why Test for Normality?

Normality is an important concept in data analysis because there are at least two problems that can result when data are not normally distributed. The first problem is that markedly non-normal data might lead to incorrect conclusions in inferential statistical analyses. Many inferential procedures are based on the assumption that the sample of observations was drawn from a normally distributed population. If this assumption is violated, the statistic can give misleading findings. For example, the independent groups *t* test assumes that both samples in the study were drawn from normally distributed populations. If this assumption is violated, then performing the analysis might cause you to incorrectly reject the null hypothesis (or incorrectly fail to reject the null hypothesis). Under these circumstances, you should analyze the data using a procedure that does not assume normality (perhaps some nonparametric procedure).

The second problem is that markedly non-normal data can have a biasing effect on correlation coefficients, as well as more sophisticated procedures that are based on correlation coefficients. For example, assume that you compute the Pearson correlation between two variables. When one or both of these variables are markedly non-normal, the correlation coefficient between them could be much larger or much smaller than the actual correlation between these variables in the population. In addition, many sophisticated data analysis procedures (such as principal component analysis) are performed on a matrix of correlation coefficients. If some or all of these correlations are distorted due to departures from normality, then the results of the analyses can again be misleading. For this reason, some experts recommend that researchers routinely check their data for major departures from normality prior to performing sophisticated analyses such as principal component analysis (Rummel, 1970).

Departures from Normality

Assume that you draw a random sample of 18 subjects from the population of persons living in retirement communities. There are a variety of ways that the data can display a departure from normality. Data for a variety of age samples, called Sample A, Sample B, and so on, are used for the following examples and in the data table called **ages.jmp**, shown in Figure 4.9.

Figure 4.9 Data Table with Six Samples of Age Distributions

	Sample A	Sample B	Sample C	Sample D	Sample E	Sample F
1	75	75	73	77	77	77
2	74	74	72	76	75	76
3	73	73	72	75	73	76
4	73	73	72	74	72	75
5	72	72	72	73	71	75
6	72	72	71	73	70	75
7	72	72	71	72	69	74
8	71	71	71	72	69	74
9	71	71	71	71	68	73
10	71	71	71	71	68	73
11	71	71	71	70	67	72
12	70	70	71	70	67	72
13	70	70	71	69	66	71
14	70	70	70	63	66	70
15	69	69	70	68	66	69
16	69	69	70	67	65	68
17	68	68	70	66	65	66
18	67	37	69	65	64	64

Columns (7/0): Sample A, Sample B, Sample C, Sample D, Sample E, Sample F, sample 3

Rows: All Rows 18, Selected 0, Excluded 0, Hidden 0, Labelled 0

Figure 4.10 shows the distribution of ages in two samples of subjects drawn from the population of retirees. This figure is somewhat different from the normal distribution shown previously in Figure 4.8 because the Distribution platform in JMP displays distribution bars horizontally so that age is plotted on the vertical axis rather than on the horizontal axis. This arrangement is similar to the stem-and-leaf plots discussed later in the chapter.

There is a display option on the menu of the response variable title bar to show the histogram with a horizontal layout, giving the shape seen in Figure 4.8. There are also options to suppress the outlier box plot that often displays by default, and to display the number of observations in each bar, as shown in this example.

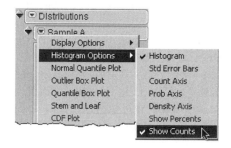

Each bar in a given distribution of Figure 4.10 represents the number of subjects in an age category. For example, in the distribution for Sample A, you can see that there is one subject at age 75, one subject at age 74, two subjects at age 73, three subjects at age 72, and so forth. The ages of the 18 subjects in Sample A range from a low of 67 to a high of 75.

Figure 4.10 Sample with Normal Distribution and Sample with an Outlier

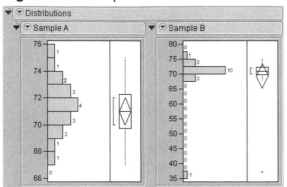

The data in Sample A form an approximately normal distribution. The distribution is called *approximately normal* because it is difficult to form a perfectly normal distribution using a small sample of just 18 cases. A statistical test (discussed later in the section "Fitting and Testing a Normal Distribution to the Sample") will show that Sample A does not demonstrate a significant departure from normality. Therefore, it would be appropriate to include the data in Sample A in an independent sample *t* test.

In contrast, there are problems with the data in Sample B. Notice that their distribution is similar to that of Sample A, except that there is an *outlier* at the lower end of the distribution. An outlier is an extreme value that differs substantially from the other values in the distribution. In this case, the outlier represents a subject whose age is only 37. This person's age is markedly different from that of the other subjects in your study. Later, you will see that this outlier causes the data set to demonstrate a significant departure from normality, making the data inappropriate for some statistical procedures. When you observe an outlier such as this, it is important to determine if an error was made when the data were entered and correct the error whenever possible. It is not usually considered statistically sound to simply delete perceived outliers from a set of data. If the data cannot be corrected, it is preferable to use some method of imputation that preserves the integrity of the data rather than delete the data.

A sample can also depart from normality because it displays *kurtosis*. Kurtosis refers to the peakedness of the distribution. The two samples displayed in Figure 4.11 demonstrate different levels of kurtosis.

Sample C in Figure 4.11 displays *positive kurtosis*, which means that the distribution is relatively peaked (tall and skinny) rather than flat. Notice that, with Sample C, there are a relatively large number of subjects that cluster around the central part of the distribution (around age 71). This is what makes the distribution peaked (relative to Sample A, for example). Distributions with positive kurtosis are also called *leptokurtic*.

In contrast, Sample D in Figure 4.11 displays *negative kurtosis*, which means that the distribution is relatively flat. Flat distributions are also described as being *platykurtic*.

Figure 4.11 Samples Displaying Positive and Negative Kurtosis

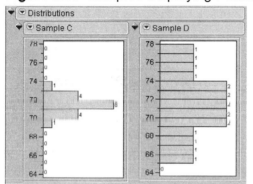

In addition to kurtosis, distributions can also demonstrate varying degrees of *skewness*, or sidedness. A distribution is skewed if the tail on one side of the distribution is longer than the tail on the other side. The distributions in Figure 4.12 show two different types of skewness.

Consider Sample E in Figure 4.12. Notice that the largest number of subjects in this distribution tends to cluster around the age of 66. The tail of the distribution that stretches above 66 (from 67 to 77) is relatively long, while the tail of the distribution that stretches below 66 (from 65 to 64) is relatively short. Clearly, this distribution is skewed. A distribution is said be *positive skewed* if the longer tail of a distribution points in the direction of *higher* values. You can see that Sample E displays positive skewness, because its longer tail points toward larger numbers such as 75, 77, and so forth.

On the other hand, if the longer tail of a distribution points in the direction of lower values, the distribution is said to be *negative skewed*. You can see that Sample F in

Figure 4.12 displays negative skewness because in that sample the longer tail points downward, in the direction of lower values (such as 66 and 64).

Figure 4.12 Samples Displaying Positive and Negative Skewness

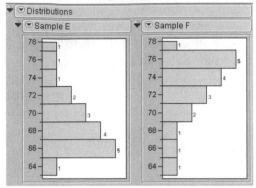

Testing for Normality from the Distribution Platform

For purposes of illustration, assume that you wish to analyze the data that are illustrated as Sample A of Figure 4.10 (the approximately normal distribution). The following sections show how to see and interpret summary statistics, produce and interpret a stem-and-leaf plot, and how to test for normality.

With the **ages** data table active (see Figure 4.9) choose **Analyze → Distribution** and select Sample A as the analysis variable in the launch dialog. By default, the results include the Histogram, Outlier Box Plot, Moments table, and Quantiles table.

Requesting More Moments

The Moments table initially shows the Mean, Standard Deviation, Standard Error of the Mean, upper and lower confidence interval for the mean, and N, the number of nonmissing observations used to compute statistics for the response variable. You can see that the analysis is based on 18 observations. The Mean and Std Dev of the age values in Sample A are 71 and 2.057, respectively.

To see additional moments, select **More Moments** from the **Display Options** menu as shown on the right in Figure 4.13. A closer look at the additional moments in Figure 4.13 shows the statistics Sum Wgt, Sum, Variance, Skewness, Kurtosis, CV, and N Missing.

Figure 4.13 Requesting More Moments

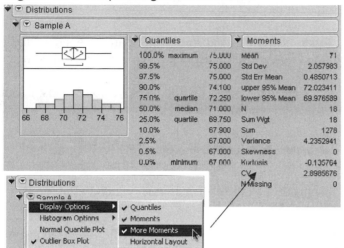

Note that the skewness statistic for the Sample A age values is zero. In interpreting the skewness statistic, keep in mind the following:

- A skewness value of zero means that the distribution is not skewed. In other words, the distribution is symmetrical, so neither tail is longer than the other.

- A positive skewness value means that the distribution is positively skewed. The longer tail points toward higher values in the distribution (as seen previously with Sample E of Figure 4.12).

- A negative skewness value means that the distribution is negatively skewed. The longer tail points toward lower values in the distribution (as with Sample F of Figure 4.12).

Because the age values in Sample A display a skewness value of zero, neither tail is markedly longer than the other in this sample.

You can see in Figure 4.13 that the kurtosis statistic for the Sample A age values is –0.135764. When interpreting this kurtosis statistic, keep in mind the following:

- A kurtosis value of zero means that the distribution displays no kurtosis; in other words, the distribution is neither relatively peaked nor relatively flat, compared to the normal distribution.

- A positive kurtosis value means that the distribution is relatively peaked, or leptokurtic.

- A negative kurtosis value means that the distribution is relatively flat, or platykurtic.

The small negative kurtosis value, approximately –0.14, indicates that Sample A is slightly flat, or platykurtic.

Fitting and Testing a Normal Distribution to the Sample

The Distribution platform has options that let you compare and test the shape of a distribution to known distributions. For example, suppose that you want to know if the age samples, Sample A and Sample B (see Figure 4.10), are approximately normal.

Use the menu accessed by the icon on the title bar next to the response variable name and select **Fit Distribution → Normal**, as shown in Figure 4.14. The Distribution platform then displays a fitted normal curve on the histogram and shows the estimated mean and standard deviation on the histogram and in the Parameter Estimates table. In this example the Outlier Box Plot, which is often displayed by default, is hidden.

Figure 4.14 Distribution Analysis for Sample A

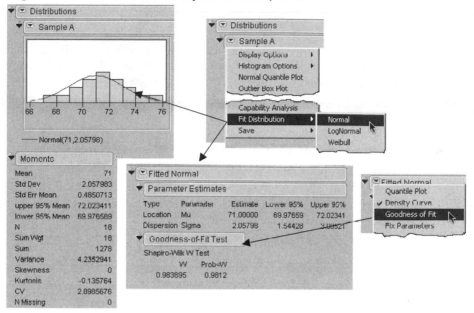

Use the menu on the Fitted Normal title bar to request additional information about the fitted distribution, including the goodness-of-fit test. This option requests a significance test for the null hypothesis that the sample data are from a normally distributed population. The results show the Shapiro-Wilk (W) statistic for samples of 2000 or less. For larger samples the Kolmogorov-Smirnov-Lillefor's (KSL) statistic is shown.

In this example, the W statistic is 0.984 with a p value of 0.9812. Remember that this statistic tests the null hypothesis that the sample data are normally distributed. This p value is very large at 0.9812, meaning that there are approximately 9,812 chances in 10,000 that you would obtain the present results if the data were drawn from a normal population. In other words, it is likely that you would obtain the present results if the sample were from a normal population. Because this statistic gives so little evidence to reject the null hypothesis, you can tentatively assume that the sample was drawn from a normally distributed population. This makes sense when you review the shape of the distribution of Sample A. The sample data clearly appear to be approximately normal. In general, you should reject the null hypothesis of normality only when the p value is less than 0.05. See Chapter 5, "Advanced Univariate Analysis," in the *JMP Statistics and Graphics Guide* (2003) for details.

Results for a Distribution with an Outlier

The data of Sample A in Figure 4.14 displays an approximately normal distribution. For purposes of contrast, now use the Distribution platform to look at the Sample B age data from Figure 4.10. Remember that Sample B was similar in shape to Sample A except for an outlier. In fact, the data for Sample B are identical to those of Sample A except for subject 18. In Sample A subject 18's age is 67 but in Sample B it is 37, which is extremely low compared to the other age values in the sample. You can turn back to Figure 4.9 to verify this.

By comparing the Moments table of Figure 4.15 (Sample B) to that of Figure 4.14 (Sample A), you can see the outlier has a considerable effect on some of the descriptive statistics for age. The mean of Sample B is now 69.33, down from the mean of 71 found for Sample A. The outlier has an even more dramatic effect on the standard deviation. With the approximately normal distribution the standard deviation was only 2.05, but with the outlier included the standard deviation is 8.27.

The skewness index for Sample B is –3.90. A negative skewness index such as this is what you expect because the outlier created a long tail that points toward the lower values in the Sample B age distribution.

The normality test (goodness-of-fit test) for Sample B gives a Shapiro-Wilk statistic of 0.458 with corresponding p value less than 0.0001. Because this p value is below 0.05, you reject the null hypothesis, and tentatively conclude that Sample B is not normally distributed. In other words, you can conclude that Sample B displays a statistically significant departure from normality (due to a single outlier, observation 18).

In Figure 4.15, the outlier box plot shows above the histogram, and the Quantiles table gives details about the skewness of the distribution. The Quantiles table shows the maximum value of age was 75 and the minimum was 37. You can quickly identify this outlier in the outlier box plot. Click on the point to see that observation 18 has the minimum value (the outlier value).

Using the outlier box plot or histogram to identify outliers might be unnecessary when working with a small data set (as in the present situation), but it can be invaluable when dealing with a large data set. For example, if there are possible outliers in a data large set of data, click on the outlier points to highlight them in the histogram and in the data table. You can then easily identify the observations that have outlying values and avoid the tedious chore of examining each observation in a data table.

Figure 4.15 Distribution Analysis for Sample B

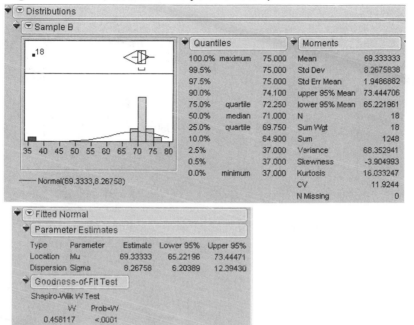

Understanding the Outlier Box Plot

Figure 4.16 shows outlier box plots for Sample A through Sample F, discussed in the previous sections. Outlier box plots are often set as the default in the preferences whenever you generate a histogram for a numeric continuous variable. All plots can be shown or hidden using the menu found on the title bar next to the variable name.

Figure 4.16 Outlier Box Plots for Six Distributions

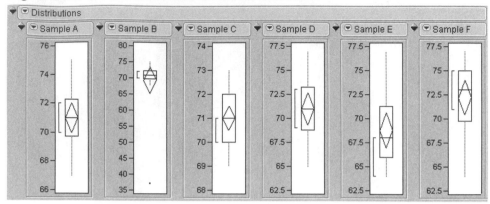

Outlier box plots are a compact display of many distribution characteristics.

- The box identifies the *interquartile range.* The ends of the box are at the 25th and 75th quantiles (also called quartiles).

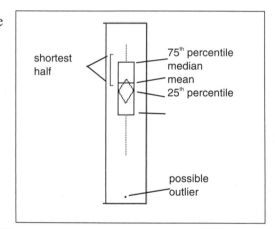

- The *means diamond* inside the box represents the mean and 95% confidence interval. The left and right diamond points are situated at the mean of the distribution, and the top and bottom points show the confidence interval.

- The line across the box indicates the median of the sample, where half the sample points are greater and half are less than that value.

- The dashed lines on the ends of the box, called whiskers, extend from the distances computed as follows:
  (upper quartile) + 1.5 * (interquartile range) and
  (lower quartile) − 1.5 * (interquartile range).

- Points falling outside this range are possible outliers.

- The bracket along the edge of the box identifies the *shortest half.* This is the most dense 50% of the sample points.

The quantile box plots in Figure 4.16 show at a glance that Sample A comes from a normally distributed sample and Sample B has a possible outlier. The box and the shortest half for Sample A are in the middle of the range of values, the mean and median are the same, and the whiskers are symmetrical. However, the outlying value in Sample B causes the distribution to be compressed at the upper end of the scale.

Sample C and Sample D were normally shaped but peaked and flat. For these kinds of distribution, the quantile box plot only shows that the shortest half (most dense) points are more confined for Sample D and more spread out for Sample E.

The positive and negative skewness in Sample E and Sample F are clearly shown. The box, the shortest half, and the median are visibly shifted in the direction of the skewness.

Understanding the Stem-and-Leaf Plot

The stem-and-leaf plot is another visual representation of data, similar to the histograms in Figure 4.10, Figure 4.11, and Figure 4.12. The stem-and-leaf plots for Sample A (the approximately normal distribution) and Sample B, the distribution with the outlier, are shown below their respective histograms in Figure 4.17.

Figure 4.17 Stem-and-Leaf Plot from Distribution Platform for Samples A and B

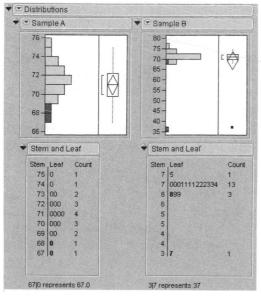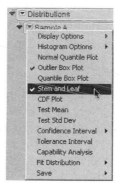

To see the stem-and-leaf plot example, run the Distribution platform again for both Sample A and Sample B. Use options from the menu on the response variable title bar to hide the Moments and Quantiles tables that are usually displayed by default. Then select the **Stem and Leaf** option from that menu to see the results shown in Figure 4.17.

> Note: To apply any option to all response variables in an analysis window at the same time, control-click (command-click on the Macintosh) and select an option from he menu on the title bar of any response variable. You only need to do this once to affect all variables.

Results for a Normal Distribution and a Distribution with an Outlier

To understand a stem-and-leaf plot, it is necessary to think of a given subject's age score as consisting of a *stem* and a *leaf*. The vertical line in the stem-and-leaf plot separates the stem from the leaf. The stem is that part of the value that appears to the left of the line, and the leaf is the part that appears to the right.

You reconstruct the data values by joining the stem values and leaf values according to the legend at the bottom of the plot. For example, subject 18 in Sample A has a value of 67. In Figure 4.17, the stem for this subject is 67 and appears to the left of the vertical line. The leaf is 0 and appears to the right of the vertical line. Subject 12 has a stem value of 68 and a leaf value of 0, which together represent the data value of 68.0.

In the stem-and-leaf plot of Figure 4.17, the numbers on the vertical axis plot the various values that could be encountered in the data table. These appear under the heading Stem. Reading from the top down, the stems are 75, 74, 73, and so forth. Notice that the values 67 and 68 have a single leaf (a single 0). This means that there is only one subject in Sample A with a stem-and-leaf value of 67.0 (an age value of 67) and one with 68.0 (an age value of 68.0). Move up an additional line, and you see the stem 69. To the right of this, two leaves appear (that is, two zeros appear). This means that there are two subjects with a stem-and-leaf value of 69.0 (two subjects with age values of 69.0). Continuing up the plot in this fashion, you can see that there are three subjects at age 70, four subjects at age 71, three at age 72, two at age 73, one at 74, and one at 75.

On the right side of the stem-and-leaf plot appears a column named Count. This column prints the number of observations that appear at each stem. Reading from the bottom up, this column again confirms that there was one subject with a score of age 67, one with a score of 68, two with a score of 69, and so forth.

The stem-and-leaf plot actively responds to clicking and to the brush tool. Clicking on the two lowest values in the stem-and-leaf plot for Sample A produces the highlighting shown in Figure 4.17, and also highlights the corresponding rows in the data table. This interactive graphic feature gives another way to quickly identify specific observations of interest.

Reviewing the stem-and-leaf plot of Figure 4.17 shows that its shape is similar to the histogram shape portrayed for Sample A in Figure 4.10. This is to be expected because both figures use similar conventions and both describe the data of Sample A. Notice in Figure 4.17 that the shape of the distribution is symmetrical, without a tail in either direction. This, too, is to be expected because Sample A demonstrates zero skewness.

Often, the stem-and-leaf plot is more complex than that for Sample A. For example, the chart on the right in Figure 4.17 is the stem-and-leaf plot produced by Sample B (the distribution with an outlier). Consider the stem-and-leaf representation at the bottom of this plot. The stem for this entry is 3, and the leaf is 7. Notice the legend at the bottom of this plot, which says "3|7 represents 37."

> **Note:** It is always important to read the explanation at the bottom of the stem-and-leaf plot that explains how to translate the stem-and-leaf values into the data values they represent.

Move up one line in the plot, and you see the stem "4." However, there are no leaves for this stem, which means that there were no subjects with a stem of 4|0 (40). Reading up the plot, no leaves appear until you reach the stem "6." The leaves on this line suggest that there was one subject with a stem-and-leaf value of 6|8, and two subjects with a stem-and-leaf value of 6|9, translating to age values of 68 and 69.

Move up an additional line, and note that there are two stems for the value 7. The first stem (moving up the plot) includes stem-and-leaf values from 7|0 through 7|4, while the next stem includes stem-and-leaf values from 7|5 through 7|9. Reviewing values in these rows, you can see that there are three subjects with a stem-and-leaf value of 7|0 (70), four with a stem-and-leaf value of 7|1 (71), and so forth.

Results for Distributions with Positive Skewness and Negative Skewness

Figure 4.18 provides some results from the Distribution analysis of Sample E, shown previously in Figure 4.12. Recall that Sample E demonstrates a positive skew and Sample F shows a negative skew.

Remember that the approximately normal distribution has a skewness index of zero. In contrast, note that the skewness index for Sample E in Figure 4.18 is 0.8698 (about 0.87). This positive skewness index is what you expect given the positive skew of the data. The skew is also reflected in the stem-and-leaf plot that appears with its histogram. Notice that the relatively long tail points in the direction of higher values for age (such as 74 and 76).

Although Sample E displays positive skewness, it does not display a significant departure from normality. In the goodness-of-fit test you can see that the Shapiro-Wilk W statistic is 0.93, with a corresponding *p* value of 0.1909. Because this *p* value is greater than 0.05, you cannot reject the null hypothesis that the data were drawn from a normal population. With small samples such as this one, the test for normality is not very powerful (that is, not very sensitive). This lack of power is why the sample was not found to display a significant departure from normality, even though it is clearly skewed.

For purposes of contrast, the results in Figure 4.19 are an analysis of Sample F, which has a negative skewness index of –0.87, shown in the Moments table. However, once again the Shapiro-Wilk test shows that the sample does not demonstrate a significant departure from normality. The stem-and-leaf plot reveals a long tail that points in the direction of lower age values for Sample F (such as 64 and 66), which is what you expect for a negatively skewed distribution. Again, the Shapiro-Wilk test for normality does not produce a significant result, even though Sample F is clearly skewed toward lower values.

The next section is a step-by-step description of how to produce the results shown in Figure 4.18 and Figure 4.19.

Figure 4.18 Sample E (Positive Skewness)

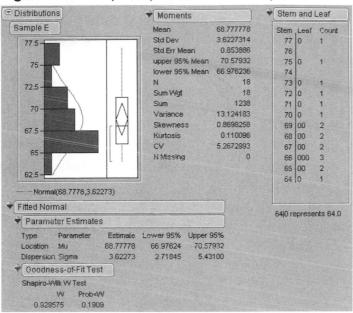

Figure 4.19 Sample F (Negative Skewness)

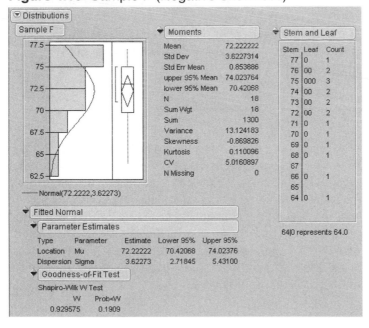

A Step-by-Step Distribution Analysis Example

To see the results in Figure 4.18, perform a distribution analysis on Sample E and Sample F in the **ages** data table and arrange the graphs and tables in a JMP layout window:

- 🖑 Do a Distribution analysis on Sample E and Sample F. That is, with the **ages** data table active, choose **Analyze → Distribution**, and select **Sample E** and **Sample F** as response variables, then click **OK** in the launch dialog.

- 🖑 Use the menu found on the title bar of the response variables to suppress the Quantiles table for both response variables (select **Display Options → Quantiles** and deselect **Quantiles**).

- 🖑 Select **Stem and Leaf** from the menu on the title bar of the response variables.

- 🖑 Select **Fit Distributions → Normal** from the menu bar for both response variables. This command overlays a fitted normal curve on the histogram, shows the estimated mean and standard deviation for the fitted normal distributions, and appends the Parameter Estimates table to the bottom of the results.

- 🖑 For both response variables, use the menu found on the Fitted Normal title bar at the bottom of the results and select **Goodness of Fit**. This command appends the Goodness of Fit table for the fitted normal distribution to the distribution analysis.

- 🖑 Finally, use the main menu bar and select **Edit → Layout**. All the results appear in a new layout window. Click anywhere in the background area to select all the results and use **Layout → Ungroup** as many times as needed to separate (ungroup) the tables. Then move the tables around on the layout surface to form any arrangement you want.

Summary

This chapter describes the importance of performing two tasks at the beginning of any investigation that involves data: screening your data to ensure better results when you analyze them, and exploring the shape of your data by performing simple, descriptive analyses to help you obtain meaningful results in later analyses. Screening and exploring your data help ensure that the data do not contain any errors that could lead to incorrect conclusions, and ultimately even to retractions of published findings. The JMP Distribution platform provides the tools you need to screen and analyze your data. Once the data undergo this initial screening you can move forward to the more sophisticated procedures described in the remainder of this book.

References

Rummel, R. J. 1970. *Applied Factor Analysis.* Evanston, IL: Northwestern
   University Press.
SAS Institute Inc. 2003. *JMP Statistics and Graphics Guide.* Cary, NC: SAS Institute Inc.

Measures of Bivariate Association

> **Overview.** This chapter discusses procedures to test the significance of the relationship between two variables. Recommendations are made for choosing the correct statistic based on the level of measurement (data type and modeling type) of the variables. The chapter shows how to use the JMP Fit Y by X platform to prepare bivariate scatterplots and perform the chi-square test of independence, and how to use the JMP Multivariate platform to compute Pearson correlations and Spearman correlations.

Significance Tests versus Measures of Association

This section defines several fundamental statistical terms. There are a large number of statistical procedures that you can use to investigate *bivariate relationships*. These procedures can provide a *test of statistical significance*, a *measure of association*, or both.

- A *bivariate relationship* involves the relationship between just two variables. For example, if you conduct an investigation in which you study the relationship between SAT verbal test scores and college grade point average (GPA), you are studying a bivariate relationship.

- A *test of statistical significance* allows you to test hypotheses about the relationship between the variables in the population. For example, the Pearson correlation coefficient tests the null hypothesis that the correlation between two continuous numeric variables is zero in the population. Suppose you draw a sample of 200 subjects and determine that the Pearson correlation between the SAT verbal test and GPA is 0.35 for this sample. You can then use this finding to test the null hypothesis that the correlation between SAT verbal and GPA is zero in the population. If the resulting test is significant at $p < 0.001$, this p value suggests that there is less than 1 chance in 1,000 of obtaining a sample correlation of 0.35 or larger if the null hypothesis is true. You therefore reject the null hypothesis, and conclude that the correlation is probably not equal to zero in the population.

- A *Pearson correlation* can also serve as a *measure of association*. A measure of association reflects the strength of the relationship between variables (regardless of the statistical significance of the relationship). For example, the absolute value of a Pearson correlation coefficient reveals the strength of the linear relationship between two variables. A Pearson correlation ranges from –1.00 through 0.00 to +1.00, with larger absolute values indicative of stronger linear relationships. For example, a correlation of 0.00 indicates no linear relationship between the variables, a correlation of 0.20 (or –0.20) indicates a weak linear relationship, and a correlation of 0.90 (or –0.90) indicates a strong linear relationship. A later section in this chapter provides more detailed guidelines for interpreting Pearson correlations.

Choosing the Correct Statistic

When investigating the relationship between two variables, there is an appropriate bivariate statistic, given the nature of the variables studied. In all analysis situations, JMP platforms use the modeling types of the variables you select for analysis and automatically compute the correct statistics. To anticipate and understand these statistics, you need to understand the level of measurement or JMP modeling type of the two variables.

Levels of Measurement and JMP Modeling Types

Chapter 1 briefly described the four levels of measurement used with variables in social science research: nominal, ordinal, interval, and ratio. These levels of measurement are equivalent to the modeling types assigned to all JMP variables.

- A variable measured on a *nominal scale* has a **nominal modeling type** in JMP and is a *classification* variable. Its values indicate the group to which a subject belongs. For example, race is a nominal variable if a subject could be classified as African American, Asian American, Caucasian American, and so forth.

- A variable measured on an *ordinal scale* has an **ordinal modeling type** in JMP and is a *ranking* variable. An ordinal variable indicates which subjects have more of the attribute being assessed and which subjects have less. For example, if all students in a classroom are ranked with respect to verbal ability so that the best student is assigned the value 1, the next best student is assigned the value 2, and so forth, this verbal ability ranking is an ordinal variable. However, ordinal scales have the characteristic that equal differences in scale value do not necessarily have equal quantitative meaning. In other words, the difference in verbal ability between student 1 and student 2 is not necessarily the same as the difference in ability between student 2 and student 3.

- An *interval scale* of measurement has a **continuous modeling type** in JMP. With an interval scale, equal differences in scale values do have equal quantitative meaning. For example, imagine that you develop a test of verbal ability. Scores on this test can range from 0 through 100, with higher scores reflecting greater ability. If the scores on this test are truly on an interval scale, then the difference between scores of 60 and 70 is equal to the difference between scores of 70 and 80. In other words, the unit of measurement is the same across the full range of the scale. Nonetheless, an interval-level scale does not have a true zero point. Among other things, this means that a score of zero on the test does not necessarily indicate a complete absence of verbal ability.

- A *ratio scale* also has a **continuous modeling type** in JMP. It has all of the properties of an interval scale and also has a true zero point. This makes it possible to make meaningful statements about the ratios between scale values. For example, body weight is assessed on a ratio scale: a score of 0 pounds indicates no body weight at all. With this variable, it is possible to state that a person who weighs 200 pounds is twice as heavy as a person who weighs 100 pounds. Other examples of ratio-level variables are age, height, and income.

Important: Remember that, in JMP, each column has a modeling type corresponding to a level of measurement. The modeling types are nominal, ordinal, and continuous, where continuous includes both the interval and ratio scales of measurement. In further discussion, levels of measurement are referred to as modeling types. The modeling types are given to the variable when it is created. You can change a variable's modeling type at any time using the Column Info dialog. Or, click on the modeling type icon next to the variable name in the Columns panel and choose a modeling type from the menu that appears. Changing a variable's modeling type can change the type of analysis generated by JMP.

A Table of Appropriate Statistics

If you know the modeling type of the variables, you can anticipate the correct statistic that JMP generates for analyzing their relationship. However, the actual situation is a bit more complex than Table 5.1 suggests because for some sets of variables, there can be more than one appropriate statistic that investigates their relationship. This chapter presents only a few of the most commonly used statistics. To learn about additional procedures and the special conditions under which they might be appropriate, consult a more comprehensive statistics textbook such as Hays (1988).

Table 5.1 identifies some of the analyses and statistics appropriate for pairs of variables with given modeling types (levels of measurement). The vertical columns of the table indicate the modeling type of the predictor (independent) variable of the pair. The horizontal rows are the modeling type of the response (predicted or criterion) variable. The appropriate test statistic for a given pair of variables is identified where a given row and column intersect, and JMP picks the correct statistic to use based on modeling types.

> **Note:** Keep in mind that for some contingency table situations and for correlations, both variables are considered as responses. Neither is a predictor.

Table 5.1 Statistics for Pairs of Variables

		Predictor Variable		
	Modeling Type	**Nominal**	**Ordinal**	**Continuous**
Response Variable *(Dependent or Criterion Variable)*	**Nominal**	contingency table chi-square test	Spearman correlation	nominal logistic chi-square test
	Ordinal	Kruskal-Wallis or contingency table chi-square test	Spearman correlation or contingency table chi-square test	ordinal logistic chi-square test
	Continuous (interval or ratio scale)	ANOVA *F* or *t* test	contingency table chi-square test	Pearson or Spearman correlation

The Chi-Square Test

When you have a nominal predictor variable and a nominal response variable, Table 5.1 indicates that a chi-square test is an appropriate statistic. To illustrate this situation, imagine that you are conducting research that investigates the relationship between geographic region of voters in the United States and political party affiliation. Assume that you have hypothesized that people who live in the Midwest are more likely to belong to the Republican Party, relative to those who live in the East or those who live in the West. To test this hypothesis, you gather data on two variables for each of 1,000 subjects. The first variable is U.S. geographic region, and each subject is coded as living either in the East, Midwest, or West. Geographic region is therefore a nominal variable that can assume one of those three values. The second variable is party membership, and each subject is coded as either being a Republican, a Democrat, or Other. Party membership is also a nominal variable that assumes one of three values.

A chi-square test with a significantly large chi-square value indicates a relationship between geographic region of voters and party affiliation. Close inspection of the two-way classification table that is produced in this analysis then indicates whether those in the Midwest are more likely to be Republicans.

The Spearman Correlation Coefficient

Table 5.1 recommends the Spearman correlation coefficient when the predictor and response variables are ordinal. An ordinal variable usually has numeric values or can be created as a numeric variable and analyzed using a continuous modeling type. As an illustration, imagine that an author has written a book that ranks the 100 largest universities in the U.S. from best to worst so that the best school ranks first (1), the second best ranks second (2), and so forth. In addition to providing an overall ranking for the institutions, the author also ranks them from best to worst with respect to a number of specific criteria such as intellectual environment, prestige, quality of athletic programs, and so forth. Assume the hypothesis that the universities' overall rankings demonstrate a strong positive correlation with the rankings of the quality of their library facilities.

To test this hypothesis, you could compute the correlation between two variables (their overall ranks and the ranks of their library facilities). Both variables are ordinal because both tell you about the numeric ordinal rankings of the universities. For example, the library facilities variable could tell you that one university has a more extensive collection of books than another university, but it does not tell you how much more extensive. Because both variables are on a numeric ordinal scale, it is appropriate to create them as numeric variables, assign them continuous modeling types, and analyze the relationship between them with the Spearman correlation coefficient.

The Pearson Correlation Coefficient

Table 5.1 shows that, when both the predictor and response are continuous variables (either the interval or ratio level of measurement), the Pearson correlation coefficient could be appropriate. For example, assume that you want to test the hypothesis that income is positively correlated with age. That is, you predict that older people tend to have more income than younger people. To test this hypothesis, suppose you obtain data from 200 subjects on both variables. The first variable is age, which is the subject's age in years. The second variable is income, which is the subject's annual income in dollars. Income can assume values such as $0, $10,000, $1,200,000, and so forth.

Both variables in this study are continuous numeric variables. In fact, they are assessed on a ratio scale because equal intervals have equal quantitative meaning and each has a true zero point—zero years of age and zero dollars of income indicate complete absence of age and income. A Pearson correlation is often the correct way to examine the relationship between these kinds of variables. The Pearson correlation is also based on other assumptions that are discussed in a later section and are also summarized at the end of this chapter.

Other Statistics

The statistics identified on the diagonal of Table 5.1 are used when both predictor and response variable have the same modeling type. For example, you should use the chi-square test when both variables are nominal or ordinal variables. The situation becomes more complex when the variables have different modeling types. For example, when the response variable is continuous and the predictor variable is nominal, the appropriate procedure is ANOVA (ANalysis Of VAriance). This chapter only discusses the Pearson correlation coefficient, the Spearman correlation coefficient, and the chi-square test. Analysis of variance (ANOVA) is covered in later chapters of this book. The Kruskal-Wallis test is not covered, but is described in other JMP documentation (see the *JMP Statistics and Graphics Guide* (2003)).

Remember that Table 5.1 presents only some of the appropriate statistics for a given combination of variables. There are many exceptions to these general guidelines. For example, Table 5.1 assumes that all nominal variables have only a relatively small number of values (perhaps two to six), that all continuous variables can take on a large number of values, and that ordinal variables might have either a small number of categorical values or a larger number of numeric values. When these conditions do not hold, the correct statistic can differ from those indicated by the table. For example, the F test produced in an ANOVA (instead of a correlation analysis) might be the appropriate statistic when both response and predictor variables are continuous. This is the case when there is an experiment in which the response variable is number of errors on a memory task and the predictor is amount of caffeine administered: 0 mg versus 100 mg versus 200 mg. The predictor variable (amount of caffeine administered) is theoretically continuous, but the correct statistic is ANOVA because this predictor has only three values. In this situation, the modeling type for caffeine administered should be assigned as nominal or ordinal for the analysis. As you read about the examples of research throughout this text, you will develop a better sense of what JMP platforms to use and how to interpret the statistics generated by JMP for a given study.

Section Summary

The purpose of this section was to provide a simple strategy for understanding measures of bivariate association under conditions often encountered in scientific research. The following section reviews information about the Pearson correlation coefficient, and shows how to use the JMP Fit Y by X platform to see scatterplots that verify the linear relationship between two variables. Subsequent sections provide a more detailed discussion of the three statistics emphasized in this chapter: the Pearson correlation, the Spearman correlation, and the chi-square test. For each of these statistics, there is an example study that shows how to do the correct JMP analysis and explains how to interpret the results.

Pearson Correlations

When to Use

You can use the Pearson product-moment correlation coefficient (symbolized as r) to examine the nature of the relationship between two variables when both variables are continuous. It is also assumed that both variables have a large number of values.

For example, it is appropriate to compute a Pearson correlation coefficient to investigate the nature of the relationship between SAT verbal test scores and grade point average (GPA). SAT Verbal is a continuous variable that can assume a wide variety of values (possible scores range from 200 through 800). Grade point average is also a continuous variable, and can also assume a wide variety of values ranging from 0.00 through 4.00. Figure 5.1 shows a partial listing of hypothetical data, **satgpa.jmp**, used in the following examples.

The Pearson correlation assumes the variables are numeric with a continuous modeling type and that the observations have been drawn from normally distributed populations. When one or both variables display a non-normal distribution (for example, when one or both variables are markedly skewed), it might be more appropriate to analyze the data with the Spearman correlation coefficient. A later section of this chapter discusses the Spearman coefficient. A summary of the assumptions of both the Pearson and Spearman correlations appears at the end of the chapter.

Figure 5.1 Partial Listing of the satgpa.jmp Data Table

	SAT Verbal	GPA	Verbal Errors	Creativity
1	211	0.44	50	5
2	234	0.98	48	89
3	241	0.57	45	69
4	241	1.20	39	112
5	243	0.90	40	100
6	250	0.82	41	125
7	300	1.15	38	165
8	310	0.89	41	190
9	320	0.99	34	175
10	326	1.34	35	210
11	370	1.27	37	260
12	380	1.54	44	285

satgpa
Columns (4/0)
 SAT Verbal
 GPA
 Verbal Errors
 Creativity

Rows
All Rows   50
Selected   0
Excluded   0
Hidden   0
Labelled   0

How to Interpret the Pearson Correlation Coefficient

To better understand the nature of the relationship between the two variables in this example, it is necessary to interpret two characteristics of a Pearson correlation coefficient.

First, the *sign* of the coefficient tells you whether there is a positive linear relationship or a negative linear relationship between the two variables.

- A *positive correlation* indicates that as values for one variable increase, values for the second variable also increase. The positive correlation illustrated on the left in Figure 5.2 shows the relationship between SAT verbal test scores and GPA in a fictitious sample of data. You can see that subjects who received low scores on the predictor variable (SAT Verbal) also received low scores on the response variable (GPA). At the same time, subjects who received high scores on SAT Verbal also received high scores on GPA. Therefore, these variables are positively correlated.

- With a *negative correlation*, as values for one variable increase, values for the second variable decrease. For example, you might expect to see a negative correlation between SAT verbal test scores and the number of errors that subjects make on a vocabulary test. The students with high SAT verbal scores tend to make few mistakes, and the students with low SAT scores tend to make many mistakes. This negative relationship is illustrated on the right in Figure 5.2.

Figure 5.2 Positive Correlation (left) and Negative Correlation (right)

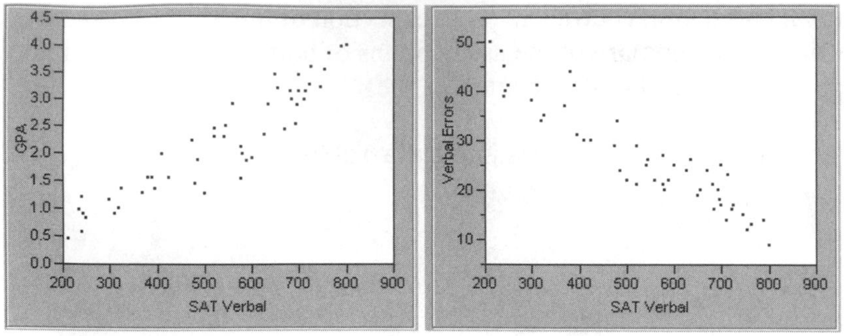

The second characteristic of a correlation coefficient is its *size*. Larger absolute values of a correlation coefficient indicate a stronger relationship between the two variables. Pearson correlation coefficients range in size from –1.00 through 0.00 to +1.00. Coefficients of 0.00 indicate no relationship between two variables. For example, if there is zero correlation between SAT scores and GPA, then knowing a person's SAT score tells you nothing about that person's GPA. In contrast, correlations of –1.00 or +1.00 indicate perfect relationships. If the correlation between SAT scores and GPA is 1.00, then knowing someone's SAT score allows you to predict that person's GPA with perfect accuracy. In the real world, SAT scores are not strongly related to GPA. The correlation between them is less than 1.00.

The following is an approximate guide for interpreting the strength of the linear relationship between two variables, based on the absolute value of the coefficient:

- ±1.00 = perfect correlation
- ±0.80 = strong correlation
- ±0.50 = moderate correlation
- ±0.20 = weak correlation
- ±0 = no correlation

You should always consider the magnitude of correlation coefficients as well as whether or not coefficients are statistically significant. This is because significance estimates are strongly influenced by sample sizes. For instance, an *r* of 0.15 (weak correlation) would be statistically significant with samples in excess of 700, whereas an *r* of 0.50 (moderate correlation) would not be statistically significant with a sample of only 15 participants.

Remember that you consider the *absolute value* of the coefficient when interpreting its size. This means that a correlation of –0.50 is just as strong as a correlation of +0.50, a correlation of –0.75 is just as strong as a correlation of +0.75, and so forth.

Linear versus Nonlinear Relationships

The Pearson correlation is appropriate only if there is a *linear relationship* between the two variables. There is a linear relationship between two variables when their scatterplot follows the form of a straight line. For example, a straight line through the points in the scatterplot on the left in Figure 5.3 fits the pattern of the data. This means that there is a linear relationship between SAT verbal test scores and GPA.

In contrast, there is a nonlinear *relationship* between two variables if their scatterplot does not follow a straight line. For example, imagine that you have a test of creativity and have administered it to a large sample of college students. With this test, higher scores reflect higher levels of creativity. Imagine further that you obtain the SAT verbal test scores for these students, plot their SAT scores against their creativity scores, and obtain the scatterplot on the right in Figure 5.3. This scatterplot shows a nonlinear relationship between SAT scores and creativity:

- Students with low SAT scores tend to have low creativity scores.

- Students with moderate SAT scores tend to have high creativity scores.

- Students with high SAT scores tend to have low creativity scores.

It is not possible to draw a straight line through the data points in the scatterplot of SAT scores and creativity shown on the right in Figure 5.3 because there is a nonlinear (or curvilinear) relationship between these variables.

Figure 5.3 A Linear Relationship (left) and a Nonlinear Relationship (right)

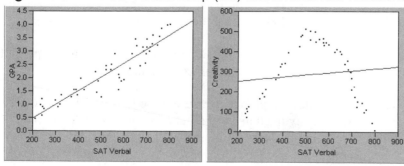

The Pearson correlation should not be used to examine the relationship between two variables involved in a nonlinear relationship because the correlation coefficient that results usually underestimates the actual strength of the relationship between the variables. The Pearson correlation between the SAT scores and creativity scores presented on the right in Figure 5.3 would indicate a very weak relationship between the two variables (the fitted straight line is nearly horizontal). However, the scatterplot shows a strong linear relationship between SAT scores and creativity, which means that for a given SAT score, you can accurately predict the creativity score.

You should verify that there is a linear relationship between two variables before computing a Pearson correlation for those variables. An easy way to verify that the relationship is linear is to prepare a scatterplot similar to those presented in Figure 5.3. This is easily done using the Fit Y by X platform in JMP, described in the next section.

Producing Scatterplots with the Fit Y by X Platform

To illustrate a bivariate analysis (an analysis between two continuous variables), imagine you have conducted a study dealing with an *investment model* that is a theory of commitment in romantic associations. The investment model identifies a number of variables that are believed to influence a person's commitment to a romantic association. Commitment refers to the subject's intention to remain in the relationship. In the following example, the variable called Commitment is the response variable. These are some of the variables that are thought to influence subject commitment (predictor variables):

> Satisfaction—the subject's affective response to the relationship

> Investment Size—the amount of time and personal resources that the subject has put into the relationship

> Alternative Value—the attractiveness of the subject's alternatives to the relationship (e.g., the attractiveness of alternative romantic partners).

Use scatterplots of the response variable plotted on the vertical axis and the predictor variables plotted on the horizontal axis to explore correlations.

Assume you have a 16-item questionnaire to measure these four variables. Twenty participants who are currently involved in a romantic relationship complete the questionnaire while thinking about their relationship. When they complete the questionnaire, their responses are used to compute four scores for each participant:

- First, each participant receives a score on the *commitment scale*. Higher values on the commitment scale reflect greater commitment to the relationship.

- Each participant also receives a score on the *satisfaction scale*, where higher scores reflect greater satisfaction with the relationship.

- Higher scores on the *investment scale* mean that the participant believes that he or she has invested a great deal of time and effort in the relationship.

- Finally, with the *alternative value scale* higher scores mean that it would be attractive to the respondent to find a different romantic partner.

Figure 5.4 lists the (fictitious) results in a JMP data table called **investment model.jmp**.

Figure 5.4 JMP Data Table of the Commitment Data

	Commitment	Satisfaction	Investment Size	Alternative Value
1	20	20	28	21
2	10	12	5	31
3	30	33	24	11
4	8	10	15	36
5	22	18	33	16
6	31	29	33	12
7	6	10	12	29
8	11	12	6	30
9	25	23	34	12
10	10	7	14	32
11	31	36	25	5
12	5	4	18	30
13	31	28	23	6
14	4	6	14	29
15	36	33	29	6
16	22	21	14	17
17	15	17	10	25
18	19	16	16	22
19	12	14	18	27
20	24	21	33	16

Columns (4/0): Commitment, Satisfaction, Investment Size, Alternative Value

Rows: All Rows 20, Selected 0, Excluded 0, Hidden 0, Labelled 0

You can use the **Fit Y by X** command on the **Analyze** menu to prepare scatterplots for various combinations of variables. For a hands-on example, open the **investment model.jmp** table and follow the mouse steps.

- To begin, choose **Analyze → Fit Y by X**, which launches the dialog in Figure 5.5.

- Select **Commitment** and click the **Y, Response** button (or drag **Commitment** from the variable list on the left of the dialog to the **Y, Response** box).

- Likewise, select two of the predictor variables, **Satisfaction** and **Alternative Value**, and click the **X, Factor** button.

- Click **OK** to see the bivariate scatterplots of Commitment by each of the X variables.

Note: There is always help available. The **Help** button in the launch dialog tells how to complete the dialog. After the results appear, choose the Help tool (question mark) from the **Tools** menu or Tools palette and click anywhere in the results.

Figure 5.5 Launch Dialog for the JMP Fit Y by X Platform

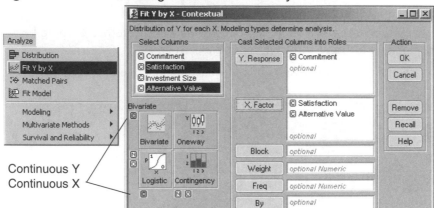

The scatterplot on the left in Figure 5.6 shows the initial results for Commitment by Satisfaction. Notice that the response variable (Commitment) is on the vertical axis and the predictor variable (Satisfaction) is on the horizontal axis. The shape of the scatterplot shows a linear relationship between Satisfaction and Commitment because it is possible to draw a good-fitting straight line through the center of the points.

The general shape of the scatterplot also suggests a fairly strong relationship between the two variables. Knowing where a subject stands on the Satisfaction variable allows you to predict, with some accuracy, where that subject stands on the Commitment variable.

Figure 5.6 also shows that the relationship between Satisfaction and Commitment is positive. Large Satisfaction values are associated with large Commitment values, and small Satisfaction values are associated with small Commitment values. This makes intuitive sense—you expect subjects who are highly satisfied with their relationships to be highly committed to those relationships.

To illustrate a negative relationship, look at the plot of Commitment by Alternative Value, shown on the right in Figure 5.6. Notice that the relationship between these two variables is negative. As you might expect, subjects who indicate that the alternatives to their current romantic partner are attractive are not terribly committed to the current partner.

Figure 5.6 Scatterplots of Commitment Scores with Satisfaction and Alternative Value

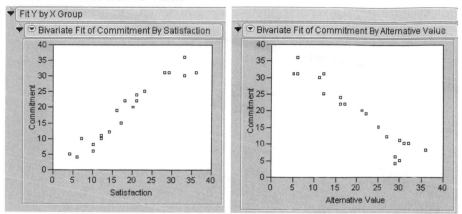

Computing Pearson Correlations

The relationships between Commitment by both Alternative Value and Satisfaction appear to be linear, so it is appropriate to assess the strength of these relationships with the Pearson correlation coefficient.

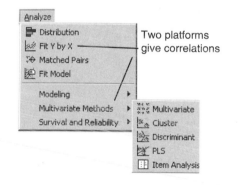

Two commands on the **Analyze** menu, **Fit Y by X** and **Multivariate Methods** → **Multivariate**, compute the Pearson Correlation coefficient. The Fit Y by X platform computes a single correlation coefficient. The Multivariate platform offers options to compute different types of correlation coefficients.

Computing a Single Correlation Coefficient

In some instances, you might want to compute the correlation between just two variables. A simple way to do this uses the Fit Y by X platform and the **Density Ellipse** option available with the bivariate scatterplot. Figure 5.7 shows the scatterplot of Commitment and Satisfaction with a 95% density ellipse overlaid on the plot. The density ellipse is computed from the sample statistics to contain an estimated 95% of the points. In addition, this option appends the Correlation table to the plot.

The names of the variables are listed first in the table, and the statistics for the variables appear to the right of the variable names. These descriptive statistics show the means and standard deviations of each variable, the Pearson correlation coefficient between them, and the number of values used to compute the descriptive statistics and correlation.

Figure 5.7 Computing the Pearson Correlation between Commitment and Satisfaction

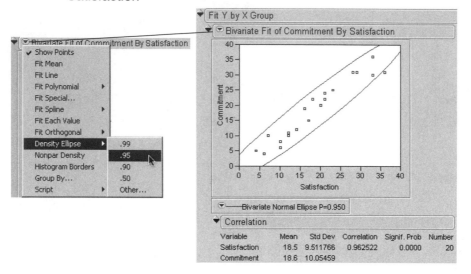

In this example the mean for Commitment is 18.6 and the standard deviation is 10.05. The mean for Satisfaction is 18.5 and its standard deviation is 9.51. There are no missing values for either variable so all 20 observations are used to compute the statistics. Observations with a missing value for either variable are not used in the analysis.

The Pearson correlation coefficient between Commitment and Satisfaction is 0.96252 (you can round this to 0.96). The *p* value associated with the correlation, labeled Signif. Prob, is next to the correlation. This is the *p* value obtained from a test of the null hypothesis that the correlation between Commitment and Satisfaction is zero in the population. Recall that a correlation value of zero means there is no relationship, or no predictive value, between two variables. More technically, the *p* value gives the probability of obtaining a sample correlation this large (or larger) if the correlation between Commitment and Satisfaction were really zero in the population.

For the correlation of 0.96, the corresponding *p* value is less than 0.0001, which shows in the Correlation table as 0.0000. This means that, given the sample size of 20, there is less than 1 chance in 10,000 of obtaining a correlation of 0.96 or larger if the population correlation is really zero. You can therefore reject the null hypothesis, and tentatively conclude that Commitment is related to Satisfaction in the population.

The alternative hypothesis for this statistical test is that the correlation is not zero in the population. This alternative hypothesis is two-sided, which means that it does not indicate that the population correlation is positive or negative, only that it is not equal to zero.

Computing All Possible Correlations for a Set of Variables

Use the Multivariate platform when you have multiple variables and want to know the correlations between each pair.

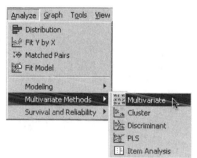

     Choose **Analyze → Multivariate Methods → Multivariate**, as shown here.

When you choose this command, a launch dialog asks you to select the variables to analyze. This platform does not distinguish between response and predictor variables. It looks only at the relationship between pairs of variables.

Figure 5.8 shows the initial results for the four variables Commitment, Satisfaction, Alternative Value, and Investment Size, beginning with the Pearson correlation coefficients for each pair. The scatterplot matrix is a visual representation of correlations. Each scatterplot has a 95% density ellipse imposed. Like the ellipse seen in the Fit Y by X platform, it is computed from the sample statistics to contain an estimated 95% of the points. The thinner, more diagonal ellipse indicates greater correlation. The menu on the Scatterplot Matrix title bar lets you change the ellipse percent (that is, 99%, 90%, or other) and change its color.

Figure 5.8 Computing All Possible Pearson Correlations

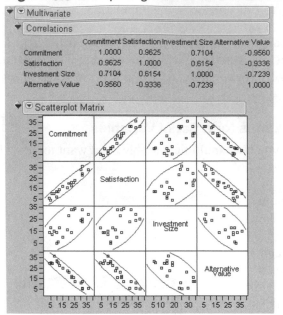

The scatterplot matrix is interactive. You can highlight points with the pointer, brush tool, or lasso tool in any plot and see them highlighted in all other plots and in the data table. The scatterplot matrix can be resized by dragging on any boundary. You can rearrange the order of the variables in the matrix by dragging any cell on the diagonal to a different position on the diagonal.

> **Note:** The Correlations table shows correlations computed only for the set of observations that have values for all variables.

🖱 Next, select the **Pairwise Correlations** option from the Multivariate title bar menu.

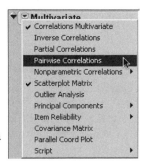

This option appends a new set of correlations at the bottom of the scatterplot matrix. The values in the Pairwise Correlations table are computed using all available values for each pair of variables. The count values among the pairs differ if any pair has a missing value for either variable. These are the same correlations given by the **Density Ellipse** option on the Fit Y by X platform. Figure 5.9

shows the Pairwise Correlations table for this example. If there had been missing values in the data, these correlations would differ from those in Figure 5.9, which use only the set of observations that have values for all variables.

Figure 5.9 Pairwise Pearson Correlations with Significance Probabilities and Bar Chart

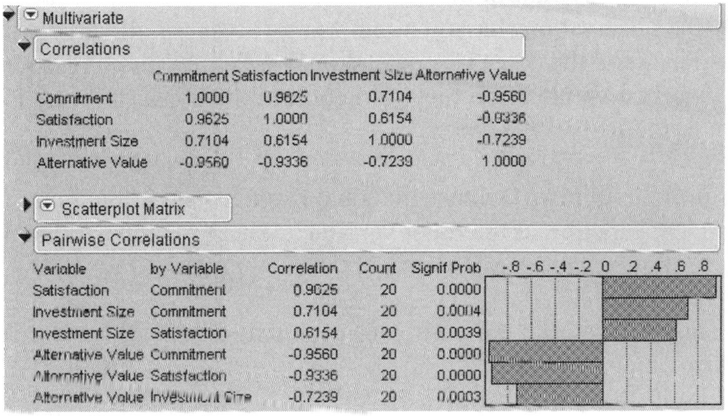

Multivariate

Correlations

	Commitment	Satisfaction	Investment Size	Alternative Value
Commitment	1.0000	0.9625	0.7104	-0.9560
Satisfaction	0.9625	1.0000	0.6154	-0.9336
Investment Size	0.7104	0.6154	1.0000	-0.7239
Alternative Value	-0.9560	-0.9336	-0.7239	1.0000

Scatterplot Matrix

Pairwise Correlations

Variable	by Variable	Correlation	Count	Signif Prob
Satisfaction	Commitment	0.9625	20	0.0000
Investment Size	Commitment	0.7104	20	0.0004
Investment Size	Satisfaction	0.6154	20	0.0039
Alternative Value	Commitment	-0.9560	20	0.0000
Alternative Value	Satisfaction	-0.9336	20	0.0000
Alternative Value	Investment Size	-0.7239	20	0.0003

You can interpret the correlations and *p* values in this output in exactly the same way as you interpret the Fit Y by X results. For example, the Pearson correlation between Investment Value and Commitment is 0.7104, with the count of 20 used for the computation. However, remember that when requesting correlations between several different pairs of variables, some correlations might be based on more subjects than others (due to missing data). In that case, the sample sizes printed for each individual correlation coefficient will be different. Finally, the *p* value of 0.0004 for these variables means that there are only 4 chances in 10,000 of observing a sample correlation this large if the population correlation is really zero. By convention, the observed correlation is said to be statistically significant if the *p* value is less than 0.05.

Notice that the bar chart to the right of the values shows the pattern of correlations, which supports some of the predictions of the investment model: Commitment is positively related to Satisfaction and Investment Size, and is negatively related to Alternative Value. With respect to strength, the correlations (for these fictitious data) range from being moderately strong to very strong.

Options Used with the Multivariate Platform

The following list summarizes the items used in the previous example, and a few of the other Multivariate platform options that you might find useful when conducting research.

Correlations Multivariate gives a matrix of Pearson (product-moment) correlation coefficients using only the observations having nonmissing values for all variables in the analysis. These correlations (and the pairwise correlations) are a measure of association that summarizes how closely a variable is a linear function of the other variables in the multivariate analysis.

Inverse Correlations provides a matrix whose diagonal elements are a function of how closely a variable is a linear function of the other variables. These diagonal elements are also called the variance inflation factor (VIF).

Partial Correlations shows the partial correlation of each pair of variables after adjusting for all other variables.

Pairwise Correlations gives a matrix of Pearson (product-moment) correlation coefficients, using all available values. The count (N) values differ if any pair has a missing value for either variable. These are the same correlations given by the **Density Ellipse** option on the Fit Y by X platform.

Nonparametric Correlations include the following:

> **Spearman's Rho** is a product-moment correlation coefficient computed on the ranks of the data values instead of the data values themselves. Spearman's correlation is covered in the next section.

> **Kendall's Tau** is based on the number of matching and nonmatching (concordant and discordant) pairs of observations, correcting for tied pairs.

> **Hoeffding's D** measure of dependence (association or correlation) is a statistical scale with a range from –0.5 to 1, where greater positive values indicate association.

See the *JMP Statistics and Graphics Guide* (2003) for more details about all the options available for the Multivariate platform.

Spearman Correlations

When to Use

Spearman's rank-order correlation coefficient (symbolized as r_s) is appropriate in a variety of circumstances. First, you can use Spearman correlations when both variables are numeric and have an ordinal modeling type (level of measurement). It is also correct when one variable is a numeric ordinal variable and the other is a continuous variable.

However, it can also be appropriate to use the Spearman correlation when both variables are continuous. The Spearman coefficient is a *distribution-free test*. A distribution-free test makes no assumption about the shape of the distribution from which the sample data are drawn. For this reason, researchers sometimes compute Spearman correlations when one or both of the variables are continuous but are markedly non-normal (such as the skewed data shown previously), which makes a Pearson correlation inappropriate. The Spearman correlation is less useful than a Pearson correlation when both variables are normal, but it is more useful when one or both variables are non-normal.

JMP computes a Spearman correlation by converting both variables to ranks and computing the correlation between the ranks. The resulting correlation coefficient can range from –1.00 to +1.00, and is interpreted in the same way as the Pearson correlation.

Computing Spearman Correlations

To illustrate this statistic, assume a teacher has administered a test of creativity to 10 students. After reviewing the results, the students are ranked from 1 to 10, with "10" representing the most creative student, and "1" representing the least creative student. Two months later, the teacher repeats the process, arriving at a slightly different set of rankings. The question now is to determine the correlation between rankings made by the two tests, given at different times. The data (rankings) are clearly on an ordinal level of measurement, so the correct statistic is the Spearman rank-order correlation coefficient.

See Figure 5.10 to follow along with this example.

- ⌐ Enter the data as shown in the data table in Figure 5.10 and name the variables test 1 and test 2.
- ⌐ Choose **Analyze → Multivariate Methods → Multivariate**.
- ⌐ In the Multivariate launch dialog, select **test 1** and **test 2** as analysis variables.
- ⌐ Click **OK** to see the initial Multivariate results (not shown in Figure 5.10).

⚓ Use the menu on the Multivariate title bar to suppress (uncheck) the default Correlations table and Scatterplot Matrix.

⚓ From the **Multivariate** menu select **Nonparametric Correlations → Spearman's Rho**.

These steps produce the Nonparametric: Spearman's Rho results seen at the bottom in Figure 5.10. The correlation coefficient between the tests is 0.9152 and the significance probability (*p* value) is 0.0002. The correlation and *p* value are interpreted as described previously for the Pearson statistics.

Figure 5.10 Computing Spearman's Rho Correlation between Variables test1 and test2

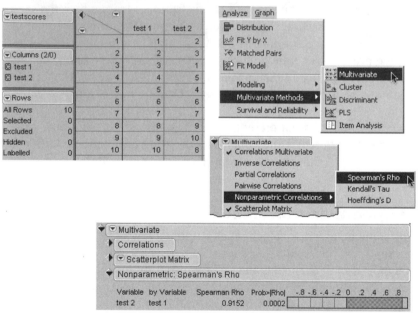

The Chi-Square Test of Independence

When to Use

The chi-square test is appropriate when both variables are classification or categorical variables, with a character data type and nominal or ordinal modeling type. Either variable can have any number of categories, but in practice the number of categories is usually small.

The Two-Way Classification Table

The nature of the relationship between two categorical variables is easiest to understand using a *two-way classification* table, also called a contingency table. A two-way classification table has rows that represent the categories (values) of one variable and columns that represent the categories of a second variable. You can review the number of observations in the various cells of a two-way table to look for a pattern that indicates some relationship between the two variables, and perform a chi-square test to determine if there is any statistical evidence that the variables are related in the population.

For example, suppose you want to prepare a table that shows one variable with two categories and a second variable with three categories. The general form for such a table appears in Table 5.2.

Table 5.2 General Form for a Two-Way Classification Table

		Column Variable		
		Column 1	Column 2	Column 3
Row Variable	Row 1	$Cell_{11}$	$Cell_{12}$	$Cell_{13}$
	Row 2	$Cell_{21}$	$Cell_{22}$	$Cell_{23}$

The point at which a row and column intersects is called a *cell*, and each cell is given a unique subscript. The first number in this subscript indicates the *row* to which the cell belongs, and the second number indicates the *column* to which the cell belongs. So the general form for the cell subscripts is $cell_{rc}$, where r = row and c = column. This means $cell_{21}$ is the intersection of row 2 and column 1, $cell_{13}$, is the intersection of row 1 and column 3, and so forth. Classification tables often show row totals and column totals, called marginal totals. The sum of all the cells is the grand total.

The chi-square test is appropriate for looking at contingency tables constructed two different ways:

- When one sample of subjects falls into cell categories and there are no predetermined marginal totals, the chi-square *test of independence* determines whether there is any association or relationship between the contingency table variables.
- When a sample is taken from each category of either the row or column variable, the marginal totals for that category in the contingency table are fixed. The chi-square *test of homogeneity* determines whether the distributions of these samples are the same for the levels of the second variable.

This section is only concerned with the chi-square test of independence.

> **Note:** The chi-square test is computed the same way whether it is testing independence or homogeneity, but the hypothesis and conclusion of a significant chi-square differ depending on the circumstances of its use.

One of the first steps in performing a chi-square test is to determine exactly how many subjects fall into each of the cells of the classification table. The pattern shown by the table cells helps you understand whether the classification variables are related.

Computing a Chi Square

To make this more concrete, assume that you are a university administrator preparing to purchase a large number of new personal computers for three of the schools in your university: the School of Arts and Sciences, the School of Business, and the School of Education. For a given school, you can purchase either IBM-compatible computers or Macintosh computers, and you need to know which type of computer the students within each school prefer.

In general terms, your research question is

> "Is there a relationship between (a) school of enrollment, and (b) computer preference?"

The chi-square test will help answer this question. If this test shows that there is a relationship between the two variables, you can review the two-way classification table to see which type of computer most students in each of the three schools prefer.

To answer this question, you draw a representative sample of 370 students from the 8,000 students that constitute the three schools. Each student is given a short questionnaire that asks just two questions:

1. In which school are you enrolled? (Circle one)

   a. School of Arts and Sciences
   b. School of Business
   c. School of Education

2. Which type of computer do you prefer that we purchase for your school?
   (Circle one)

   a. IBM compatible
   b. Macintosh

These two questions constitute the two nominal-level variables for your study.

- Question 1 defines a school of enrollment variable that can take on one of three values (Arts and Sciences, Business, or Education)
- Question 2 defines a computer preference variable that can take on one of two values (IBM compatible or Macintosh).

These are nominal variables because they only indicate group membership and provide no quantitative information.

You can now prepare the two-way classification table shown in Table 5.3. Computer preference is the row variable—row 1 represents students who preferred IBM compatibles and row 2 represents those who preferred Macintosh. School of enrollment is the column variable.

Table 5.3 shows the number of students that appear in each cell of the classification table. For example, the first row of the table shows that, among those students who preferred IBM compatibles, 30 were Arts and Sciences students, 100 were Business students, and 20 were Education students.

Table 5.3 Two-Way Table of Computer Preference by School of Enrollment

		School of Enrollment		
		Arts and Sciences	Business	Education
Computer	IBM Compatible	$n = 30$	$n = 100$	$n = 20$
Preference	Macintosh	$n = 60$	$n = 40$	$n = 120$

Remember that the purpose of the study is to determine whether there is any relationship between the two variables—that is, to determine whether school of enrollment is related to computer preference. This is just another way of saying, "If you know what school a student is enrolled, does that help you predict what type of computer that student is likely to prefer?" When you ask this kind of question about a set of nominal variables, the statistic to address the question is the chi-square test of independence. You want to know if computer preference is dependent (or not) on school of enrollment.

In this case, the answer to the question is easiest to find if you review the table one column at a time. For example, review just the Arts and Sciences column of the table. Notice that most of the students ($n = 60$) preferred Macintosh computers, while fewer ($n = 30$) preferred IBM compatibles. The column for the Business students shows the opposite trend—most business students ($n = 100$) preferred IBM compatibles. Finally, the pattern for the Education students was similar to that of the Arts and Sciences students in that the majority ($n = 120$) preferred Macintosh computers.

In short, there appears to be a relationship between school of enrollment and computer preference. Business students appear to prefer IBM compatibles, and Arts and Sciences and Education students appear to prefer Macintoshes. But this is just a trend that you observed in the *sample*. Is this trend strong enough to allow you to conclude that the variables are probably related in the *population* of 8,000 students? To determine this you must conduct a chi-square test of independence.

Tabular versus Raw Data

You can use JMP to compute the chi-square test of independence regardless of whether you are dealing with tabular data or raw data. When you are working with *raw data*, you are working with data that have not been summarized or tabulated in any way. For example, imagine that you have administered your questionnaire to 370 students and have not yet tabulated their responses—you merely have 370 completed questionnaires. In this situation, you are working with raw data.

On the other hand, *tabular data* are data that have been summarized in a table. For example, imagine that it was actually another researcher who administered this questionnaire, then summarized subject responses in a two-way classification table similar to Table 5.3. In this case you are dealing with tabular data.

In computing the chi-square statistic, there is no advantage to using one form of data or the other. The following section shows how to request the chi-square statistic in JMP for tabular data. A subsequent section shows how to analyze raw data.

Computing Chi-Square Values from Tabular Data

Often, the data to be analyzed with a chi-square test of independence have already been summarized and entered into a JMP data table like the one shown in Figure 5.11.

This JMP data table, **computerprefs.jmp**, contains three variables for each line of data. The first variable is a character variable named Preference that codes student computer preferences. The second is a character variable named School that gives students' school enrollment. The third variable is a numeric variable called Number that indicates how many students are in a given preference by school cell.

By now, you might wonder why there is so much emphasis on preparing a two-way classification table when you want to perform a chi-square test of independence. This is necessary because computing a chi-square statistic involves comparing the *observed frequencies* in each cell of the classification table (the number of observations that actually appear in each cell) and the *expected frequencies* in each cell of the table (the number of observations that you would expect to appear in each cell if the row variable and the column variable were completely unrelated).

However, the JMP analysis finds these expected frequencies needed for the chi-square computation regardless of whether the data are in raw form or tabular form.

Figure 5.11 Tabular Data and Launch Dialog to Generate a Contingency Table

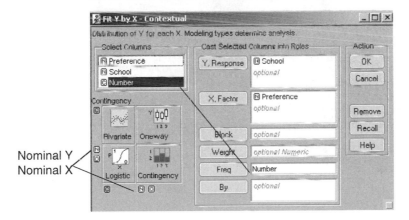

Now that your two-way classification data are in a JMP data table, you can request the chi-square statistic as follows.

 With the **computerprefs.jmp** table active, choose **Analyze → Fit Y by X** and complete the launch dialog as shown above in Figure 5.11.

🖰 Note that the Number variable must be assigned as a **Freq** variable because the data are already in tabular form in the data table. The Number value tells how many observations are in each cell of the two-way contingency table, as illustrated previously in Table 5.3.

🖰 Click **OK** to see the initial analysis results shown in Figure 5.12.

The Contingency Table

In the 2 by 3 classification table, called the Contingency Table in the analysis results, the name of the row variable (Preference) is on the far left of the table. The label for each row taken from the data table appears on the left side of the appropriate row. The first row (labeled "IBM") represents the students who preferred IBM-compatible computers, and the second row (labeled "MAC") represents students who preferred Macintoshes.

The name of the column variable (School) appears above the three columns. Each column is headed with its label— "Arts" represents the Arts and Sciences students, "Business" students are tabulated in the second column, and "Education" is the third column.

Figure 5.12 The Contingency Table and Tests for the Computer Preferences Data

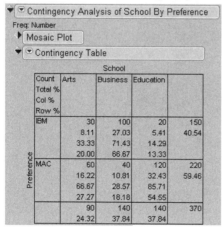

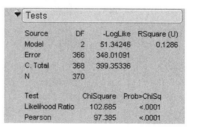

Where a given row and column intersect, information regarding the subjects in that cell is provided. The upper-left cell in the table is a legend that lists the quantities in the cells. The menu on the Contingency Table title bar (shown to the right) gives you options to show or hide these quantities in the contingency table cells by checking or unchecking them.

Count is the count or the raw number of subjects in the cell.

Total % is the total percentage or the percentage of subjects in that cell, relative to the total number of subjects (the number of subjects in the cell divided by the total number of subjects).

Col % is the column percentage or the percentage of subjects in that cell, relative to the number of subjects in that column. For example, there are 30 subjects in the IBM Arts cell, and 90 subjects in the Arts column. Therefore, the column percent for this cell is 30/90 = 33.33%.

Row % is the row percentage or the percentage of subjects in that cell, relative to the number of subjects in that row. For example, there are 30 subjects in the IBM Arts cell, and 150 subjects in the IBM row. Therefore, the row percent for this cell is 30/150 = 20%.

Expected shows the expected cell frequencies (E), which are the cell frequencies that would be expected if the two variables were in fact independent (unrelated). This is a very useful option for determining the nature of the relationship between the two variables. It is computed as the product of the corresponding row total and column total, divided by the grand total.

Deviation is the observed (actual) cell frequency (O) minus the expected cell frequency (E).

Cell Chi Square is the chi-square value for the individual cell, computed as $(O-E)^2 / E$.

In the present study, it is particularly revealing to review the classification table one column at a time, and to pay particular attention to the column percent. First, consider the Arts column. The column percent entries show that only 33.33% of the Arts and Sciences students preferred IBM compatibles, while 66.67% preferred Macintoshes. Next, consider the Business column, which shows the reverse trend: 71.43% of the Business students preferred IBM compatibles while only 28.57% preferred Macintoshes. The trend of the Education students is similar to that for the Arts and Sciences students: only 14.29% preferred IBM compatibles, while 85.71% preferred Macintoshes.

These percentages reinforce the suspicion that there is a relationship between school of enrollment and computer preference. To find out if the relationship is statistically significant, you must consult the chi-square test of independence shown below and in Figure 5.12.

The Tests Report

The upper part of the Tests report is arranged like an Analysis of Variance table for continuous data. This table assumes there is a linear model where one variable is the fixed predictor (X) and the other is the response (Y). The negative log-likelihood (–LogLike) plays the same role for categorical data as the sum of squares does for continuous data. It measures the fit and uncertainty of the model,

Tests			
Source	DF	-LogLike	RSquare (U)
Model	2	51.34246	0.1286
Error	366	348.01091	
C. Total	368	399.35336	
N	370		

Test	ChiSquare	Prob>ChiSq
Likelihood Ratio	102.685	<.0001
Pearson	97.385	<.0001

showing quantities for the Corrected Total (C. Total) partitioned into Model and Error. Likewise, the total number of observations is shown, with total degrees of freedom partitioned into those for Model and for Error. The Rsquare (U) is the proportion of uncertainty (–LogLike) explained by the model.

The lower part of the Tests report shows two chi-square tests of independence, the Likelihood Ratio test (computed as twice the negative log-likelihood for Model in the Tests table) and the familiar Pearson chi-square test. These statistics test the null hypothesis that, in the population, the two variables (Preference and School) are independent, or unrelated. When the null hypothesis is true, expect the value of chi square to be relatively small—the stronger the relationship between the two variables in the sample, the larger chi square will be.

The Tests report shows that the obtained value of the Pearson chi-square test is 97.385 with 2 degrees of freedom. The Model DF toward the top of the Tests report shows the degrees of freedom. The degrees of freedom for the chi-square test are calculated as

$$df = (r - 1)(c - 1)$$

where

r = number of categories for the row variable

c = number of categories for the column variable.

For the current analysis, the row variable (Preference) has two categories, and the column variable (School) has three categories, so the degrees of freedom are calculated as

$$df = (2-1)(3-1) = (1)(2) = 2$$

The obtained (computed) chi-square value, 97.385, is quite large for the given degrees of freedom. The probability value, or p value, for this chi-square statistic is printed below the heading Prob>ChiSq in the Tests report. This p value is less than 0.0001, which means that there is less than one chance in 10,000 of obtaining a chi-square value of this size (or larger) if the variables were independent in the population. You can therefore reject the null hypothesis, and tentatively conclude that school of enrollment is related to computer preferences.

Computing Chi-Square Values from Raw Data

In many cases the data are not presummarized and don't have a Number column, as in the previous example (see Figure 5.11). Instead there is one row for each respondent. If the data in the example above had not been tabulated, the data table would have had 30 rows with School value listed as "Arts" and Preference value listed as "IBM," 100 rows for a School value of "Business" and Preference value of "IBM," and so forth. The data table would have had 370 rows and only two variables. To analyze data in this raw form, choose **Analyze → Fit Y by X** and proceed as described in Figure 5.11 but without a **Freq** variable. The results are identical to those described for tabular data.

Fisher's Exact Test for 2 by 2 Tables

A 2 by 2 table contains just two rows and two columns. Said another way, a two-way classification table for a chi-square study is a 2 by 2 table where there are two values for the row variable and two values for the column variable. For example, imagine that you modified the preceding computer preference study so that there were only two values for the school of enrollment variable ("Arts and Sciences" and "Business"). The resulting table is called a 2 by 2 table because it has only two rows ("IBM" and "Macintosh") and two columns ("Arts and Sciences" and "Business").

When analyzing a 2 by 2 classification table, it is best to use *Fisher's exact test* instead of the standard chi-square test. This test is automatically appended to the Fit Y by X contingency table analysis whenever the classification table is a 2 by 2 table (see Figure 5.13). In the JMP analysis results, consult the *p* values listed in the Tests report as Prob Alternative Hypothesis, for the Left, Right, and 2-Tail hypotheses. These estimate the probability of observing a table that gives at least as much evidence of a relationship as the one actually observed, given that the null hypothesis is true. In other words, when the *p* value for Fisher's exact test is less than 0.05, you can reject the null hypothesis that the two nominal-scale variables are independent in the population, and conclude that they are related.

Figure 5.13 Contingency Table Analysis with Fisher's Exact Test for 2 by 2 Table

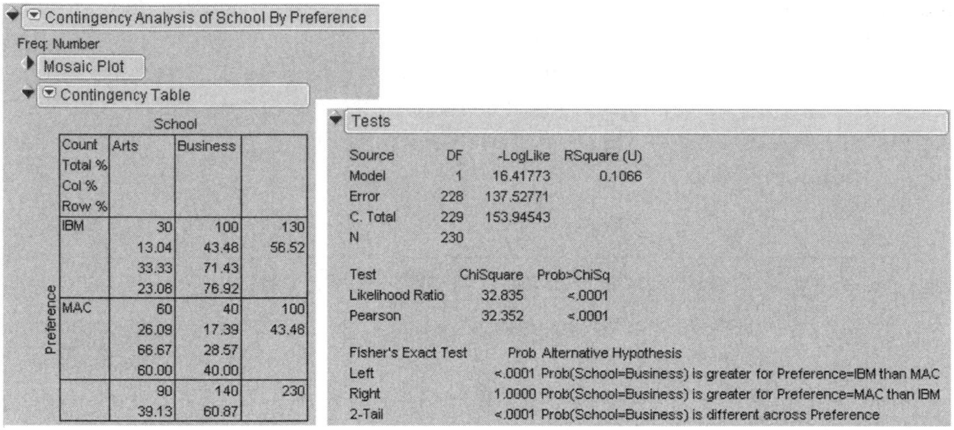

Note: The chi-square test might not be valid if the observed frequency in any of the cells is zero, or if the expected frequency in any of the cells is less than five (use the **Expected** option on the **Contingency Table** menu to see expected cell frequencies). JMP displays a warning when this situation occurs. When these minimums are not met, consider gathering additional data or combining similar categories of subjects in order to increase cell frequencies.

Summary

Bivariate associations are the simplest types of associations studied, and the statistics presented here (the Pearson correlation, the Spearman correlation, and the chi-square test of independence) are appropriate for investigating most types of bivariate relationships found in the social sciences and many other areas of research. With these relatively simple procedures behind you, you are now ready to proceed to the *t* test, the one-way analysis of variance, and other tests of group differences.

Assumptions Underlying the Tests

Assumptions Underlying the Pearson Correlation Coefficient

Continuous variables

Both the predictor and response variables have a continuous modeling type, with continuous values measures on an interval or ratio scale.

Random sampling

Each subject in the sample contributes one score for the predictor variable and one score for the response variable. These pairs of scores represent a random sample drawn from the population of interest.

Linearity

The relationship between the response variable and the predictor variable should be linear. This means that, in the population, the mean response scores at each value of the predictor variable fall in a straight line. The Pearson correlation coefficient is not appropriate for assessing the strength of the relationship between two variables involved in a curvilinear relationship.

Bivariate normal distribution

The pairs of scores should follow a bivariate normal distribution. That is, scores for the response variable should form a normal distribution at each value of the predictor variable. Similarly, scores for the predictor variable should form a normal distribution at each value of the response variable. When scores represent a bivariate normal distribution, they form an *elliptical* scatterplot when plotted (their scatterplot is shaped like a football—fat in the middle and tapered on the ends). However, the Pearson correlation coefficient is robust (works well) with respect to the normality assumption when the sample size is greater than 25.

Assumptions Underlying the Spearman Correlation Coefficient

Ordinal modeling type

Both the predictor and response variables should have numeric values and an ordinal modeling type. However, continuous numeric variables are sometimes analyzed with the Spearman correlation coefficient when one or both variables are markedly skewed.

Assumptions Underlying the Chi-Square Test of Independence

Nominal or ordinal modeling type

Both variables must have either a nominal or ordinal modeling type. The Fit Y by X platform treats both nominal and ordinal variables as classification variables and analyzes them the same way.

Random sampling

Subjects contributing data should represent a random sample drawn from the population of interest.

Independent cell entries

Each subject must appear in only one cell. The fact that a given subject appears in one cell should not affect the probability of another subject appearing in any other cell.

Expected frequencies of 5 or more

For tables with expected cell frequencies less than 5, the chi-square test might not be reliable. A standard rule of thumb (Cochran, 1954) is to avoid using the chi-square test for tables with expected cell frequencies less than 1, or when more than 20% of the table cells have expected cell frequencies less than 5. For 2 by 2 tables, use Fisher's exact test whenever possible.

References

Cochran, W. G. 1954. "Some Methods of Strengthening the Common Chi-Square Tests." *Biometrics* 10:417–451.

Hays, W. L. 1988. *Statistics*. 4th ed. New York: Holt, Rinehart and Winston.

Assessing Scale Reliability with Coefficient Alpha

Overview. This chapter shows how to use the Multivariate platform in JMP to compute the coefficient alpha reliability index (Cronbach's alpha) for a multiple-item scale. It reviews basic issues regarding the assessment of reliability, and describes the circumstances under which a measure of internal consistency is likely to be high. Fictitious questionnaire data are analyzed to demonstrate how you can perform an item analysis to improve the reliability of scale responses.

Introduction: The Basics of Scale Reliability

You can compute coefficient alpha, also called Cronbach's alpha, when you have administered a multiple-item rating scale to a group of participants and want to determine the internal consistency of responses to the scale. The items constituting the scale can be scored dichotomously, such as "right" or "wrong." Or, items can use a multiple-point rating format—participants can respond to each item using a 7-point "agree-disagree" rating scale.

This chapter shows how to use the Multivariate platform in JMP to compute the coefficient alpha for the types of scales often used in social science research and surveys in general. However, this chapter will not show how to actually *develop* a multiple-item scale for use in research. To learn about recommended approaches for creating summated rating scales, see Spector (1992).

Example of a Summated Rating Scale

A *summated rating scale* consists of a short list of statements, questions, or other items to which participants respond. Often, the items that constitute the scale are statements, and participants indicate the extent to which they agree or disagree with each statement by selecting some response on a rating scale, such as a 7-point rating scale in which 1 = "Strongly Disagree" and 7 = "Strongly Agree." The scale is called a *summated* scale because the researcher can sum responses to all selected responses to create an overall score on the scale. These scales are often referred to as *Likert-type* scales.

> **Note:** Likert scale items can be averaged to give a response instead of summed. Using the mean response eliminates the problem of non-response and is sometimes easier to interpret than the sum of the responses.

As an example, imagine that you are interested in measuring job satisfaction in a sample of employees. To do this, you might develop a 10-item scale that includes items such as

"In general, I am satisfied with my job." Employees respond to these items using a 7-point response format in which 1 = "Strongly Disagree" and 7 = "Strongly Agree."

You administer this scale to 200 employees and compute a job satisfaction score by summing each employee's responses to the 10 items. Scores can range from a low of 10 (if the employee circled "Strongly Disagree" for each item) to a high of 70 (if the employee circled "Strongly Agree" for each item). You hope to use job satisfaction scores as a predictor variable in research. However, the people who later read about your research have questions about the psychometric properties of scale responses. At the very least, you need to show empirical evidence that responses to the scale are reliable. This chapter discusses the meaning of scale reliability and shows how to use JMP to obtain an index of internal consistency for summated rating scales.

True Scores and Measurement Error

Most observed variables measured in the social sciences (such as scores on your job satisfaction scale) consist of two components:

- A *true score* that indicates where the participant actually stands on the variable of interest

- *Measurement error*, which is a part of almost all observed variables, even those variables that seem to be objectively measured.

Imagine that you assess an observed variable, age, in a group of participants by asking them to indicate their age in years. To a large extent, this observed variable (what the participants wrote down) is influenced by the true score component. That is, the participant's response is influenced by how old they actually are. Unfortunately, this observed variable is also influenced by measurement error. Some participants write down the wrong age because they don't know how old they are, some write the wrong age because they don't want the researcher to know how old they are, and other participants write the wrong age because they didn't understand the question. In short, it is likely that there will not be a perfect correlation between the observed variable (what the participants write down) and their true age scores.

Measurement error can occur even though the age variable is relatively objective and straightforward. If this simple question can be influenced by measurement error, imagine how much more error results when more subjective constructs, such as the items that constitute your job satisfaction scale, are measured.

Underlying Constructs versus Observed Variables

In applied research, it is useful to draw a distinction between *underlying constructs* and *observed variables*.

- An underlying construct is the hypothetical variable that you want to measure. In the job satisfaction study, for example, you want to measure the underlying construct of job satisfaction within a group of employees.

- The observed variable consists of the responses you actually obtain. In this example, the observed variable consisted of scores on the 10-item measure of job satisfaction. These scores might or might not be a good measure of the underlying construct.

Reliability Defined

With this understanding, it is now possible to provide some definitions. A *reliability coefficient* can be defined as the percent of variance in an observed variable that is accounted for by true scores on the underlying construct. For example, imagine that, in the study just described, you were able to obtain two scores for the 200 employees in the sample—their observed scores on the job satisfaction questionnaire and their true scores on the underlying construct of job satisfaction. Assume that you compute the correlation between these two variables. The square of this correlation coefficient represents the reliability responses to your job satisfaction scale. It is the percent of variance in observed job satisfaction scores accounted for by true scores on the underlying construct of job satisfaction.

The preceding is a technical definition for reliability, but this definition is of little use in practice because it is usually not possible to obtain true scores for a variable. For this reason, reliability is often defined in terms of the *consistency* of the scores obtained on the observed variable. An instrument is said to be reliable if it is shown to provide consistent scores upon repeated administration, upon administration by alternate forms, and so forth. A variety of methods to estimate scale reliability are used in practice.

Test-Retest Reliability

Assume that you administer your measure of job satisfaction to a group of 200 employees, first in January and again in March. If the instrument is reliable, you expect that the participants who provided high scores in January will tend to provide high scores again in March, and that those with low scores in January will also have low scores in March. These results support the test-retest reliability of responses to the scale. You assess test-retest reliability by administering the same instrument to the same sample of participants at two points in time, and then computing the correlation between sets of scores.

But what is an appropriate interval over which questionnaires should be administered? Unfortunately, there is no hard-and-fast rule here. The interval depends on what is being measured. For enduring constructs such as personality variables, test-retest reliability has been assessed over several decades. For other constructs such as depressive symptomatology, the interval tends to be much shorter due to the fluctuating course of depression and its symptoms. Generally speaking, the test-retest interval should not be so short that respondents recall their responses to specific items (less than a week) but not so long as to measure natural variability in the construct (real change in depressive symptoms). The former leads to an overstatement of test-retest reliability, whereas the latter leads to understatement of test-retest reliability.

Internal Consistency

Another important aspect of reliability is *internal consistency*. Internal consistency is the extent to which the individual items that constitute a test correlate with one another or with the test total.

Internal consistency and test-retest procedures measure different aspects of scale reliability estimation. That is, internal consistency evaluation is not an alternative to test-retest measures. For example, some constructs, such as mood measures, do not typically exhibit test-retest reliability, but should show internal consistency.

In the social sciences, a widely used index of internal consistency is the coefficient alpha or Cronbach's alpha (Cronbach, 1951), symbolized by the Greek letter α.

Cronbach's Alpha

Cronbach's alpha is a general formula for scale reliability based on internal consistency It gives the lowest estimate of reliability that can be expected for an instrument.

The formula for coefficient alpha is

$$r_{xx} = \left(\frac{N}{N-1} \right) \left(\frac{S^2 - \Sigma S^2_i}{S^2} \right)$$

where

r_{xx} = Cronbach's alpha.

N = number of items in the instrument.

S^2 = variance of the summated scale scores. That is, assume that you compute a total score for each participant by summing responses to the scale. The variance of this total score variable is S^2.

ΣS^2_i = the sum of the variances of the individual items that constitute this scale.

Other factors held constant, Cronbach's alpha is high when there are many highly correlated items in the scale. To understand why Cronbach's alpha is high when the items are highly correlated with one another, consider the second term in the preceding formula:

$$\left(\frac{S^2 - \Sigma S^2_i}{S^2} \right)$$

This term shows that the variance of the summated scales scores, S^2, is divided by itself to compute Cronbach's alpha. However, the combined variance of the individual items is first subtracted from this variance before the division is performed. This part of the equation shows that as the combined variance of the individual items becomes small, Cronbach's alpha becomes larger.

This is important because (with other factors held constant) stronger correlations between the individual items give a smaller ΣS^2 term. Thus, Cronbach's alpha for responses to a given scale is likely to be large to the extent that the variables constituting that scale are strongly correlated.

Computing Cronbach's Alpha

Imagine that you have conducted research in the area of prosocial behavior and have developed an instrument designed to measure two separate underlying constructs: *helping others* and *financial giving*.

- Helping others refers to prosocial activities performed to help coworkers, relatives, and friends.
- Financial giving refers to giving money to charities or the homeless.

See Chapter 14, "Principal Component Analysis," for a more detailed description of these constructs. In the following questionnaire, items 1 to 3 are designed to assess helping others and items 4 to 6 are designed to assess financial giving.

```
Instructions: Below are a number of activities in which people sometimes
engage. For each item, please indicate how frequently you have engaged in this
activity over the past six months. Provide your response by circling the
appropriate number to the left of the item, and use the following response
format:
7 = Very Frequently
6 = Frequently
5 = Somewhat Frequently
4 = Occasionally
3 = Seldom
2 = Almost Never
1 = Never
_____
1 2 3 4 5 6 7      1.   Went out of my way to do a favor for a coworker
1 2 3 4 5 6 7      2.   Went out of my way to do a favor for a relative.
1 2 3 4 5 6 7      3.   Went out of my way to do a favor for a friend.
1 2 3 4 5 6 7      4.   Gave money to a religious charity.
1 2 3 4 5 6 7      5.   Gave money to a charity not associated with a religion.
1 2 3 4 5 6 7      6.   Gave money to a panhandler.
```

Assume that you have administered this 6-item questionnaire to 50 participants. For the moment, consider only the reliability of the scale that includes items 1 to 3 (the items that assess helping others).

Further assume that you made a mistake in assessing the reliability of this scale. Suppose you erroneously believed that the "helping others" construct was assessed by items 1 to 4 (whereas, in reality, that construct was assessed by items 1 to 3). It is instructive to see what happens when you mistakenly include item 4 in the analysis. The next section looks at this 4-item scale.

This example uses the JMP data table called **helpfulness.jmp**. Figure 6.1 shows a partial listing of the data. Ordinarily, you would not compute Cronbach's alpha in this case because internal consistency is often underestimated with so few items. Cronbach's alpha also tends to overestimate the internal consistency of responses to scales with 40 or more items (Cortina, 1993; Henson, 2001).

The JMP Multivariate Platform

In JMP, the Multivariate platform gives simple statistics and correlations and, optionally, can compute Cronbach's alpha (internal consistency) for a summated rating scale.

- Open the **helpfulness** data table. Note that the items are called Q1–Q6, and each row is the set of responses given by the 50 participants.

Figure 6.1 Partial Listing of the Helpfulness Survey Data Table

	Q1	Q2	Q3	Q4	Q5	Q6
1	5	5	6	7	5	4
2	5	6	7	3	4	3
3	7	7	7	2	2	2
4	6	6	5	2	4	3
5	6	6	6	6	6	5
6	3	5	3	3	2	4
7	7	6	7	1	5	3
8	6	6	6	6	5	6
9	3	3	4	3	3	3
10	5	6	7	2	3	2

- On the **Analyze** main menu, the **Multivariate Methods** command has several platforms. For this example, choose **Analyze → Multivariate Methods → Multivariate**.
- When the launch dialog appears, select **Q1**, **Q2**, **Q3**, and **Q4** as analysis (Y) variables, as shown in Figure 6.2.

Figure 6.2 Multivariate Command and Launch Dialog

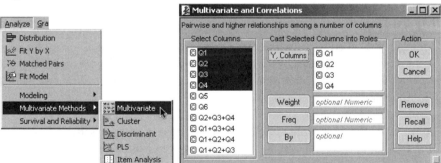

Click **OK** in the launch dialog to see the initial results. By default, the initial Multivariate platform results show a correlations table and a bivariate scatterplot matrix with a density ellipse for each pair of Y variables.

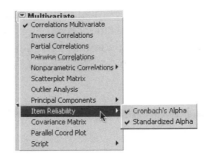 Use options on the Multivariate platform to see the results in Figure 6.3. That is, turn off (deselect) **Scatterplot Matrix**, and select both **Cronbach's Alpha** and **Standardized Alpha** from the **Item Reliability** submenu, as shown by the menu to the right.

Cronbach's Alpha for the 4-Item Scale

Figure 6.3 shows the Cronbach's alpha computations. Cronbach's alpha for the nonstandardized set, noted as Entire Set on the report, is only 0.4904. This is the reliability coefficient for items Q1 to Q4. This reliability coefficient is often reported in published articles.

How Large Is an Acceptable Reliability Coefficient?

Nunnally (1978) suggests that 0.70 is an acceptable reliability coefficient. Reliability coefficients less than 0.70 are generally seen as inadequate. However, you should remember that this is only a rule of thumb and scientists sometimes report a coefficient alpha (Cronbach's alpha) under 0.70 (and sometimes even under 0.60).

Also, a larger alpha coefficient is not always necessarily better than a smaller one. An ideal estimate of internal consistency is thought to be between 0.80 and 0.90 (Clark and Watson, 1995; DeVellis, 1991) because estimates in excess of 0.90 suggest item redundancy or inordinate scale length.

Alpha Comparison for Item Selection

The coefficient alpha of 0.49 reported in Figure 6.3 is not acceptable. In some situations, the reliability of responses to a multiple-item scale improves by dropping a redundant item. One useful way to evaluate an item and determine if it should be deleted is to compare its α with the α for the entire set. The rule of thumb is that if an item α is greater than the overall α that includes the item, then scale reliability improves when that item is dropped from the set.

For example, the Excluded Col section of the Cronbach's Alpha reports show the Q4 alpha value to be 0.7766, which is considerably larger than the Entire Set value of 0.4904. Likewise, the standardized alpha for Q4 is 0.7733 and the overall standardized alpha is 0.5797. These comparisons indicate that Q4 should be dropped from the analysis. Coefficient alpha computations reveal that Q4 was accidentally included in this initial example.

Figure 6.3 Multivariate Results Showing Correlations and Cronbach's Alpha

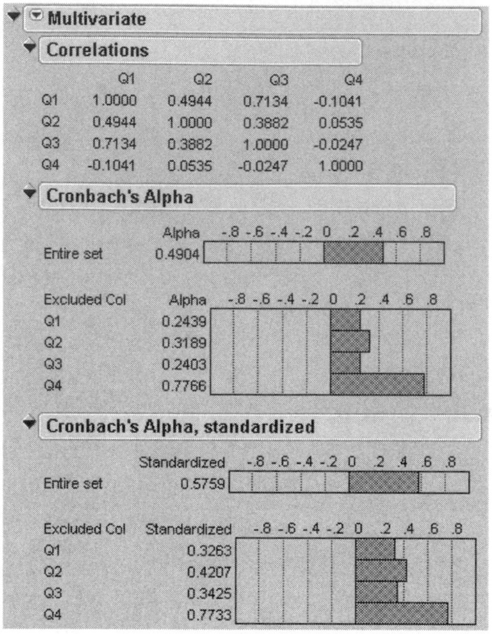

Item-Total Correlations

In some situations, the reliability of responses to a multiple-item scale improves by deleting those items with poor *item-total correlations*. An item-total correlation is the correlation between an individual item and the sum of the remaining items that constitute the scale. A small item-total correlation is evidence that the item is not measuring the same construct measured by the other scale items. This means that you could choose to discard items that show small item-total correlations. The next section shows how to compute item-total correlations.

Method to Compute Item-Total Correlation

JMP version 5 does not automatically compute item-total correlations. However, there are several ways to construct them using JMP features. This section shows how to use the data table and formulas, followed by the Multivariate platform to look at item-total correlations.

Begin by constructing four new variables in the **helpfulness** data table. Each new variable is the sum of three original variables. There is a sum column for each combination of three original variables.

- Use the **Add Multiple Columns** command from the **Cols** main menu to add four numeric columns to the data table.
- Give the columns convenient names that identify their formulas, such as Q2+Q3+Q4, Q1+Q3+Q4, Q1+Q2+Q4, and Q1+Q2+Q3.
- For each new column, use its formula editor to compute column values. To do this, right-click in the column name area and select **Formula** from the popup menu, as shown in Figure 6.4. Or, highlight the column and choose **Formula** from the **Cols** main menu. When the formula editor appears, enter the formula identified by the column name.

See the section "Computing Column Values with a Formula" in Chapter 3, "Working with JMP Data," for a detailed example of how to use the JMP Formula Editor.

Figure 6.4 Create Columns and Compute Sums

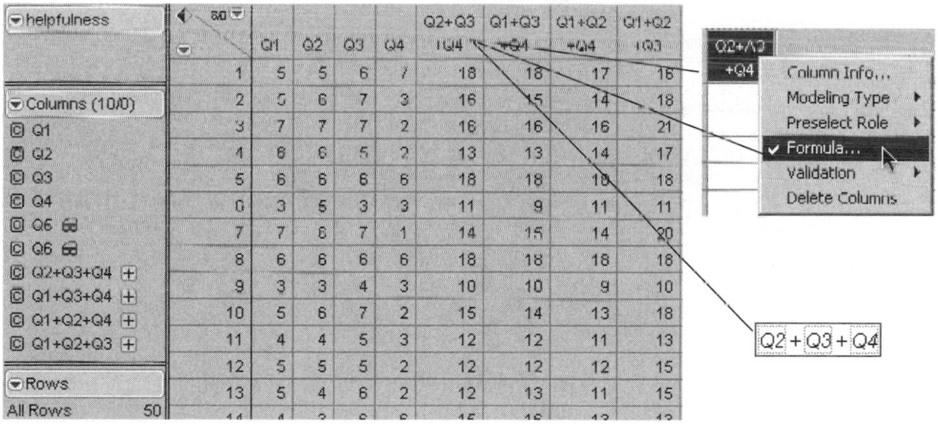

Caution: The **helpfulness** data table includes these computed variables. In Figure 6.4, the Q5 and Q6 variables are hidden and are not shown in the data table.

Use the Multivariate platform to compute correlations between the original variables and the constructed sums.

- Choose **Multivariate Methods** → **Multivariate** from the **Analyze** menu.
- In the Multivariate launch dialog, select **Q1** through **Q4** and the four sum variables in the Select Columns list and click **Y, Columns** in the dialog. The completed dialog should look like the one shown at the top in Figure 6.5.
- Click **OK** in the launch dialog to see the results shown at the bottom of Figure 6.5.

Figure 6.5 Item-Total Correlations for Each Item

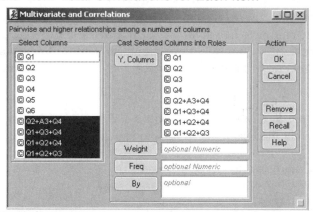

The item-total correlations are now shown in the Correlations table. You only need to look at the portion of the table shown boxed in Figure 6.5. The item-total correlations are the diagonal quantities. For example, to see the correlation between Q1 and the sum of Q2, Q3, and Q4, look at the intersection of the row labeled Q1 and the column labeled Q2+Q3+Q4.

You can see in that items 1 to 3 demonstrate reasonably strong correlations with the sum of the remaining items on the scale. However, item Q4 demonstrates an item-total correlation of approximately –0.037, which suggests that item Q4 is not measuring the same construct as items Q1 to Q3.

Cronbach's Alpha for the 3-Item Scale

Now look at the alpha computations for the 3-item scale with Q4 removed.

- As before, choose **Multivariate Methods → Multivariate** from the **Analyze** menu.
- In the Multivariate launch dialog, select **Q1** through **Q3** from the Select Columns list and click **Y, Columns** in the dialog. Click **OK** to see the results shown in Figure 6.6.

The results for three variables, Q1, Q2, and Q3, show a raw-variable coefficient alpha of 0.7766. This coefficient exceeds the recommended minimum value of 0.70 (Nunnally, 1978). Clearly, responses to the Helping Others subscale demonstrate a much higher level of reliability with item Q4 deleted.

The alpha computation of 0.7766 that occurs if you remove Q4 from the analysis is expected because Q4 was mistakenly included in the analysis. This mistake was clear because Q4 demonstrates a correlation with the remaining scale items of only –0.037 (see Figure 6.5). You can substantially improve this scale by removing the item that is not measuring the same construct assessed by the other items.

Figure 6.6 Correlations and Cronbach's Alpha for the 3-Item Scale

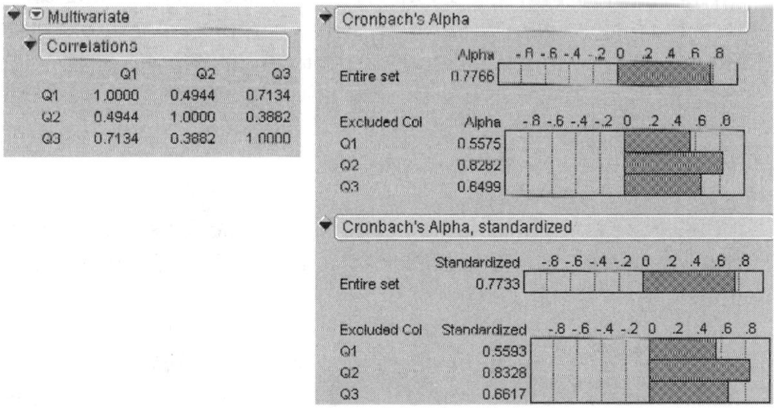

Summarizing the Results

Researchers typically report the reliability of a scale in a table that reports simple descriptive statistics for the study's variables such as means, standard deviations, and intercorrelations. In these tables, Cronbach's alpha (coefficient alpha) estimates are usually reported on the diagonal of the correlation matrix, within parentheses. Table 6.1 shows an example of this approach to summarize items for three constructs (scales).

Note: The summary data in Table 6.1 is an example only, and not for the data used in this chapter.

Table 6.1 Means, Standard Deviations, Correlations, and Coefficient Alpha Reliability Estimates for the Study's Variables

Variables	Mean	SD	Scale 1	Scale 2	Scale 3
Scale 1	13.56	2.54	(0.90)		
Scale 2	15.60	3.22	0.37	(0.78)	
Scale 3	12.55	1.32	0.25	0.53	(0.77)

Note: N = 200. Reliability estimates appear on the diagonal.

In Table 6.1, information for the first scale variable is at the intersection of the Scale 1 row and Scale 1 column. Where the row and column intersect, the coefficient alpha index for the first scale item is 0.90. Likewise, you see the coefficient alpha for Scale 2 where row 2 intersects with column 2 (alpha = 0.78), and the coefficient alpha for Scale 3 where row 3 intersects with column 3 (alpha = 0.77).

When reliability estimates are computed for a relatively large number of scales, it is common to report them in a table (such as Table 6.1) and make only passing reference to them within the text of the paper when within acceptable parameters. For example, within a section on instrumentation, you might indicate the following:

"Estimates of internal consistency as measured by Cronbach's alpha all exceeded 0.70 and are reported on the diagonal of Table 6.1."

When reliability estimates are computed for only a small number of scales, it is possible to instead report these estimates within the body of the text itself. Here is an example of how this might be done:

"Internal consistency of scale responses was assessed by Cronbach's alpha. Reliability estimates were 0.90, 0.78, and 0.77 for responses to Scale 1, Scale 2, and Scale 3, respectively."

Summary

Assessing scale reliability with Cronbach's alpha (or some other reliability index) should be one of the first tasks you undertake when conducting questionnaire research. If responses to selected scales are not reliable, there is no point performing additional analyses. You can often improve suboptimal reliability estimates by deleting items with poor item-total correlations in keeping with the procedures discussed in this chapter. When several subscales on a questionnaire display poor reliability, it may be advisable to perform a principal component analysis or an exploratory factor analysis on responses to all questionnaire items to determine which tend to group together empirically. If many items load on each retained factor and if the factor pattern obtained from such an analysis displays a simple structure, chances are good that responses to the resulting scales will demonstrate adequate internal consistency.

References

Clark, L. A., and D. Watson. 1995. "Constructing Validity: Basic Issues in Objective Scale Development." *Psychological Assessment* 7:309–319.

Cortina, J. M. 1993. "What Is Coefficient Alpha? An Examination of Theory and Applications." *Journal of Applied Psychology* 78.98–104.

Cronbach, L. J. 1951. "Coefficient Alpha and the Internal Structure of Tests." *Psychometrika* 16:297–334.

DeVellis, R. F. 1991. *Scale Development: Theory and Applications*. Newbury Park, CA: Sage Publications

Henson, R. K. 2001. "Understanding Internal Consistency Estimates: A Conceptual Primer on Coefficient Alpha." *Measurement and Evaluation in Counseling and Development* 34;177–189.

Nunnally, J. 1978. *Psychometric Theory*. New York: McGraw-Hill.

Spector, P. E. 1992. *Summated Rating Scale Construction: An Introduction*. Newbury Park, CA: Sage Publications.

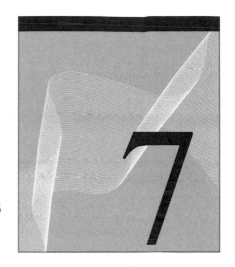

t Tests: Independent Samples and Paired Samples

> **Overview.** This chapter describes the differences between the independent-samples *t* test and the paired-samples *t* test, and shows how to perform both types of analyses. An example of a research design is developed that provides data appropriate for each type of *t* test. With respect to the independent-samples test, this chapter shows how to use JMP to determine whether the equal-variances or unequal-variances *t* test is appropriate, and how to interpret the results. There are analyses of data for paired-samples research designs, with discussion of problems that can occur with paired data.

Introduction: Two Types of *t* Tests

A *t* test is appropriate when an analysis involves a single nominal or ordinal predictor that assumes only two values (often called treatment conditions), and a single continuous response variable. A *t* test helps you determine if there is a significant difference in mean response between the two conditions. There are two types of *t* tests that are appropriate for different experimental designs.

First, the *independent-samples t test* is appropriate if the observations obtained under one treatment condition are independent of (unrelated to) the observations obtained under the other treatment condition. For example, imagine you draw a random sample of subjects and randomly assign each subject to either Condition 1 or Condition 2 in your experiment. You then determine scores on an attitude scale for subjects in both conditions, and use an independent-samples *t* test to determine whether the mean attitude score is significantly higher for the subjects in Condition 1 than for the subjects in Condition 2. The independent-samples *t* test is appropriate because the observations (attitude scores) in Condition 1 are unrelated to the observations in Condition 2. Condition 1 consists of one group of people and Condition 2 consists of a different group of people who are not related to, or affected by, the people in Condition 1.

The second type of test is the *paired-samples t test*. This statistic is appropriate if each observation in Condition 1 is paired in some meaningful way with a corresponding observation in Condition 2. There are a number of ways that this pairing happens. For example, imagine you draw a random sample of subjects and decide that each subject is to provide two attitude scores—one score after being exposed to Condition 1 and a second score after being exposed to Condition 2. You still have two samples of observations (the sample from Condition 1 and that from Condition 2), but the observations from the two samples are now *related*. If a given subject has a relatively high score on the attitude scale under Condition 1, that subject might also score relatively high under Condition 2. In analyzing the data, it makes sense to pair each subject's scores from Condition 1 and Condition 2. Because of this pairing, a paired-samples *t* statistic is calculated differently than an independent-samples *t* statistic.

This chapter is divided into two major sections. The first deals with the independent-samples *t* test, and the second deals with the paired-samples test. These sections describe additional examples of situations in which the two procedures might be appropriate.

Earlier, you read that a *t* test is appropriate when the analysis involves a nominal or ordinal predictor variable and a continuous response. A number of additional assumptions should also be met for the test to be valid and these assumptions are summarized at the end of this chapter. When these assumptions are violated, consider using a nonparametric statistic instead. See the *JMP Statistics and Graphics Guide* (2003) for examples of nonparametric statistics.

The Independent-Samples *t* Test

Example: A Test of the Investment Model

The investment model of emotional commitment (Rusbult, 1980) illustrates the hypothesis tested by the independent-samples *t* test. As discussed in earlier chapters, the investment model identifies a number of variables expected to affect a subject's commitment to romantic relationships (as well as to some other types of relationships). *Commitment* can be defined as the subject's intention to remain in the relationship and to maintain the relationship. One version of the investment model predicts that commitment will be affected by four variables—rewards, costs, investment size, and alternative value. These variables are defined as follows

Rewards are the number of "good things" that the subject associates with the relationship (the positive aspects of the relationship).

Costs are the number of "bad things" or hardships associated with the relationship.

Investment size is the amount of time and personal resources that the subject has "put into" the relationship.

Alternative value is the attractiveness of the subject's alternatives to the relationship (the attractiveness of alternative romantic partners).

At least four testable hypotheses can be derived from the investment model as it is described here.

- Rewards have a causal effect on commitment.
- Costs have a causal effect on commitment.
- Investment size has a causal effect on commitment.
- Alternative value has a causal effect on commitment.

This chapter focuses on testing only the first hypothesis: the prediction that the level of rewards affects commitment.

Rewards refer to the positive aspects of the relationship. Your relationship would score high on rewards if your partner were physically attractive, intelligent, kind, fun, rich, and so forth. Your relationship would score low on rewards if your partner were unattractive, unintelligent, unfeeling, dull, and so forth. It can be seen that the hypothesized relationship between rewards and commitment makes good intuitive sense: an increase in rewards should result in an increase in commitment. The predicted relationship between these variables is illustrated in Figure 7.1.

Figure 7.1 Hypothesized Causal Relationship between Rewards and Commitment

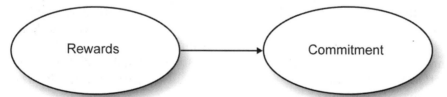

There are a number of ways that you could test the hypothesis that rewards have a causal effect on commitment. One approach involves an experimental procedure in which subjects are given a written description of different fictitious romantic partners and asked to rate their likely commitment to these partners. The descriptions are written so that a given fictitious partner can be described as a "high-reward" partner to one group of subjects, and as a "low-reward" partner to a second group of subjects. If the hypothesis about the relationship between rewards and commitment is correct, you expect to see higher commitment scores for the high-reward partner. This part of the chapter describes a fictitious study that utilizes just such a procedure, and tests the relevant null hypothesis using an independent-samples *t* test.

The Commitment Study

Assume that you have drawn a sample of 20 subjects, and have randomly assigned 10 subjects to a high-reward condition and 10 to a low-reward condition. All subjects are given a packet of materials, and the following instructions appear on the first page:

> In this study, you are asked to imagine that you are single and not involved in any romantic relationship. You will read descriptions of 10 different "partners" with whom you might be involved in a romantic relationship. For each description, imagine that you are involved in a romantic relationship with that person. Think about what it would be like to date that person, given his/her positive features, negative features, and other considerations. After you have thought about it, rate how committed you would be to maintaining your romantic relationship with that person. Each "partner" is described on a separate sheet of paper, and at the bottom of each sheet there are four items with which you can rate your commitment to that particular relationship.

The paragraph that describes a given partner provides information about the extent to which the relationship with that person is rewarding and costly. It also provides information relevant to the investment size and alternative value associated with the relationship.

The Dependent Variable

The dependent variable in this study is the subject's commitment to a specific romantic partner. It would be ideal if you could arrive at a single score that indicates how committed a given subject is to a given partner. High scores would reveal that the subject is highly committed to the partner, and low scores would indicate the opposite. This section describes one way that you could use rating scales to arrive at such a score.

At the bottom of the sheet that describes a given partner, the subject is provided with four items that use a 9-point Likert-type rating format. Participants are asked to respond to these items to indicate the strength of their commitment to the partner described on that page. The following items are used in making these ratings.

PLEASE RATE YOUR COMMITMENT TO THIS PARTNER BY
CIRCLING YOUR RESPONSE TO EACH OF THE FOLLOWING ITEMS:

How committed are you to remaining in this relationship?
Not at All Committed   1 2 3 4 5 6 7 8 9   Extremely Committed

How likely is it that you will maintain this relationship?
Definitely Plan Not to   1 2 3 4 5 6 7 8 9   Definitely Plan to Maintain
Maintain

How likely is it that you will break up with this partner soon?
Extremely Likely   1 2 3 4 5 6 7 8 9   Extremely Unlikely

"I feel totally committed to this partner."
Disagree Strongly   1 2 3 4 5 6 7 8 9   Agree Strongly

Notice that, with each of the preceding items, circling a higher response number (closer to "9") reveals a higher level of commitment to the relationship. For a given partner, the subject's responses to these four items were summed to arrive at a final commitment score for that partner. This score could range from a low of 4 (if the subject had circled the "1" on each item) to a high of 36 (if the subject had circled the "9" on each item). These scores serve as the dependent variable in your study.

Manipulating the Independent Variable

The independent variable in this study is "level of rewards associated with a specific romantic partner." This independent variable is manipulated by varying the descriptions of the partners shown to the two treatment groups.

The partner descriptions given to the high-reward group are identical to those given to the low-reward group, but this is true only for the first nine descriptions. For partner 10, there is an important difference between the descriptions provided to the two groups. The sheet given to the high-reward group describes a relationship with a relatively high level of rewards, but the one given to the low-reward group describes a relationship with a relatively low level of rewards. Here is the description seen by subjects in the high-reward condition:

PARTNER 10: Imagine that you have been dating partner 10 for about a year, and you have put a great deal of time and effort into this relationship. There are not very many attractive members of the opposite sex where you live, so it would be difficult to replace this person with someone else. Partner 10 lives in the same neighborhood as you, so it is easy to see him/her as often as you like. This person enjoys the same recreational activities that you enjoy, and is also very good-looking.

Notice how the preceding description provides information relevant to the four investment model variables discussed earlier. The first sentence provides information dealing with investment size ("you have put a great deal of time and effort into this relationship"), and the second sentence deals with alternative value ("There are not very many attractive members of the opposite sex where you live"). The third sentence indicates that this is a low-cost relationship because "it is easy to see him or her as often as you like." In other words, there are no hardships associated with seeing this partner. If the descriptions said that the partner lives in a distant city, this would be a high-cost relationship.

However, you are most interested in the last sentence in this description, because it describes the level of rewards associated with the relationship. The relevant sentence is "This person enjoys the same recreational activities that you enjoy, and is also very good-looking." This statement establishes partner 10 as a high-reward partner for the subjects in the high-reward group.

In contrast, consider the description of partner 10 given to the low-reward group. Notice that it is identical to the description given to the high-reward group with regard to the first three sentences. The last sentence, however, deals with rewards, so this last sentence is different for the low-reward group. It describes a low-reward relationship:

PARTNER 10: Imagine that you have been dating partner 10 for about 1 year, and you have put a great deal of time and effort into this relationship. There are not very many attractive members of the opposite sex where you live, so it would be difficult to replace this person with someone else. Partner 10 lives in the same neighborhood as you, so it is easy to see him/her as often as you like. This person does not enjoy the same recreational activities that you enjoy, and is not very good-looking.

For this study, the vignette for partner 10 is the only scenario of interest. The analysis is only for the subjects' ratings of their commitment to partner 10, and disregards their responses to the first nine partners. The first nine were included to give the subjects some practice at evaluating commitment before encountering item 10.

Also notice the logic behind these experimental procedures: both groups of subjects are treated in exactly the same way with respect to everything except the independent variable. Descriptions of the first nine partners are identical in the two groups. Even the description of partner 10 is identical with respect to everything except the level of rewards associated with the relationship. Therefore, if the subjects in the high-reward group are significantly more committed to partner 10 than the subjects in the low-reward group, you can be reasonably confident that it is the level of reward manipulation that affected their commitment ratings. It would be difficult to explain the results in any other way.

In summary, you began your investigation with 10 of 20 subjects randomly assigned to the high-reward condition and the other 10 subjects assigned to the low-reward condition. After the subjects complete their task, you disregard their responses to the first nine scenarios, but record their responses to partner 10 and analyze these responses.

Entering the Data into a JMP Data Table

Remember that an independent-samples *t* test is appropriate for comparing two samples of observations. It allows you to determine whether there is a significant difference between the two samples with respect to the mean scores on their responses. More technically, it allows you to test the null hypothesis that, in the population, there is no difference between the two groups with respect to their mean scores on the response criterion. This section shows how to use the JMP Fit Y by X bivariate platform to test this null hypothesis for the current fictitious study.

The predictor variable in the study is "level of reward." This variable can assume one of two values: subjects were either in the high-reward group or in the low-reward group. Because this variable simply codes group membership, you know that it is measured on a nominal scale. In coding the data, you can give subjects a score of "High" if they were in the high-reward condition, and a score of "Low" if they were in the low-reward condition. You need a name for this variable, so call it Reward Group.

The response variable in this study is commitment: the subjects' ratings of how committed they would be to a relationship with partner 10. When entering the data, the response is the sum of the rating numbers that have been circled by the subject in responding to partner 10. This variable can assume values from 4 through 36, and is a continuous numeric variable. Call this variable Commitment in the JMP data table.

Figure 7.2 shows the JMP table, **commitment difference.jmp**, with this hypothetical data. Each line of data contains the group and the commitment response for one subject. Data from the 10 high-reward subjects were keyed first, followed by data from the 10 low-reward subjects. It is not necessary to enter the data sorted this way—data from low-reward and high-reward subjects could have been keyed in a random sequence.

Figure 7.2 Listing of the Commitment Difference JMP Data Table

commitment difference

Columns (2/0)
Reward Group
Commitment

Rows

Rows	
All Rows	20
Selected	0
Excluded	0
Hidden	0
Labelled	0

	Reward Group	Commitment
1	High	25
2	High	22
3	High	27
4	High	24
5	High	22
6	High	20
7	High	24
8	High	23
9	High	22
10	High	24
11	Low	12
12	Low	10
13	Low	15
14	Low	13
15	Low	16
16	Low	9
17	Low	13
18	Low	14
19	Low	15
20	Low	13

Performing a *t* Test in JMP

In the previous chapter the **Fit Y by X** command in JMP was used to look at measures of association between two continuous numeric variables. Now this platform is used to test the relationship between a continuous numeric response variable and a nominal classification (predictor) variable.

- To begin the analysis, choose **Fit Y by X** from the **Analyze** menu.
- Select **Commitment** from the **Select Columns** list and click the **Y, Response** button.
- Select **Reward Group** from the **Select Columns** list and click the **X, Factor** button.

Figure 7.3 shows the completed launch dialog.

Figure 7.3 Launch Dialog for Oneway Platform

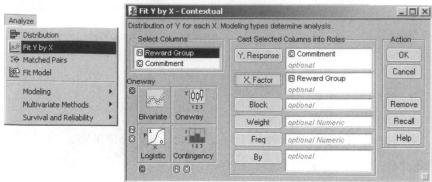

> **Note:** At any time, click the **Help** button to see help for the Oneway
> (Fit Y by X) platform. Or, choose the question mark (?) tool from the **Tools**
> menu or Tools palette and click on the analysis results.

Results from the JMP Analysis

Click **OK** in the launch dialog to see the initial results. Figure 7.4 presents the results
obtained from the JMP Fit Y by X platform for a continuous Y variable and a nominal or
ordinal X variable. Initially, only the scatterplot of Commitment by Reward Group
shows. To see a *t* test, select **Means/Anova/Pooled t** from the menu found on the
analysis title bar.

Notice that the title of the analysis is Oneway Analysis of Commitment By Reward Group. The Fit Y by X platform always performs a one-way analysis when the Y (dependent) variable is numeric and the X (independent or predictor) variable is nominal. When the X variable only has two levels, there are two types of independent *t* test:

- The **Means/Anova/Pooled t** gives the *t* test using the pooled standard error. This option also performs an analysis of variance, which is appropriate if the X variable has more than two levels.

- The **t Test** menu option tests the difference between two independent groups assuming unequal variances, and therefore uses an unpooled standard error.

The analysis results are divided into sections. The scatterplot is shown by default, with the Y variable (Commitment) on the Y axis and the predictor X variable (Reward Group) on the X axis. The **Means/Anova/Pooled t** option overlays *means diamonds*, illustrated to the right, on the groups in the scatterplot and appends several additional tables to the analysis results.

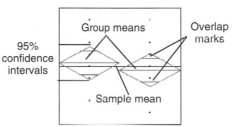

The means diamonds are a graphical illustration of the *t* test. If the overlap marks do not vertically separate the groups, as in this example, the groups are probably not significantly different. The groups appear separated if there is vertical space between the top overlap mark of one diamond and the bottom overlap of the other.

The t Test report gives the results of the *t* test for the null hypothesis that the means of the two groups do not differ in the population. The following section describes a systematic approach for interpreting the *t* test.

Figure 7.4 Results of *t* Test Analysis

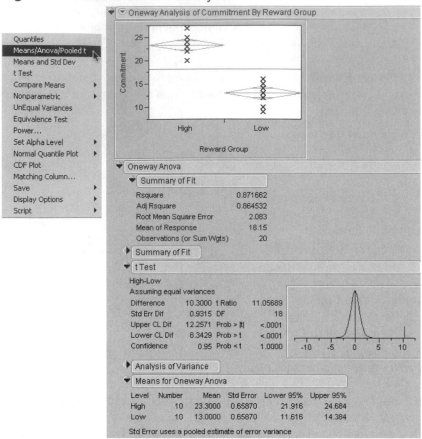

To change the markers in the scatterplot from the default small dots:

- 🖱 Select all rows in the data table.
- 🖱 Choose the **Markers** command from the **Rows** main menu and select a different marker from the markers palette.
- 🖱 Right-click in the scatterplot area and choose the **Marker Size** command. Select the size marker you want from the marker size palette.

Steps to Interpret the Results

1. **Make sure that everything looks reasonable.** As stated previously, the name of the nominal-level predictor variable appears on the X axis of the scatterplot, which shows the group values "Low" and "High." Statistics in the Means for Oneway

Anova table show the sample size (Number) for both high- and low-reward groups is 10. The mean Commitment score for the high-reward group is 23.3 and the mean score for the low-reward group is 13. The pooled standard error (Std Error) is 0.6587.

You should carefully review each of these figures to verify that they are within the expected range. For example, in this case you know there were no missing values so you want to verify that data for 10 subjects were observed for each group. In addition, you know that the Commitment variable was constructed such that it could assume possible values between 4 and 36. The observed group mean values are within these bounds, so there is no obvious evidence that an error was made keying the data.

2. **Review the *t* statistic and its associated probability value.** The **Means/Anova/Pooled t** option in this example lets you review the equal-variances *t* statistic, as noted at the beginning of the t Test table. This *t* statistic assumes that the two samples have equal variances. In other words, the distribution of scores around the means for both samples is similar.

Descriptive statistics for the difference in the group means are listed on the left in the t Test table in Figure 7.4. The information of interest, on the right in this table, is the obtained *t* statistic, its corresponding degrees of freedom, and probability values for both one-tailed and two-tailed tests.

- The obtained *t* statistic (*t* ratio) is 11.057 (which is quite large).
- This *t* statistic is associated with 18 degrees of freedom.
- The next item, Prob > |t|, shows that the *p* value associated with this *t* is less than 0.0001. This is the two-sided *t* test, which tests the hypothesis that the true difference in means (difference in the population) is neither significantly greater than nor significantly less than the observed difference of 10.3.

But what does this *p* value ($p < 0.0001$) really mean?

This p value is the probability that you would obtain a t statistic as large as 11.057 or larger (in absolute magnitude) if the null hypothesis were true; that is, you would expect to observe an absolute *t* value greater than 11.056 by chance alone in only 1 of 10,000 random samples of size 20 if there were no difference in the population means. If this null hypothesis were true, you expect to obtain a *t* statistic close to zero.

You can state the null hypothesis tested in this study as follows:

"In the population, there is no difference between the low-reward group and the high-reward group with respect to their mean scores on the commitment variable."

Symbolically, the null hypothesis can be represented

$$\mu_1 = \mu_2 \text{ or } \mu_1 - \mu_2 = 0$$

where μ_1 is the mean commitment score for the population of people in the high-reward condition, and μ_2 is the mean commitment score for the population of people in the low-reward condition.

Remember that anytime you obtain a *p* value less than 0.05, you reject the null hypothesis, and because your obtained *p* value is so small in this case, you can reject the null hypothesis of no commitment difference between groups. You can therefore conclude that there is probably a difference in mean commitment in the population between people in the high-reward condition and those in the low-reward condition.

The two remaining items in the t Test table are the one-tailed probabilities for the observed *t* value, which tests not only that there is a difference, but also the direction of the difference.

- Prob > t is the probability (< 0.0001 in this example) that the difference in population group means is greater than the observed difference.

- Prob < t is the probability (1.0000 in this example) that the difference in population group means is less than the observed difference.

3. **Review the graphic for *t* test**. The plot to the right in the t Test table in Figure 7.4 illustrates the *t* test. The plot is for the *t* density with 18 degrees of freedom. The obtained *t* value appears as a red line on the plot. In this case, the *t* value of 11.057 falls far into the right tail of the distribution, making it easy to see why it is so unlikely that independent samples would produce a higher *t* than the one shown if the difference between the groups in the population is close to zero.

4. **Review the sample means.** The significant *t* statistic indicates that the two populations are probably different from each other. The low-reward group has a mean score of 13.0 on the commitment scale and the high-reward group has a mean score of 23.3. It is therefore clear that, as you expected, the high-reward group demonstrates a higher level of commitment than the low-reward group.

5. **Review the confidence interval for the difference between the means.** A confidence interval extends from a lower confidence limit to an upper confidence limit. The t Test table in Figure 7.4 (also shown to the right) gives the upper confidence limit of the difference, Upper CL Dif, of 12.2571 and the lower confidence limit, Lower CL Dif, of 8.3429. Thus, the 95% confidence interval for the difference between means extends from 12.2571 to 8.3429.

High-Low			
Assuming equal variances			
Difference	10.3000	t Ratio	11.05689
Std Err Dif	0.9315	DF	18
Upper CL Dif	12.2571	Prob > |t|	<.0001
Lower CL Dif	8.3429	Prob > t	<.0001
Confidence	0.95	Prob < t	1.0000

This means you can estimate with a 95% probability that in the population, the actual difference between the mean of the low-reward condition and the mean of the high-reward condition is somewhere between 12.2571 and 8.3429. Notice that this interval does not contain the value of zero (difference). This is consistent with your rejection of the null hypothesis, which states "in the population, there is no difference between the low- and high-reward groups with respect to their mean scores on the commitment variable."

6. **Compute the index of effect size.** In this example the p value is less than the standard criterion of .05 so you reject the null hypothesis. You know that there is a statistically significant difference between the observed commitment levels for the high- and low-reward conditions. But is it a relatively large difference? The null hypothesis test alone does not tell whether the difference is large or small. In fact, when the sample is very large, you can obtain statistically significant results even if the difference is relatively trivial.

Because of this limitation of null hypothesis testing, many researchers now supplement statistics such as t tests with measures of effect size. The exact definition of effect size varies depending on the type of analysis. For an independent-samples t test, *effect size* can be defined as the degree to which one sample mean differs from a second sample mean stated in terms of standard deviation units. That is, it is the absolute value of the difference between the group means divided by the pooled estimate of the population standard deviation.

The formula for effect size, denoted d, is

$$d = \frac{\left|\overline{X}_1 - \overline{X}_2\right|}{S_p}$$

where

$\overline{X}_1$ = the observed mean of sample 1 (the participants in treatment condition 1)

$\overline{X}_2$ = the observed mean of sample 2 (the participants in treatment condition 2)

S_p = the pooled estimate of the population standard deviation

To compute the formula, use the sample means from the Means for Oneway Anova table, discussed previously. The estimate of the population standard deviation is the Root Mean Square Error found in the Summary of Fit table of the *t* test analysis, which gives the pooled estimate of the population standard deviation (see Figure 7.4).

$$d = \frac{|23.3 - 13.0|}{2.083} = 4.9448$$

This result tells you that the sample mean for the low-reward condition differs from the sample mean of the high-reward condition by 4.9448 standard deviations. To determine whether this is a relatively large or small difference, you can consult the guidelines provided by Cohen (1992), shown in Table 7.1.

Table 7.1 Guidelines for Interpreting *t* Test Effect Sizes

Effect Size	Computed d Statistic
Small	$d = 0.20$
Medium	$d = 0.50$
Large	$d = 0.80$

The computed d statistic of 4.9448 for the commitment study is larger than the large-effect value in Table 7.1. This means that the differences between the low- and high-reward participants in commitment levels for partner 10 produced both a statistically significant and a very large effect.

A General Outline for Summarizing Analysis Results

In performing an independent-samples *t* test (and other analyses), the following format can be used to summarize the research problem and results:

A) Statement of the problem
B) Nature of the variables
C) Statistical test
D) Null hypothesis (H_0)

E) Alternative hypothesis (H_1)
F) Obtained statistic
G) Obtained probability (p) value
H) Conclusion regarding the null hypothesis
I) Sample means and confidence interval of the difference
J) Effect size
K) Figure representing the results
L) Formal description of results for a paper

As an illustration, here is a summary of the preceding example analysis, according to this format.

A) **Statement of the problem:**
The purpose of this study was to determine whether there was a difference between people in a high-reward relationship and those in a low-reward relationship with respect to their mean commitment to the relationship.

B) **Nature of the variables:**
This analysis involved two variables. The predictor variable was level of rewards, which was measured on a nominal scale and could assume two values: a low-reward condition (coded as "Low") and a high-reward condition (coded as "High"). The response variable was commitment, which was a numeric continuous variable constructed from responses to a survey, with values ranging from 4 through 36.

C) **Statistical test:**
Independent-samples t test, assuming equal variances.

D) **Null hypothesis (H_0):**
$\mu_1 = \mu_2$. In the population, there is no difference between people in a high-reward relationship and those in a low-reward relationship with respect to their mean levels of commitment.

E) **Alternative hypothesis (H_1):**
$\mu_1 \neq \mu_2$. In the population, there is a difference between people in a high-reward relationship and those in a low-reward relationship with respect to their mean levels of commitment.

F) **Obtained statistic:**
$t = 11.057$.

G) **Obtained probability (p) value:**
$p < .0001$.

H) Conclusion regarding the null hypothesis:
Reject the null hypothesis.

I) Sample means and confidence interval of the difference:
The difference between the high-reward and the low-reward means was
23.3 – 13 = 10.3. The 95% confidence interval for this difference extended from
8.3429 to 12.2571.

J) Effect size:
d = 4.94 (large effect size).

K) Figure representing the results:
To produce the chart shown here:
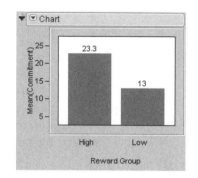
   🖰 Choose **Chart** from the **Graph** menu.
   🖰 Select **Reward Group** as X. Select
   Commitment and choose **Mean** from the
   statistic menu as Y.
   🖰 Shift-click both bars to highlight them.
   🖰 Select the **Label/Unlabel** command from
   the **Rows** menu to label the bars.
   Highlighting the bars selects all rows in
   the data table.

L) Formal description of results for a paper:
Most chapters of this text show you how to summarize the results of an analysis in a
way that would be appropriate if you were preparing a paper to be submitted for
publication in a scholarly research journal. These summaries generally follow the
format recommended in the *Publication Manual of the American Psychological
Association* (2001), which is required by many journals in the social sciences. Here is
an example of how the current results could be summarized according to this format:

Results were analyzed using an independent-samples t test. This analysis revealed a
significant difference between the two groups, $t(18)$ = 11.05689; $p < 0.0001$. The sample
means are displayed with a bar chart, which illustrates that subjects in the high-reward
condition scored significantly higher on commitment than did subjects in the low-reward
condition (for high-reward group, Mean = 23.30, SD = 1.95; for low-reward group, Mean =
13.00, SD = 2.21). The observed difference between means was 10.30 and the 95%
confidence interval for the difference between means extended from 8.34 to 12.26. The
effect size was computed as d = 4.94. According to Cohen's (1992) guidelines for t tests,
this represents a large effect.

An Example with Nonsignificant Differences

Researchers do not always obtain significant results when performing investigations such as the one described in the previous section. This section repeats the analyses, this time using fictitious data that result in a nonsignificant *t* test. The conventions for summarizing nonsignificant results are then presented.

The data table for the following example is **commitment no difference.jmp**. The data have been modified so that the two groups do not differ significantly on mean levels of commitment. Figure 7.5 shows this JMP table. Simply "eyeballing" the data reveals that similar commitment scores seem to be displayed by subjects in the two conditions. Nonetheless, a formal statistical test is required to determine whether significant differences exist.

Figure 7.5 Example Data for Nonsignificant *t* Test

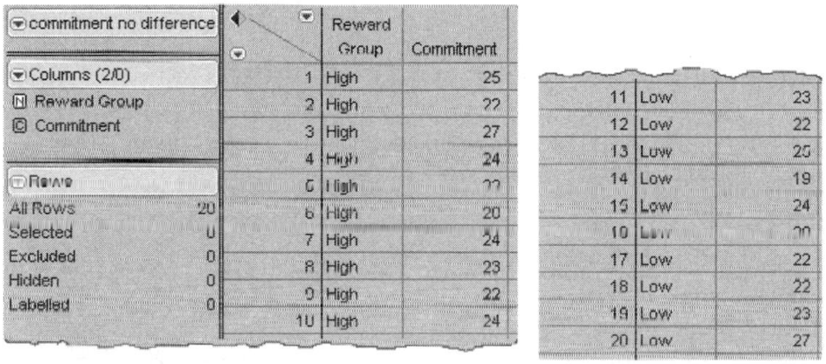

Proceed as before.

- ◌ Use the **Fit Y by X** command on the **Analyze** menu.
- ◌ When the results appear, select **Means/Anova/Pooled t** from the menu on the analysis title bar.
- ◌ Also select the **Means and Std Dev** option, as illustrated in Figure 7.6. This option shows the pooled standard deviation used to compute *t* values in this example and in the previous example.

Figure 7.6 Results of Analysis with Nonsignificant *t* Test

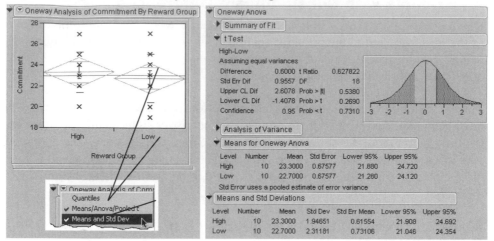

The t Test table in Figure 7.6 shows the *t* statistic, assuming equal variances, to be small at 0.627 and the *p* value for this *t* statistic to be large at 0.5380. Because this *p* value is greater than the standard cutoff of 0.05, you can say that the *t* statistic is nonsignificant. These results mean that you don't reject the null hypothesis of equal population means on Commitment. In other words, you conclude that there is not a significant difference between mean levels of Commitment in the two samples.

This analysis shows the Means and Std Deviations table for the data, which gives the Std Dev and Std Error Mean for each level of the Commitment variable. Note that, as before, the Std Error value in the Means for Oneway Anova table is the same for both levels because it is the pooled error that is used in the *t* test computations. The mean Commitment score is 23.3 for the high-reward group and 22.7 for the low-reward group. The t Test table shows the difference between the means is 0.6 with lower confidence limit of –1.4078 and upper confidence limit of 2.6078. Notice that this interval includes zero, which is consistent with your finding that the difference between means is nonsignificant.

Compute the effect size of the difference as the difference between the means divided by the estimate of the standard deviation of the population (Root Mean Square Error in the Summary of Fit table):

$$d = \frac{\overline{X}_1 - \overline{X}_2}{S_p}$$

$$d = \frac{23.3 - 22.7}{2.137} = 0.2808$$

Thus the index of effect size for the current analysis is 0.2802. According to Cohen's guidelines in Table 7.1, this value falls between a small and medium effect.

For this analysis, the statistical interpretation format appears as follows. This is the same study as shown previously, so you can complete items A through E in the same way.

F) Obtained statistic:
$t = 0.6278$.

G) Obtained probability (*p*) value:
$p = 0.5380$.

H) Conclusion regarding the null hypothesis:
Fail to reject the null hypothesis.

I) Sample means and confidence interval of the difference:
The difference between the high-reward and the low-reward means was 23.3   22.7 = 0.6. The 95% confidence interval for this difference extended from   1.4078 to 2.6078.

J) Effect size:
$d = 0.2808$ (small to medium effect size).

K) Figure representing the results:

To produce the chart shown here:

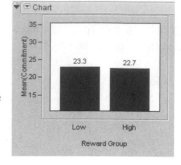

- 🖱 Choose the **Chart** command from the **Graph** menu.
- 🖱 Select **Reward Group** as X. Select **Commitment** and choose **Mean** from the statistic menu as Y.
- 🖱 Shift-click both bars to highlight them. Highlighting the bars selects all rows in the data table.
- 🖱 Select the **Label/Unlabel** command from the **Rows** menu to label the bars. Highlighting the bars selects all rows in the data table.

L) Formal description of results for a paper:

The following is an example of a formal description of the results:

> Results were analyzed using an independent-samples t test. This analysis failed to reveal a significant difference between the two groups, $t(18) = 0.628$; $p = 0.538$. The bar chart of the sample means illustrates that subjects in the high-reward condition demonstrated scores on commitment that were similar to those shown by subjects in the low-reward condition (for high-reward group, Mean = 23.30, SD = 1.95; for low-reward group, Mean = 22.70, SD = 2.31). The observed difference between means was 0.6 and the 95% confidence interval for the difference between means extended from −1.41 to 2.61. The effect size was computed as $d = 0.28$. According to Cohen's (1992) guidelines for t tests, this represents a small to medium effect.

The Paired-Samples *t* Test

The paired-samples t test (sometimes called the *correlated-samples t* test or *matched-samples t* test) is similar to the independent-samples test in that both procedures compare two samples of observations, and determine whether the mean of one sample significantly differs from the mean of the other. With the independent-samples procedure, the two groups of scores are completely independent. That is, an observation in one sample is not related to any observation in the other sample. Independence is achieved in experimental research by drawing a sample of subjects and randomly assigning each subject to either Condition 1 or Condition 2. Because each subject contributes data under only one condition, the two samples are empirically independent.

In contrast, each score in one sample of the paired-samples procedure is *paired* in some meaningful way with a score in the other sample. There are several ways that this can happen. The following examples illustrate some of the most common paired situations.

Examples of Paired-Samples Research Designs

Be aware that the following fictitious studies illustrate paired-sample designs, but might not represent sound research methodology from the perspective of internal or external validity. Problems with some of these designs are reviewed later.

Each Subject Is Exposed to Both Treatment Conditions

Earlier sections described an experiment in which level of reward was manipulated to see how it affected subjects' level of commitment to a romantic relationship. The study required that each subject review 10 people and rate commitment to each fictitious romantic partner. The dependent variable is the rated amount of commitment the subjects displayed toward partner 10. The independent variable is manipulated by varying the description of partner 10: subjects in the "high-reward" condition read that partner 10 had positive attributes, while subjects in the "low-reward" condition read that partner 10 did not have these attributes. This study is an independent-samples study because each subject was assigned to either a high-reward condition or a low-reward condition (but no subject was ever assigned to both conditions).

You can modify this (fictitious) investigation so that it follows a paired-samples research design by conducting the study with only one group of subjects instead of two groups. Each subject rates partner 10 twice, once after reading the low-reward version of partner 10, and a second time after reading the high-reward version of partner 10.

It would be appropriate to analyze the data resulting from such a study using the paired-samples *t* test because it is possible to meaningfully pair observations under both conditions. For example, subject 1's rating of partner 10 under the low-reward condition could be paired with his or her rating of partner 10 under the high-reward condition, subject 2's rating of partner 10 under the low-reward condition could be paired with his or her rating of partner 10 under the high-reward condition, and so forth. Table 7.2 shows how the resulting data could be arranged in tabular form.

Remember that the dependent variable is still the commitment ratings for partner 10. Subject 1 (John) has two scores on this dependent variable—a score of 11 obtained in the low-reward condition, and a score of 19 obtained in the high-reward condition. John's score from the low-reward condition is paired with his score from the high-reward condition. The same is true for the remaining participants.

Table 7.2 Fictitious Data from a Study Using
a Paired-Samples Procedure

	Low-Reward Condition	High-Reward Condition
John	11	19
Mary	9	22
Tim	10	23
Susan	9	18
Maria	12	21
Fred	12	25
Frank	17	22
Edie	15	25
Jack	14	24
Shirley	19	31

Matching Subjects

The preceding study used a type of *repeated-measures* approach. There is only one sample of participants, and repeated measurements on the dependent variable (commitment) are taken from each participant. That is, each person contributes one score under the low-reward condition and a second score under the high-reward condition.

A different approach could have used a type of *matching procedure*. With a matching procedure, a given participant provides data under only one experimental condition. However, each person is matched with certain conditions to a different person who provides data under the other experimental condition:

- The participants are matched on some variable that is expected to be related to the dependent variable.

- The matching is done prior to the manipulation of the independent variable.

For example, imagine that it is possible to administer an "emotionality scale" to subjects. Further, prior research has shown that scores on this scale are strongly correlated with scores on romantic commitment (the dependent variable in your study). You could administer this emotionality scale to 20 participants, and use their scores on that scale to match them. That is, you could place them in pairs according to their similarity on the emotionality scale.

For example, suppose scores on the emotionality scale range from a low of 100 to a high of 500. Assume that John scores 111 on this scale, and William scores 112. Because their scores are very similar, you pair them together, and they become subject pair 1. Tim scores 150 on this scale, and Fred scores 149. Because their scores are very similar, you also pair them together as subject pair 2. Table 7.3 shows how you could arrange these fictitious pairs of subjects.

Within a subject pair, one participant is randomly assigned to the low-reward condition, and one is assigned to the high-reward condition. Assume that, for each of the pairs in Table 7.3, the person listed first was randomly assigned to the low-reward condition, and the person listed second was assigned to the high-reward condition. The study then proceeds in the usual way, with subjects rating the various pairs described in the questionnaire.

Table 7.3 Fictitious Data from a Study Using a Matching Procedure

| | Commitment Ratings of Partner 10 | |
Subject Pairs	Low-Reward Condition	High-Reward Condition
Subject pair 1 (John and William)	8	19
Subject pair 2 (Tim and Fred)	9	21
Subject pair 3 (Frank and Jack)	10	21
Subject pair 4 (Howie and Jim)	10	23
Subject pair 5 (Andy and Floyd)	11	24
Subject pair 6 (Walter and Rich)	13	26
Subject pair 7 (James and Denny)	14	27
Subject pair 8 (Reuben and Joe)	14	28
Subject pair 9 (Mike and Peter)	16	30
Subject pair 10 (George and Dave)	18	32

Table 7.3 shows that, for subject pair 1, John gave a commitment rating of 8 to partner 10 in the low-reward condition; William gave a commitment score of 19 to partner 10 in the high-reward condition. When analyzing the data, you pair John's score on the commitment variable with William's score on commitment. The same will be true for the remaining subject pairs. A later section shows how to analyze the data using JMP.

> **Note:** Remember that subjects are placed together in pairs on the basis of some matching variable *before the independent variable is manipulated.* The subjects are *not* placed together in pairs on the basis of their scores on the dependent variable. In the present case, subjects are paired based on the similarity of their scores on the emotionality scale, administered previously. Later, the independent variable is manipulated and the subjects' commitment scores are recorded. Although they are not paired on the basis of their scores on the dependent variable, you hope that their scores on the dependent variable will be correlated. There is more discussion on this in a later section.

Taking Pretest and Posttest Measures

Consider now a different type of research problem. Assume that an educator believes that taking a foreign language course causes an improvement in critical-thinking skills among college students. To test the hypothesis, the educator administers a test of critical-thinking skills to a single group of college students at two points in time:

1. A pretest is administered at the beginning of the semester (prior to taking the language course).

2. A posttest is administered at the end of the semester (after completing the course).

The data obtained from the two test administrations appear in Table 7.4.

Table 7.4 Fictitious Data from Study Using
a Pretest-Posttest Procedure

| Subject | Scores on Test of Critical-Thinking Skills | |
	Pretest	Posttest
John	34	55
Mary	35	49
Tim	39	59
Susan	41	63
Maria	43	62
Fred	44	68
Frank	44	69
Edie	52	72
Jack	55	75
Shirley	57	78

You can analyze these data using the paired-samples *t* test because it is meaningful to pair the same subject's pretest and posttest scores. When the data are analyzed, the results indicate whether there was a significant increase in critical-thinking scores over the course of the semester.

Problems with the Paired-Samples Approach

Some of the studies described in the preceding section use fairly weak experimental designs. This means that, even if you had conducted the studies, you might not have been able to draw firm conclusions from the results because alternative explanations could be offered for those results.

Order Effects

Consider the first investigation that exposes each subject to both the low-reward version of partner 10 and the high-reward version of partner 10. If you designed this study poorly, it might suffer from confoundings that make it impossible to interpret the results. For example, suppose you design a study so that each subject rates the low-reward version first and the high-reward version second. If you then analyze the data and find that higher commitment ratings were observed for the high-reward condition, you would not know whether to attribute this finding to the manipulation of the independent variable (level of rewards) or to *order effects*. Order effects are the possibility that the order in which the treatments were presented influences scores on the dependent variable. For example, it is possible that subjects tend to give higher ratings to partners that are rated later in serial order. If this is the case, the higher ratings observed for the high-reward partner may simply reflect such an order effect.

Alternative Explanations

The third study, which investigated the effects of a language course on critical-thinking skills, also displays a weak experimental design. The single-group, pretest-posttest design assumes you administered the test of critical-thinking skills to the students at the beginning and again at the end of the semester. It further assumes that you observe a significant increase in their skills over this period as would be consistent with your hypothesis that the foreign language course helps develop critical-thinking skills.

However, there are other reasonable explanations for the findings. Perhaps the improvement was simply due to the process of *maturation*—changes that naturally take place as people age. Perhaps the change is due to the general effects of being in college, independent of the effects of the foreign language course. Because of the weak design used in this study, you will probably never be able to draw firm conclusions about what was really responsible for the students' improvement.

This is not to argue that researchers should never obtain the type of data that can be analyzed using the paired-samples *t* test. For example, the second study described previously (the matching procedure) was reasonably sound and might have provided interpretable results. The point is that research involving paired samples must be designed very carefully to avoid the sorts of problems discussed here. You can deal with most of these difficulties through the appropriate use of counterbalancing, control groups, and other strategies. The problems inherent in repeated measures and matching designs, along with the procedures that can be used to handle these problems, are discussed in Chapter 11, "One-Way ANOVA with One Repeated-Measures Factor," and Chapter 12, "Factorial ANOVA with Repeated-Measures Factors and Between-Subjects Factors."

When to Use the Paired-Samples Approach

When conducting a study with only two treatment conditions, you often have the choice of using either the independent-samples approach or the paired-samples approach. One of the most important considerations is the extent to which the paired-samples analysis can result in a more sensitive test. That is, to what extent is the paired-samples approach more likely to detect significant differences when they actually do exist?

It is important to understand that the paired-samples *t* test has one important weakness in regard to sensitivity—it has only *half* the degrees of freedom as the equivalent independent-samples test. Because the paired-samples approach has fewer degrees of freedom, it must display a larger *t* value than the independent-samples *t* test to attain statistical significance.

However, under the right circumstances, the paired-samples approach results in a smaller standard error of the mean (the denominator in the formula used to compute the *t* statistic), and a smaller standard error usually results in a more sensitive test. The exception is that the paired-samples approach results in a smaller standard error only if scores on the two sets of observations are positively correlated with one another. This concept is easiest to understand with reference to the pretest-posttest study shown in Table 7.4.

Notice that scores on the pretest appear to be positively correlated with scores on the posttest. That is, subjects who obtained relatively low scores on the pretest (such as John) also tended to obtain relatively low scores on the posttest. Similarly, subjects who obtained relatively high scores on the pretest (such as Shirley) also tended to obtain relatively high scores on the posttest. This shows that although the subjects might have displayed a general improvement in critical-thinking skills over the course of the semester, their ranking relative to one another remained relatively constant. The subjects with the lowest scores at the beginning of the term still tended to have the lowest scores at the end of the term.

The situation described here is the type of situation that makes the paired samples *t* test the optimal procedure. Because pretest scores are correlated with posttest scores, the paired-samples approach should yield a fairly sensitive test.

The same logic applies to the other studies described previously. For example, look at the values in Table 7.2, from the study in which subjects were assigned to pairs based on matching criteria. There appears to be a correlation between scores obtained in the low-reward condition and those obtained in the high-reward condition. This could be because subjects were first placed into pairs based on the similarity of their scores on the emotionality scale, and the emotionality scale is predictive of how subjects respond to the commitment scale. For example, both John and William (subject pair 1) display relatively low scores on commitment, presumably because they both scored low on the emotionality scale that was initially used to match them. Similarly, both George and Dave (subject pair 10) scored relatively high on commitment, presumably because they both scored high on emotionality.

This illustrates why it is so important to select *relevant* matching variables when using a matching procedure. There is a correlation between the two commitment variables above because (presumably) emotionality is related to commitment. If you had instead assigned subjects to pairs based on some variable that is not related to commitment (such as subject shoe size), the two commitment variables would not be correlated, and the paired-samples *t* test would not provide a more sensitive test. Under those circumstances, you achieve more power by instead using the independent-samples *t* test and capitalizing on the greater degrees of freedom.

An Alternative Test of the Investment Model

The remainder of this chapter shows how to use JMP to perform paired-sample *t* tests, and describes how to interpret the results. The first example is based on the fictitious study that investigates the effect of levels of reward on commitment to a romantic relationship. The investigation included 10 subjects, and each subject rated partner 10 after reviewing the low-reward version of partner 10, and again after reviewing the high-reward version. Figure 7.7 shows the data (shown previously in Table 7.1) keyed into a JMP data table, called **commitment paired.jmp**.

Figure 7.7 Paired Commitment Data

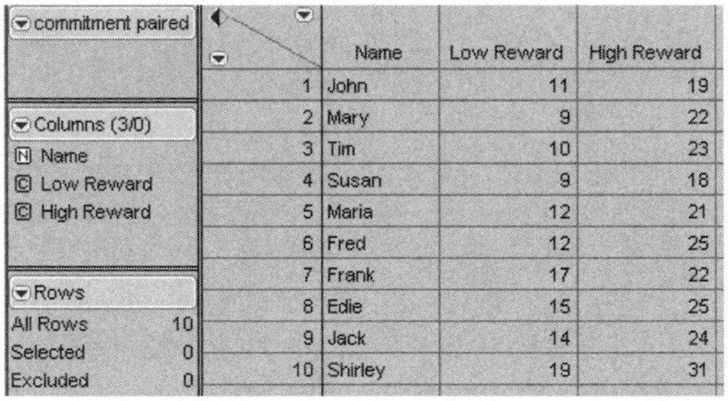

	Name	Low Reward	High Reward
1	John	11	19
2	Mary	9	22
3	Tim	10	23
4	Susan	9	18
5	Maria	12	21
6	Fred	12	25
7	Frank	17	22
8	Edie	15	25
9	Jack	14	24
10	Shirley	19	31

commitment paired

Columns (3/0)
N Name
C Low Reward
C High Reward

Rows
All Rows 10
Selected 0
Excluded 0

Notice in the JMP table that the data are arranged exactly as they were presented in Table 7.2. The two score variables list commitment ratings obtained when subjects reviewed the low-reward version of partner 10 (Low Reward) and when they reviewed the high-reward version (High Reward).

It is easy to do a paired *t* test in JMP.

🖰 Choose the **Matched Pairs** command from the **Analyze** menu.

🖰 When the launch dialog appears, select the Low Reward and the High Reward variables (the set of paired responses) and click the **Y, Paired Response** button to enter them as the variables to be analyzed.

🖰 Click **OK** to see the results in Figure 7.8.

Figure 7.8 Results from Paired *t* Analysis

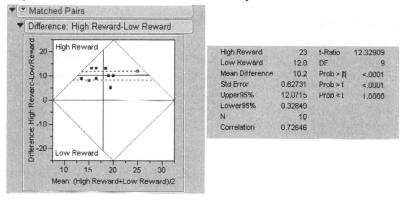

Interpreting the Paired *t* Plot

As in most JMP analyses, the results start with a graphic representation of the analysis. The illustration here describes the paired *t* test plot, using y1 and y2 as the paired variables. The vertical axis is the difference between the group means, with the zero line that represents zero difference between means. If the

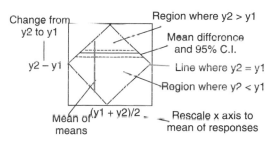

dotted 95% confidence lines (around the plotted difference) also encompass the zero reference, then you visually conclude there is no difference between group means. In Figure 7.8, the mean difference and its confidence lines are far away from the zero reference so you can visually conclude there is a difference between groups.

Note: The diamond-shaped rectangle in the plot results from the **Reference Frame** option found on the menu on the Matched Pairs title bar.

Interpreting the Paired *t* Statistics

The Matched Pairs report lists basic summary statistics. You can see that the mean commitment score in the low-reward condition is 12.8 and the mean score in the high-reward condition is 23.0. Subjects displayed higher levels of commitment for the high-reward version of partner 10.

Also note in the Matched Pairs report that the lower 95% confidence limit (Lower95%) is 8.3265 and the upper limit (Upper95%) is 12.0715. This lets you estimate with 95% probability that the actual difference between the mean of the low-reward condition and the mean of the high-reward condition (in the population) is between these two confidence limits. Also, as shown in the plot, this interval does not contain zero, which indicates you will be able to reject the null hypotheses. If there was no difference between group means, you expect the confidence interval to include zero (a difference score of zero).

Note that the Mean Difference is 10.2 and the standard error of the difference is 0.82731. The paired *t* analysis determines whether this mean difference is significantly different from zero. Given the way that this variable was created, a positive value on this difference indicates that, on the average, scores for High Reward tended to be higher than scores for Low Reward. The direction of this difference is consistent with your prediction that higher rewards are associated with greater levels of commitment.

Next, review the results of the *t* test to determine whether this mean difference score is significantly different from zero. The *t* statistic in a paired-samples *t* test is computed using the following formula:

$$t = M_d / SE_d$$

where

M_d = the mean difference score

SE_d = the standard error of the mean for the difference scores (the standard deviation of the sampling distribution of means of difference scores).

This *t* value in this example is obtained by dividing the mean difference score of 10.2 by the standard error of the difference (0.82731 shown in the results table), giving $t =$ 12.32909. Your hypothesis is one-sided—you expect high-reward groups to have significantly higher commitment scores. Therefore, the probability of getting a greater positive *t* value shows as Prob > t and is less than 0.0001. This *p* value is much lower than the standard cutoff of 0.05, which indicates that the mean difference score of 10.2 is significantly greater than zero. Therefore you can reject the null hypothesis that the population difference score was zero, and conclude that the mean commitment score of 23.0 observed with the high-reward version of partner 10 is significantly higher than the mean score of 12.8 observed with low-reward version of partner 10. In other words, you tentatively conclude that the level of reward manipulation had an effect on rated commitment.

The degrees of freedom associated with this *t* test are $N-1$, where N is the number of pairs of observations in the study. This is analogous to saying that N is equal to the number of difference scores that are analyzed. If the study involves taking repeated measures from a single sample of subjects, N will be equal to the number of subjects. However, if the study involves two sets of subjects who are matched to form subject pairs, N will be equal to the number of subject *pairs*, which is one-half the total number of subjects.

The present study involved taking repeated measures from a single sample of 10 participants. Therefore, $N = 10$ in this study, and the degrees of freedom are $10 - 1 = 9$, as in the *t* test results shown in Figure 7.8.

Effect Size of the Result

The previous example in this chapter defined effect size, *d*, as the degree to which a mean score obtained under one condition differs from the mean score obtained under a second condition. For the paired *t* test, the *d* statistic is computed by dividing the difference between the means by the estimated standard deviation of the population of difference scores. That is,

$$d = \frac{\overline{X}_1 - \overline{X}_2}{S_d}$$

where

$\overline{X}_1$ = the observed mean of sample 1 (the participants in treatment condition 1)

$\overline{X}_2$ = the observed mean of sample 2 (the participants in treatment condition 2)

S_d = the estimated standard deviation of the population of difference scores

The paired *t* test report does not show s_d but it is computed as the standard error (Std Err), shown in the *t* test analysis report, multiplied by the square root of the sample size (N). That is,

$$s_d = (\text{Std Err} * \sqrt{N}) = (0.82731 * \sqrt{10}) = (0.82731 * 3.162) = 2.616$$

Then *d* is computed:

$$d = \frac{|\overline{X}_1 - \overline{X}_2|}{S_d} = \frac{|23 - 12.8|}{2.616} = \frac{10.2}{2.616} = 3.999$$

Thus, the obtained index of effect size for the current study is 3.999, which means the commitment score under the low-reward condition differs from the mean commitment score under the high-reward condition by almost four standard deviations. To determine whether this is a large or small difference, refer to the guidelines provided by Cohen (1992), shown in Table 7.1. Your obtained *d* statistic of 3.999 is much larger than the "large effect" value of 0.80. This means that the manipulation in your study produced a very large effect.

Summarizing the Results of the Analysis

One way to produce a summary bar chart is to stack the paired commitment scores in the two reward columns into a single column, and use the Chart platform to plot the means of the reward groups. The example in Chapter 3, "Working with JMP Data," uses these paired data to show how to stack columns. With the **Commitment Paired** data table active,

- Choose **Stack** from the **Tables** menu.
- Select the **Low Reward** and **High Reward** variables and click **Add** to add them to the **Stack Columns** list, shown in Figure 7.9.
- Clear the **Stack By Row** box in the dialog, which is checked by default.

Figure 7.9 Stack Dialog to Stack Commitment Scores
into a Single Column

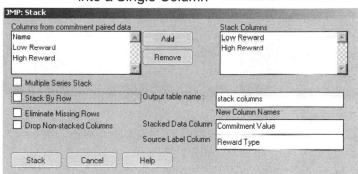

🖱 Click the **Stack** button to see the data table shown on the left in Figure 7.11. This data table with stacked columns is now in the same form as the one shown previously in Figure 7.5, used to illustrate the independent-samples *t* test.

🖱 Choose the **Chart** command from the **Graph** menu and complete the Chart dialog as in Figure 7.10.

Figure 7.10 Completed Chart Dialog to Chart Mean
Commitment Scores

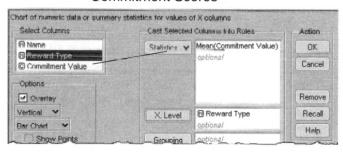

🖰 Click **OK** in the Chart dialog to see the bar chart on the right in Figure 7.11.

Figure 7.11 Stacked Table and Bar Chart of Means

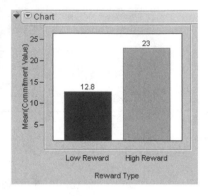

	Name	Reward Type	Commitment Value
1	John	Low Reward	11
2	Mary	Low Reward	9
3	Tim	Low Reward	10
4	Susan	Low Reward	9
5	Maria	Low Reward	12
6	Fred	Low Reward	12
7	Frank	Low Reward	17
8	Edie	Low Reward	15
9	Jack	Low Reward	14
10	Shirley	Low Reward	19
11	John	High Reward	19
12	Mary	High Reward	22
13	Tim	High Reward	23
14	Susan	High Reward	18
15	Maria	High Reward	21
16	Fred	High Reward	25
17	Frank	High Reward	22
18	Edie	High Reward	25
19	Jack	High Reward	24
20	Shirley	High Reward	31

Shift-click both bars to select them.
Then choose **Rows → Label/Unlabel**
to label the bars with the mean values.

You can summarize the results of the present analysis following the same format used with the independent groups *t* test, as presented earlier in this chapter.

Results were analyzed using a paired-samples *t* test. This analysis revealed a significant difference between mean levels of commitment observed in the two conditions, $t(9) = 12.329; p < 0.0001$. The sample means are displayed as a bar chart in Figure 7.11, which shows that mean commitment scores appear significantly higher in the high-reward condition (mean = 23) than in the low-reward condition (mean = 12.8). The observed difference between these scores was 10.2, and the 95% confidence interval for the difference extended from 8.3285 to 12.0715. The effect size was computed as $d = 3.999$. According to Cohen's (1992) guidelines for *t* tests, this represents a very large effect.

A Pretest-Posttest Study

An earlier section presented the hypothesis that taking a foreign language course leads to an improvement in critical thinking among college students. To test this hypothesis, assume that you conducted a study in which a single group of college students took a test of critical-thinking skills both before and after completing a semester-long foreign language course. The first administration of the test constituted the study's pretest, and the second administration constituted the posttest. The JMP data table **languagetext.jmp**, shown on the left in Figure 7.12, has the results of the study.

Analyze the data using the same approach as shown in the previous example.

- ⌐⊕ Choose the **Matched Pairs** command from the **Analyze** menu.
- ⌐⊕ When the launch dialog appears, select **Pretest** (each participant's score on the pretest) and **Posttest** (each participant's score on the posttest) and click the **Y, Paired Response** button to enter them as the pair of variables to be analyzed.
- ⌐⊕ Click **OK** to see the results in Figure 7.8.

Figure 7.12 Data for Pretest and Posttest Language Scores

	Name	Pretest	Posttest
1	Paul	34	55
2	Vilem	35	49
3	Pavel	39	59
4	Sunil	41	63
5	Maria	43	62
6	Fred	44	68
7	Jirka	44	69
8	Eduardo	52	72
9	Ashur	66	75
10	Shirley	57	78

Matched Pairs
Difference: Posttest-Pretest

Posttest	65	t-Ratio	21.79518
Pretest	44.4	DF	9
Mean Difference	20.6	Prob > |t|	<.0001
Std Error	0.94516	Prob > t	<.0001
Upper95%	22.7381	Prob < t	1.0000
Lower95%	18.4619		
N	10		
Correlation	0.94779		

The positive mean difference in score (Posttest – Pretest) is consistent with the hypothesis that taking a foreign language course causes an improvement in critical thinking. You can interpret the results in the same manner as in the previous example. This analysis revealed a significant difference between mean levels of pretest scores and posttest scores, with $t(9) = 21.7951$ and $p < 0.0001$.

Summary

Many simple investigations involve the comparison of responses from only two treatment conditions. The *t* test is one of the most commonly used statistical procedures to compare the mean responses from two treatment conditions. This chapter described the independent-samples *t* test and the paired-samples *t* test, and showed how to perform both types of analyses.

The JMP Fit by Y platform was used to perform the independent-samples test. You saw how to determine whether the equal-variances or unequal-variances *t* test is appropriate, and how to interpret the results. Examples of various kinds of paired data were discussed, with explanation of problems that can occur with paired data. The Matched Pairs platform was used to analyze paired data with a paired *t* test.

You were given an example of a structured outline for summarizing analysis results, and saw how to prepare a formal description of the results for publication.

When an investigation involves more than two conditions, however, the *t* test is no longer appropriate, and you usually replace it with the *F* test obtained from an analysis of variance (ANOVA). The simplest ANOVA procedure—the one-way ANOVA with one between-groups factor—is the topic of the next chapter.

Assumptions Underlying the *t* Test

Assumptions Underlying the Independent-Samples *t* Test

Level of measurement

The response variable should be assessed on an interval or ratio level of measurement. The predictor variable should be a nominal-level variable that must include two categories (groups).

Independent observations

A given observation should not be dependent on any other observation in either group. In an experiment, you normally achieve this by drawing a random sample and randomly assigning each subject to only one of the two treatment conditions. This assumption would be violated if a given subject contributed scores on the response variable under both treatment conditions. The independence assumption is also violated when one subject's behavior influences another subject's behavior within the same condition. The texts discussed in this chapter rely on the assumption of independent observations. If this assumption is not met, inferences about the population (results of hypothesis tests) can be misleading or incorrect.

Random sampling

Scores on the response variable should represent a random sample drawn from the populations of interest.

Normal distributions

Each sample should be drawn from a normally distributed population. If each sample contains over 30 subjects, the test is robust against moderate departures from normality.

Homogeneity of variance

To use the equal-variances *t* test, you should draw the samples from populations with equal variances on the response variable. If the null hypothesis of equal population variances is rejected, you should use the unequal-variances *t* test.

Assumptions Underlying the Paired-Samples *t* Test

Level of measurement

The response variable should be assessed on an interval or ratio level of measurement. The predictor variable should be a nominal-level variable that must include just two categories.

Paired observations

A given observation appearing in one condition must be paired in some meaningful way with a corresponding observation appearing in the other condition. You can accomplish this by having each subject contribute one score under Condition 1 and a separate score under Condition 2. Observations could also be paired by using a matching procedure to create the sample.

Independent observations

A given subject's score in one condition should not be affected by any other subject's score in either of the two conditions. It is acceptable for a given subject's score in one condition to be dependent upon his or her *own score* in the other condition. This is another way of saying that it is acceptable for subjects' scores in Condition 1 to be correlated with their scores in Condition 2.

Random sampling

Subjects contributing data should represent a random sample drawn from the populations of interest.

Normal distribution for difference scores

The differences in paired scores should be normally distributed. These difference scores are normally created by beginning with a given subject's score on the dependent variable obtained under one treatment condition, and subtracting from it that subject's score on the dependent variable obtained under the other treatment condition. It is not necessary that the individual dependent variables be normally distributed, as long as the distribution of difference scores is normally distributed.

Homogeneity of variance

The populations represented by the two conditions should have equal variances on the response criterion.

References

American Psychological Association. 2001. *Publication Manual of the American Psychological Association.* 5th ed. Washington: American Psychological Association.

Cohen, J. 1992. "A Power Primer." *Psychological Bulletin* 112:155–159.

Hartigan, J. A., and B. Kleiner. 1981. "Mosaics for Contingency Tables." *Proceedings of the Thirteenth Symposium on the Interface between Computer Science and Statistics*, ed. W. F. Eddy, 268–273. New York: Springer-Verlag.

Rusbult, C. E. 1980. "Commitment and Satisfaction in Romantic Associations: A Test of the Investment Model." *Journal of Experimental Social Psychology* 16:172–186.

SAS Institute Inc. 2003. *JMP Statistics and Graphics Guide.* Cary, NC: SAS Institute Inc.

One-Way ANOVA with One Between-Subjects Factor

> **Overview.** In this chapter you learn how to enter data and use JMP to perform a one-way analysis of variance. This chapter focuses on the *between-subjects* design, in which each subject is exposed to only one condition under the independent variable. The use of a multiple comparison procedure (Tukey's HSD test) is described, and guidelines for summarizing the results of the analysis in tables and text are provided.

Introduction: One-Way ANOVA Between-Subjects Design

One-way analysis of variance (ANOVA) is appropriate when an analysis involves

- a single nominal or ordinal predictor (independent) variable that can assume two or more values
- a single continuous numeric response variable.

In the preceding chapter (Chapter 7, "*t* Tests: Independent Samples and Paired Samples"), you learned about the independent samples *t* test, which you can use to determine whether there is a significant difference between two groups of subjects with respect to their scores on a continuous numeric response variable. But what if you are conducting a study in which you need to compare more than just two groups? In that situation, it is often appropriate to analyze your data using a one-way ANOVA.

The analysis that this chapter describes is called *one-way* ANOVA because you use it to analyze data from studies in which there is only one predictor variable (or independent variable). In contrast, Chapter 9, "Factorial ANOVA with Two Between-Subjects Factors," presents a statistical procedure that is appropriate for studies with two predictor variables.

The Aggression Study

To illustrate a situation for which one-way ANOVA might be appropriate, imagine you are conducting research on aggression in children. Assume that a review of prior research has led you to believe that consuming sugar causes children to behave more aggressively. You therefore wish to conduct a study that will test the following hypothesis:

> "The amount of sugar consumed by eight-year-old children increases the levels of aggression that they subsequently display."

To test your hypothesis, you conduct an investigation in which each child in a group of 60 children is randomly assigned to one of three treatment conditions:

- 20 children are assigned to the "0 grams of sugar" treatment condition.

- 20 children are assigned to the "20 grams of sugar" treatment condition.

- 20 children are assigned to the "40 grams of sugar" treatment condition.

The independent variable in the study is the amount of sugar consumed. You manipulate this variable by controlling the amount of sugar that is contained in the lunch each child receives. In this way, you ensure that the children in the "0 grams of sugar" group are actually consuming 0 grams of sugar, that the children in the "20 grams of sugar" group are actually consuming 20 grams, and so forth.

The dependent variable in the study is the level of aggression displayed by each child. To measure this variable, a pair of observers watches each child for a set period of time each day after lunch. These observers tabulate the number of aggressive acts performed by each child during this time. The total number of aggressive acts performed by a child over a two-week period serves as that child's score on the dependent response variable. You can see that the data from this investigation are appropriate for a one-way ANOVA because

- the study involves a single predictor variable that is measured on a nominal scale (amount of sugar consumed)

- the predictor variable assumes more than two values (the 0-gram group, the 20-gram group, and the 40-gram group)

- the study involves a single response variable (number of aggressive acts) that is treated as a continuous numeric variable.

Between-Subjects Designs versus Repeated-Measures Designs

The research design that this chapter discusses is referred to as a *between-subjects* design because each subject appears in only one group, and comparisons are made *between* different groups of subjects. For example, in the experiment just described, a given subject is assigned to just one treatment condition (such as the 20-gram group), and provides data on the dependent variable from only that specific treatment condition.

A distinction, therefore, is made between a between-subjects design and a repeated-measures design. With a *repeated-measures* design, a given subject provides data under each of the treatment conditions used in the study. It is called a repeated-measures design because each subject provides repeated measurements on the dependent variable.

A one-way ANOVA with one between-subjects factor is directly comparable to the independent-samples *t* test from the last chapter. The difference is that you can use a *t* test to compare just two groups, but you use a one-way ANOVA to compare two or more groups. In the same way, a one-way ANOVA with one repeated-measures factor is very similar to the paired-samples *t* test from the last chapter. Again, the main difference is that you use a *t* test to analyze data from just two treatment conditions but use a repeated-measures ANOVA with data from two or more treatment conditions. The repeated-measures ANOVA is covered in Chapter 11, "One-Way ANOVA with One Repeated-Measures Factor."

Multiple Comparison Procedures

When you analyze data from an experiment with a between-subjects ANOVA, you can state the null hypothesis like this:

> "In the population, there is no difference between the various treatment conditions with respect to their mean scores on the dependent variable."

For example, for the study on aggression, you might state a null hypothesis that, in the population, there is no difference between the 0-gram group, the 20-gram group, and the 40-gram group with respect to the mean number of aggressive acts performed. This null hypothesis could be represented symbolically as

$$\mu_1 = \mu_2 = \mu_3$$

where μ_1 represents the mean level of aggression shown by the 0-gram group, μ_2 represents mean aggression shown by the 20-gram group, and μ_3 represents mean aggression shown by the 40-gram group.

When you analyze your data, JMP tests the null hypothesis by computing an F statistic. If the F statistic is sufficiently large (and the p value associated with the F statistic is sufficiently small), you can reject the null hypothesis. In rejecting the null, you tentatively conclude that, in the population, at least one of the three treatment conditions differs from at least one other treatment condition on the measure of aggression.

However, this leads to a problem: which pairs of treatment groups are significantly different from one another? Perhaps the 0-gram group is different from the 40-gram group, but is not different from the 20-gram group. Perhaps the 20-gram group is different from the 40-gram group, but is not different from the 0-gram group. Perhaps all three groups are significantly different from one another.

Faced with this problem, researchers routinely rely on *multiple comparison procedures*. Multiple comparison procedures are statistical tests used in studies with more than two groups to help determine which pairs of groups are significantly different from one another. JMP offers several different multiple comparison procedures:

- a simple *t* test to compare each pair of groups
- Hsu's MCB, which compares each group to the best (one group you choose)
- Tukey's HSD (Honestly Significant Difference), which compares each pair of groups with an adjusted probability
- Dunnett's test to compare each group to a control group you specify.

This chapter shows how to request and interpret Tukey's HSD test, also called Tukey's studentized range test or the Tukey-Kramer HSD test. The Tukey test is especially useful when the various treatment groups in the study have unequal numbers of subjects, which is often the case.

Statistical Significance versus Magnitude of the Treatment Effect

This chapter also discusses the R^2 statistic from the results of an analysis of variance. In an ANOVA, R^2 represents the percent of variance in the response that is *accounted for* or *explained* by variability in the predictor variable. In a true experiment, you can view R^2 as an index of the magnitude of the treatment effect. It is a measure of the strength of the relationship between the predictor and the response. Values of R^2 range from 0 through 1. Values closer to 0 indicate a weaker relationship between the predictor and response, and values closer to 1 indicate a stronger relationship.

For example, assume that you conduct the preceding study on aggression in children. If your independent variable (amount of sugar consumed by the children) has a very weak effect on the level of aggression displayed by the children, R^2 will be a small value, perhaps 0.02 or 0.04. On the other hand, if your independent variable has a very strong effect on their level of aggression, R^2 will be a larger value, perhaps 0.20 or 0.40. Exactly how large R^2 must be to be considered "large" depends on a number of factors that are beyond the scope of this chapter.

It is good practice to report R^2 or some other measure of the magnitude of the effect in a published paper because researchers like to draw a distinction between results that are merely statistically significant and those that are truly meaningful. The problem is that very often researchers obtain results that are statistically significant, but not meaningful in terms of the magnitude of the treatment effect. This is especially likely to happen when conducting research with a very large sample. When the sample is very large (say, several hundred subjects), you might obtain results that are statistically significant even though your independent variable has a very weak effect on the dependent variable (this is because many statistical tests become very sensitive to minor group differences when the sample is large).

For example, imagine that you conduct the aggression study with 500 children in the 0-gram group, 500 children in the 20-gram group, and 500 children in the 40-gram group. It is possible that you would analyze your data with a one-way ANOVA, and obtain an *F* value that is significant at $p < 0.05$. Normally, this might lead you to rejoice. But imagine that you then calculate R^2 for this effect, and learn that R^2 is only 0.03. This means that only 3% of the variance in aggression is accounted for by the amount of sugar consumed. Obviously, your manipulation has had a very weak effect. Even though your independent variable is statistically significant, most researchers would argue that it does not account for a meaningful amount of variance in children's aggression.

This is why it is helpful to always provide a measure of the magnitude of the treatment effect (such as R^2) along with your test of statistical significance. In this way, your readers will always be able to assess whether your results are truly meaningful in terms of the strength of the relationship between the predictor variable and the response.

Example with Significant Differences between Experimental Conditions

To illustrate one-way ANOVA, imagine that you replicate the study that investigated the effect of rewards on commitment in romantic relationships, described previously in Chapter 7. However, in this investigation you use three experimental conditions instead of the two experimental conditions described previously.

Recall that the preceding chapter hypothesized that the rewards people experience in a romantic relationship have a causal effect on their commitment to those relationships. You tested this prediction by conducting an experiment with 20 participants. The participants were asked to read the descriptions of 10 potential romantic partners. For each partner, the subjects imagined what it would be like to date this person, and rated how committed they would be to a relationship with that person. For the first 9 partners, every subject saw exactly the same description. However, there were some important differences with respect to partner 10:

- Half of the subjects had been assigned to a *high-reward* condition, and these subjects were told that partner 10 "enjoys the same recreational activities that you enjoy, and is very good-looking."

- The other half of the subjects were assigned to the *low-reward* condition, and were told that partner 10 "does not enjoy the same recreational activities that you enjoy, and is not very good-looking."

You are now going to repeat this experiment using the same procedure as before, but this time add a third experimental condition called the *mixed-reward* condition. Here is the description of partner 10 to be read by subjects assigned to the mixed reward group.

> PARTNER 10: Imagine that you have been dating partner 10 for about 1 year, and you have put a great deal of time and effort into this relationship. There are not very many attractive members of the opposite sex where you live, so it would be difficult to replace this person with someone else. Partner 10 lives in the same neighborhood as you, so it is easy to be together as often as you like. Sometimes this person seems to enjoy the same recreational activities that you enjoy, and sometimes not. Sometimes partner 10 seems to be very good-looking, and sometimes not.

Notice that the first three sentences of the preceding description are identical to the descriptions read by subjects in the low-reward and high-reward conditions in the previous experiment. These three sentences deal with investment size, alternative value, and costs, respectively. However, the last two sentences deal with rewards, and they are different from the sentences dealing with rewards in the other two conditions. In this mixed-reward condition, partner 10 is portrayed as being something of a mixed bag. Sometimes this partner is rewarding to be around, and sometimes not.

The purpose of this new study is to determine how the mixed-reward version of partner 10 will be rated. Will this version be rated more positively than the version provided in the low-reward condition? Will it be rated more negatively than the version provided in the high-reward condition?

To conduct this study, you begin with a total sample of 18 subjects. You randomly assign 6 subjects to the high-reward condition, and they read the high-reward version of partner 10, as it was described in the chapter on *t* tests. You randomly assign 6 subjects to the low-reward condition, and they read the low-reward version of partner 10. Finally, you randomly assign 6 subjects to the mixed-reward condition, and they read the mixed-reward version just presented. As was the case in the previous investigation, you analyze subjects' ratings of how committed they would be to a relationship with partner 10 (and ignore their ratings of the other 9 partners).

You can see that this study is appropriate for analysis with a one-way ANOVA, between-groups design because

- it involves a single predictor variable assessed on a nominal scale (type of rewards)
- it involves a single response variable assessed on an interval scale (rated commitment)
- it involves three treatment conditions (low-reward, mixed-reward, and high-reward), making it inappropriate for analysis with an independent-groups *t* test.

The following section shows how to analyze fictitious data using JMP to do a one-way ANOVA with one between-subjects factor.

Using JMP for Analysis of Variance

The data table used in this study is called **mixed reward.jmp** (see Figure 8.1). There is one observation for each subject. The predictor variable is type of reward, called Reward Group, with values "low," "mixed," and "high." The response variable is Commitment. It is the subjects' rating of how committed they would be to a relationship with partner 10. The value of this variable is based on the sum of subjects' responses to four questionnaire items described in the last chapter and can assume values from 4 through 36. It is a numeric variable with a continuous modeling type.

Figure 8.1 Mixed Reward Data Table

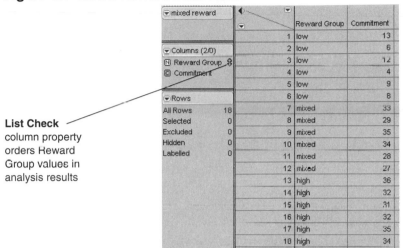

List Check
column property
orders Reward
Group values in
analysis results

Ordering Values in Analysis Results

The downward arrow icon to the right of the Reward Group name in the Columns panel indicates that column is assigned the List Check column property. The List Check property lets you specify the order in which you want variable values to appear in analysis results. In this example, you want to see Reward Group values listed in the order, "low," "mixed," and "high." Unless otherwise specified, values in reports appear in alphabetic order.

One way to order values in analysis results is to use the Column Info dialog and assign a special property to the column called List Check (see Figure 8.2). To do this:

- Right-click in the Reward Group column name area or on the column name in the Columns panel, and select **Column Info** from the menu that appears.
- When the Column Info dialog appears, select **List Check** from the **New Property** menu. Notice that the list of values is in the default (alphabetic) order.
- Highlight values in the **List Check** list and use the **Move Up** or **Move Down** button to rearrange the values in the list, as shown in Figure 8.2.
- Click **Apply**, then **OK**.

Figure 8.2 Column Info Dialog with List Check Property to Order Values in Results

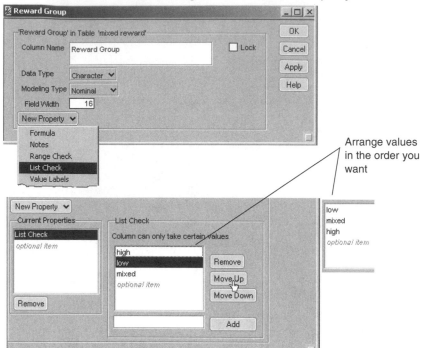

The Fit Y by X Platform for One-Way ANOVA

The easiest way to perform a one-way ANOVA in JMP uses the Fit Y by X platform, as described in the previous chapter on *t* tests. Earlier, you learned that you can use a multiple comparison procedure when the ANOVA reveals significant results and you want to determine which pairs of groups are significantly different. The following example shows a one-way ANOVA, followed by Tukey's HSD test. Remember that you only need to request a multiple comparisons test if the overall *F* test for the analysis is significant.

- 🖱 Choose **Fit Y by X** from the **Analyze** menu.
- 🖱 When the launch dialog appears, select **Reward Group** and click **X, Factor**.
- 🖱 Select **Commitment** and click **Y, Response**.

Your completed dialog should look like the one in Figure 8.3. Note that the modeling type of the variables, shown by the icon to the left of the variable name, determines the type of analysis. The legend in the lower-left corner of the dialog describes the type of analysis appropriate for all combinations of Y and X variable modeling types. It is important to note that you don't have to specifically request a one-way ANOVA—the platform is constructed to perform the correct analysis based on modeling types. The type of analysis shows at the top left of the legend and in the legend cell that represents the modeling types of the two variables.

Figure 8.3 Fit Y by X Dialog for One-Way ANOVA

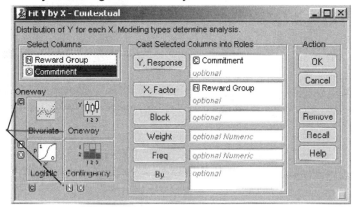

Y, Response is
continuous;
X, Factor is
nominal

Results from the JMP Analysis

- ⟡ Click **OK** to see the initial scatterplot, shown on the left in Figure 8.4. You suspect from looking at the scatterplot that the "high" and "mixed" groups are the same and that both of these groups are significantly different from the "low" group.
- ⟡ Choose **Means/Anova** from the menu on the Oneway Analysis title bar to see the scatter plot on the right in Figure 8.4.

The means diamonds visually support the conjecture about group differences. The diamonds for the "high" and "mixed" groups do not separate sufficiently to indicate a difference, but both are visibly above the "low" group.

Figure 8.4 Graphical Results of One-Way ANOVA

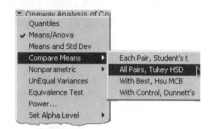

🖑 To confirm the visual group comparison, select **Compare Means** from the menu on the title bar, and select **All Pairs, Tukey HSD** from its submenu, as shown here. Figure 8.5 shows the ANOVA results and the comparison of group means.

Figure 8.5 Results of ANOVA with Means Comparisons

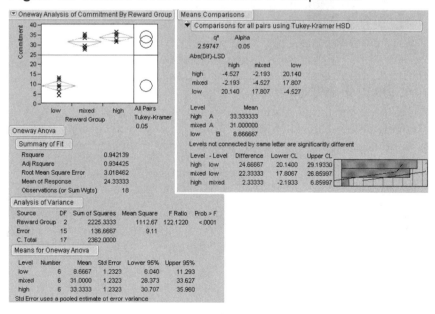

Steps to Interpret the Results

1. **Make sure that everything looks reasonable**. Check the analysis title and the scatterplot to verify you have completed the intended analysis. The title bar should state "Oneway Analysis of Commitment (*response*) by Reward Group (*predictor*)." The name of the response appears on the Y axis and the predictor on the X axis.

   The Summary of Fit table shows N = 18, the number of observations used in the analysis. This N is also the total number of participants for whom you have a complete set of data. A complete set of data means you have scores on both the predictor variable and the response variable. The degrees of freedom are 2 for Reward Group in the Analysis of Variance table, which is the number of groups (3) minus 1. The total corrected degrees of freedom (C. Total) are always equal to N – 1 = 18 – 1 = 17. If any of these basic data do not look correct, you might have made an error in entering or modifying the data.

   The means for the three experimental groups are listed in the Means for Oneway Anova table. Notice they range from 8.67 through 33.33. Remember that your participants used a scale that could range from 4 through 36, so these group means appear reasonable. If those means had fallen outside this range, there was a probable error when entering the data.

2. **Review the appropriate *F* statistic and its associated *p* value.** Once you verify there are no obvious errors, continue to review the results. Look at the *F* statistic in the Analysis of Variance table, shown here, to see if you can reject the null hypothesis. You can state the null hypothesis for the present study as follows:

   > "In the population, there is no difference between the high-reward group, the mixed-reward group, and the low-reward group with respect to mean commitment scores."

   Symbolically, you can represent the null hypothesis this way:

   $H_0: \mu_1 = \mu_2 = \mu_3$

where μ_1 is the mean commitment score for the population of people in the low-reward condition, μ_2 is the mean commitment score for the population of people in the mixed-reward condition, and μ_3 is the mean commitment score for the population of people in the high-reward condition.

Note: The Analysis of Variance table shown above uses the Beveled option. To change the appearance of any results table, right-click on the table and choose the option you want.

The line in the analysis of Variance table denoted by the predictor variable name, Reward Group, shows the statistical information appropriate for the null hypothesis, $\mu_1 = \mu_2 = \mu_3$.

- The heading called Source stands for *source of variation*. The first item under Source is the predictor (or model) variable, Reward Group. Its degrees of freedom, DF, are 2. The degrees of freedom associated with a predictor are equal to $k - 1$, where k represents the number of experimental groups.

- The Sum of Squares (SS for short) associated with the predictor variable, Reward Group, is 2225.33. It is found by subtracting the Error SS from the total SS found in the Analysis of Variance table. Compare the predictor (Reward Group) SS to the total SS (C. Total) to see how much of the variation in the Commitment data is accounted for by the predictor variable.

  Note: The ratio of the predictor SS to the corrected SS gives the R^2 (Rsquare) statistic found in the Summary of Fit table. This is the proportion of variation explained by the predictor variable, Reward Group. In this example the R^2 is 2225.33 / 2362.00 = .942. This means that 94.2% of the total variation in the Commitment data is accounted for by the *between-groups* variation of the Reward Group variable. This is a very high value of R^2, much higher than you are likely to obtain in an actual investigation (remember that the present data are fictitious).

- The Mean Square for Reward Group, also called the *mean square between groups*, is the sum of squares (2225.33) divided by its associated degrees of freedom (2), giving 1112.67. This quantity is used to compute the F ratio needed to test the null hypothesis.

- The F statistic (F Ratio) to test the null hypothesis is 122.12, which is very large. It is computed as the ratio of the model (Reward Group) mean square to the Error mean square, 1112.6667 / 9.1111 = 122.12.

- The probability (Pr > F) associated with the F ratio is less than 0.0001. This is the probability of obtaining (by chance alone) an F statistic as large or larger than the one in this analysis if the null hypothesis were true. When a p value is less than 0.05, you often choose to reject the null hypothesis. In this example you reject the null hypothesis of no population differences. In other words, you conclude that there is a significant effect for the type of rewards independent variable.

  Because the F statistic is significant, you reject the null hypothesis of no differences in population means. Instead, you tentatively conclude that at least one of the population group means is different from at least one of the other population group means. However, because there are three experimental conditions, you now have a new problem: which of these groups is significantly different from the others?

3. **Prepare your own version of the ANOVA summary table.** Before moving on to interpret the sample means for the three experimental groups, you should prepare your own version of the ANOVA summary table (see Table 8.1). All of the information you need is presented in the Analysis of Variance table shown in Figure 8.5 except for the R^2 value. The reward group information is copied directly from the Analysis of Variance table. The within-groups information deals with the error variance in your sample and is denoted as Error in the Analysis of Variance table. The total is denoted as C. Total in the JMP results. You do not need to include the p value in the table. Instead, mark the significant F value with an asterisk (*). At the bottom of the table place a note that indicates the level of significance.

You can easily calculate R^2 by hand as

Reward Group SS / C. Total SS = 2225.33 / 2362 = 0.9421

The R^2 of 0.94 indicates that 94% of the variance in commitment is accounted for by the Reward Group independent variable. This is a very high value of R^2, much higher than you are likely to obtain in an actual investigation.

Table 8.1 An Alternative ANOVA Summary Table

ANOVA Summary Table for Study Investigating the Relationship between Type of Rewards and Commitment				
Source	df	SS	MS	F	R^2
Reward Group	2	2225.33	1112.67	122.12*	0.94
Within Groups	15	136.67	9.11		
Total	17	2362.00			

Note: N = 18; $p < .0001$.

4. **Review the sample means and multiple comparison tests**. When you request any means comparison test, means comparison circles for that test appear next to the

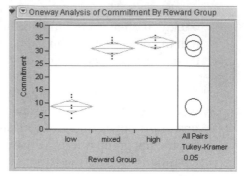

scatterplot, as shown here and in Figure 8.5. The horizontal radius of a circle aligns with its respective group mean on the scatterplot. You can quickly compare each pair of group means visually by examining how the circles intersect. Means comparison circles that do not overlap or overlap only slightly represent group means that *are* significantly different. Circles with significant overlapping represent group means that *are not* different.
See the *JMP Statistics and Graphics Guide* (2003) and Sall (1992) for a detailed discussion of means comparison circles.

5. **Review the Means Comparison reports.** The results of the means comparisons test were shown previously in Figure 8.5. At the top of the Means Comparisons report you see an alpha level of 0.05. This alpha level means that if any of the groups are significantly different from the others, they are different with a significance level of $p < 0.05$. The results give the following information:

   - To the left of the alpha at the top of the report, you see a $q*$ value of 2.59747. The q statistic is the quantile used to scale the Least Significant Difference (LSD) values. This value plays the same role in the computations as the t quantile in t tests that compare group means. Using a t test to compare more than two group means gives significance values that are too liberal. The q value is constructed to give more accurate probabilities. If the Tukey-Kramer test finds significant differences between any pair of means, they are said to be "Honestly Significant Differences." Thus, this test is called the Tukey HSD test.

   - Next you see a table of the actual absolute difference in the means minus the LSD (shown here). The Least Significant Difference (LSD) is the difference that would be significant. Pairs with a positive value are significantly different.

$q*$	Alpha	
2.59747	0.05	

Abs(Dif)-LSD	high	mixed	low
high	-4.527	-2.193	20.140
mixed	-2.193	-4.527	17.807
low	20.140	17.807	-4.527

- The Tukey letter grouping shows the mean scores for the three groups on the response variable (the Commitment variable in this study) and whether or not these group means are different. You can see that the high-reward group has a mean Commitment score of 33.3, the mixed-reward group has a mean score of 31.0, and the low-reward group has a mean score of 8.7. The letter A identifies the high-reward group and the letter B identifies the low-reward group. You conclude these two groups have significantly different mean Commitment scores.

Level		Mean
high	A	33.333333
mixed	A	31.000000
low	B	8.666667

Levels not connected by same letter are significantly different

Similarly, different letters also identify the mixed-reward and low-reward groups, so their means are significantly different. However, the letter A identifies both the high- and mixed-reward groups. Therefore, the means of these two groups are not significantly different.

- The difference between each pair of means shows at the bottom of this report, with lower and upper confidence limits for each pair. The bar chart of the differences, ordered from high to low, gives another quick visual comparison.

Examining the means comparison circles for Tukey's HSD test and the Means Comparisons report, you can conclude there was no significant difference in the mean response between the high-reward group and the mixed-reward group. However, there is a significant difference between low-reward group means and the other two group means.

> **Note:** Remember that you should review the results of a multiple comparison procedure only if the *F* value from the preceding Analysis of Variance table is statistically significant. If the *F* for the predictor variable is not significant, then you don't need to request a multiple comparison test.

Summarizing the Results of the Analysis

When performing a one-factor ANOVA, you can use the following format to summarize the results of your analysis:

A) Statement of the problem

B) Nature of the variables

C) Statistical test

D) Null hypothesis (H_0)

E) Alternative hypothesis (H_1)

F) Obtained statistic (F)

G) Obtained probability (p) value

H) Conclusion regarding the null hypothesis

I) Magnitude of treatment effect (R^2)

J) Analysis of Variance summary table

K) Figure representing the results

An example summary follows:

A) **Statement of the problem:**

The purpose of this study was to determine if there was a difference between people in a high-reward relationship, a mixed-reward relationship, and a low-reward relationship, with respect to their commitment to the relationship.

B) **Nature of the variables:**

This analysis involved two variables. The predictor variable was types of reward, which was measured on a nominal scale and could assume one of three values: a low-reward condition, a mixed-reward condition, and a high-reward condition. The numeric continuous response variable represented subjects' commitment.

C) **Statistical test:**

One-way ANOVA, between-subjects design.

D) **Null hypothesis (H_0):**

$\mu_1 = \mu_2 = \mu_3$; in the population, there is no difference between subjects in a high-reward relationship, subjects in a mixed-reward relationship, and subjects in a low-reward relationship with respect to their mean commitment scores.

E) **Alternative hypothesis (H_1):**

In the population, there is a difference between at least two of the following three groups with respect to their mean commitment scores: subjects in a high-reward relationship, subjects in a mixed-reward relationship, and subjects in a low-reward relationship.

F) **Obtained statistic:**

$F(2,15) = 122.12$.

G) **Obtained probability (p) value:**

$p < 0.0001$.

H) **Conclusion regarding the null hypothesis:**

Reject the null hypothesis.

I) Magnitude of the treatment effect:
$R^2 = 0.94$ (large treatment effect).

J) ANOVA Summary Table:
See Table 8.1.

K) Figure representing the results:
To produce the bar chart shown here, do the following.

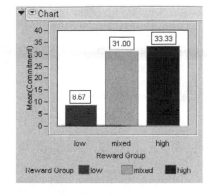

- Choose the **Chart** command from the **Graph** menu.
- In the launch dialog, select **Reward Group** as X. Select **Commitment** and choose **Mean** from the **Statistics** menu in the dialog, as shown in Figure 8.6, and click OK to see the bar chart.
- Use the annotate tool (🅰) found on the **Tools** menu or tools palette to label the bars with their respective mean values given in the analysis results.

Figure 8.6 Completed Chart Launch Dialog to Chart Group Means

Formal Description of Results for a Paper

You could use the following approach to summarize this analysis for a paper in a professional journal:

> Results were analyzed using a one-way ANOVA, between-subjects design. This analysis revealed a significant effect for type of rewards, $F(2,15) = 122.12$; $p < 0.001$. The sample means are displayed in Figure 8.5. Tukey's HSD test showed that subjects in the high-reward and mixed-reward conditions scored with significant difference on commitment versus subjects in the low-reward condition ($p < 0.05$). There were no significant differences between subjects in the high-reward condition and subjects in the mixed-reward condition.

Example with Nonsignificant Differences between Experimental Conditions

In this section you repeat the preceding analysis on a data table designed to provide nonsignificant results. This will help you learn how to interpret and summarize results when there are no significant differences between group differences. The data table for this example is **mixed reward nodiff.jmp**, shown to the right.

	Reward Group	Commitment
1	low	15
2	low	17
3	low	10
4	low	19
5	low	16
6	low	18
7	mixed	14
8	mixed	17
9	mixed	16
10	mixed	18
11	mixed	11
12	mixed	12
13	high	17
14	high	20
15	high	14
16	high	16
17	high	16
18	high	17

Use the same steps described for the previous example to analyze the data.

- 🖱 Choose **Fit Y by X** from the **Analyze** menu.
- 🖱 When the launch dialog appears, select **Reward Group** and click **X, Factor**.
- 🖱 Select **Commitment** and click **Y, Response**.
- 🖱 Click **OK** in the launch dialog to see the results in Figure 8.7.

Again, you are most interested in the Analysis of Variance table. The F statistic (F Ratio) for this analysis is 0.8295, with associated probability (Prob > F) of 0.4553. Because this p value is greater than .05, you fail to reject the null hypothesis and conclude that the types of reward independent variable did not have a significant effect on rated commitment.

The Means for Oneway Anova table shows the mean scores on Commitment for the three groups. You can see that there is little difference between these means as all three groups display an average score on Commitment between 14 and 17. You can use the scatterplot with means diamonds to graphically illustrate results, or use the mean values from the analysis to prepare a bar chart like the one described in the previous section.

Figure 8.7 Results of ANOVA for Example with No Differences

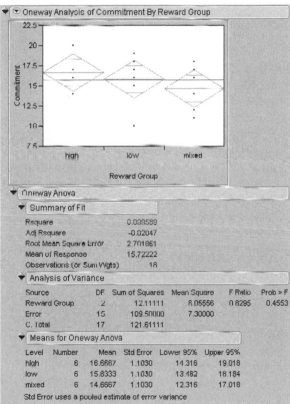

Summarizing the Results of the Analysis

This section summarizes the present results according to the statistical interpretation format presented earlier. Because the same hypothesis is being tested, items A through E are completed in the same way as before. Therefore, only items F through J are presented here. Refer to the results shown in Figure 8.7.

F) Obtained statistic:

$F(2,15) = 0.8395$

G) Obtained probability (*p*) value:

$p = 0.4553$ (not significant)

H) Conclusion regarding the null hypothesis:

Fail to reject the null hypothesis.

I) Magnitude of treatments effect:

R^2 = Reward Group SS / C. Total SS = 12.11 / 121.61 = 0.0996

J) ANOVA Summary Table:

See Table 8.2.

K) Figure representing results:

See means diamonds plot in Figure 8.7, or prepare bar chart as shown for previous example.

Table 8.2 An Alternative ANOVA Summary Table

ANOVA Summary Table for Study Investigating the Relationship between Type of Rewards and Commitment					
Source	**df**	**SS**	**MS**	**F**	**R^2**
Reward Group	2	12.11	6.06	0.8259	0.0996
Within Groups	15	109.50	7.30		
Total	17	121.61			

Note: N = 18.

Formal Description of Results for a Paper

You could summarize the results from the present analysis in the following way for a published paper:

> Results were analyzed using a one-way ANOVA, between-subjects design. This analysis failed to reveal a significant effect for type of rewards, $F(2,15) = 0.83$, not significant with negligible treatment effect of $R^2 = 0.10$. The sample means are displayed in Figure 8.7, which shows that the three experimental groups demonstrated similar commitment scores.

Understanding the Meaning of the *F* Statistic

An earlier section mentioned that you obtain significant results in an analysis of variance when the *F* statistic produced in the Analysis of Variance assumes a relatively large value. This section explains why this is so.

The meaning of the *F* statistic might be easier to understand if you think about it in terms of what results would be expected if the null hypothesis were true. If there really were no differences between groups with respect to their means, you expect to obtain an *F* statistic close to 1.00. In the previous analysis, the obtained *F* value was much larger than 1.00 and, although it is possible to obtain an *F* of 122.12 when the population means are equal, it is extremely unlikely. In fact, the calculated probability (the *p* value) in the analysis was less than or equal to 0.0001 (less than 1 in 10,000). Therefore, you rejected the null hypothesis of no population differences.

But why do you expect an *F* value of 1.00 when the population means are equal? And why do you expect an *F* value greater than 1.00 when the population means are not equal? To understand this, you need to understand how the *F* statistic is calculated. The formula for the *F* statistic is as follows:

$$F = \frac{\text{MS}_{\text{between groups}}}{\text{MS}_{\text{within groups}}}$$

- The numerator in this ratio is $MS_{between\ groups}$, which represents the *mean square between groups*. This between-subjects variability is influenced by two sources of variation—error variability and variability due to the differences in the population means.

- The denominator in this ratio is $MS_{within\ groups}$, which represents the mean square *within groups*. The within-groups variability is the sum of the variability measures that occur within each single treatment group. This is a measure of variability that is influenced by only one source of variation—error variability. For this reason, the $MS_{within\ groups}$ is often referred to as the MS_{error}.

Now you can see why the *F* statistic should be larger than 1.00 when there are differences in population means (when the null hypothesis is incorrect). Consider writing the formula for the *F* statistic as follows:

$$F = \frac{MS_{between\ groups}}{MS_{within\ groups}} = \frac{error\ variability\ +\ variability\ due\ to\ differences\ in\ population\ means}{error\ variability}$$

If the means are different, the predictor variable (types of reward) has an effect on commitment. Therefore, the numerator of the *F* ratio contains two sources of variation—both error variability and variability due to the differences in the population means. However, the denominator ($MS_{within\ groups}$) is influenced by only one source of variance—error variability. Note that both the numerator and the denominator are influenced by error variability, but the numerator is affected by an additional source of variance, so the numerator should be larger than the denominator. Whenever the numerator is larger than the denominator in a ratio, the resulting quotient is greater than 1.00. In other words, you expect to see an *F* value greater than 1.00 when there are differences between means. It follows that you reject the null hypothesis of no population differences when you obtain a sufficiently large *F* statistic.

In the same way, you can also see why the *F* statistic should be approximately equal to 1.00 when there are no differences in means values (when the null hypothesis is correct). Under these circumstances, the predictor variable (level of rewards) has no effect on commitment. Therefore, the $MS_{between\ groups}$ is not influenced by variability due to differences between the means. Instead, it is influenced only by error variability. So the formula for the *F* statistic reduces to

$$F = \frac{MS_{between\ groups}}{MS_{within\ groups}} = \frac{error\ variability}{error\ variability}$$

The number in the numerator in this formula is fairly close to the number in the denominator. Whenever a numerator in a ratio is approximately equal to the denominator in that ratio, the resulting quotient is close to 1.00. Hence, when a predictor variable has no effect on the response variable, you expect an F statistic close to 1.00.

Summary

The one-way analysis of variance is one of the most flexible and widely used procedures in the social sciences and other areas of research. It allows you to test for significant differences between groups, and its applicability is not limited to studies with just two groups (as the t test is limited).

This chapter also discussed

- the R^2 statistic from the results of an analysis of variance, which represents the percentage of variance in the response that is *accounted for* or *explained* by variability in the predictor variable
- how to interpret the graphical results produced by JMP for a one-way ANOVA
- Tukey's HSD multiple comparison test, for comparing group means
- a systematic format to use when summarizing the results of an analysis
- the construction and meaning of the F statistic used in ANOVA.

The following chapter introduces a more flexible procedure called *factorial* analysis of variance. Factorial ANOVA lets you analyze data from studies in which more than one independent variable is manipulated, and this allows you to test more than one hypothesis in a single study, as well as to test a type of effect called an "interaction."

Assumptions Underlying One-Way ANOVA with One Between-Subjects Factor

Modeling type
The response variable should be assessed on an interval or ratio level of measurement. The predictor variable should be a nominal-level variable (a categorical variable) that includes two or more categories.

Independent observations
A given observation should not be dependent on any other observation in any group (for a more detailed explanation of this assumption, see Chapter 7).

Random sampling

Scores on the response variable should represent a random sample drawn from the populations of interest.

Normal distributions

Each group should be drawn from a normally distributed population. If each group has 30 or more subjects, the test is robust against moderate departures from normality.

Homogeneity of variance

The populations represented by the various groups should have equal variances on the response. If the number of subjects in the largest group is no more than 1.5 times greater than the number of subjects in the smallest group, the test is robust against violations of the homogeneity assumption (Stevens, 2002).

References

Sall, J. 1992. "Graphical Comparison of Means." *Statistical Computing and Statistical Graphics Newsletter* (American Statistical Association) 3:27–32.

SAS Institute Inc. 2003. *JMP Statistics and Graphics Guide.* Cary, NC: SAS Institute Inc.

Stevens, J. 2002. *Applied Multivariate Statistics for the Social Sciences.* 4th ed. Mahwah, NJ: Lawrence Erlbaum Associates.

Factorial ANOVA with Two Between-Subjects Factors

> **Overview.** This chapter shows how to use JMP to perform a two-way analysis of variance. The focus is on factorial designs with two *between-groups* factors, meaning that each subject is exposed to only one condition under each independent variable. Guidelines are provided for interpreting results that do not display a significant interaction, and separate guidelines are provided for interpreting results that do display a significant interaction. For significant interactions, you see how to graphically display the interaction and how to perform tests for simple effects.

Introduction to Factorial Designs

The preceding chapter described a simple experiment in which you manipulated a single independent variable, Type of Rewards. Because there was a single independent variable in that study, it was analyzed using a one-way analysis of variance.

But imagine there are actually two independent variables that you want to manipulate. You might think that it is necessary to conduct two separate experiments, one for each independent variable, but in many cases it is possible to manipulate both independent variables in a single study.

The research design used in these studies is called a *factorial design*. In a factorial design, two or more independent variables are manipulated in a single study so that the treatment conditions represent all possible combinations of the various levels of the independent variables.

In theory, a factorial design can include any number of independent variables. This chapter illustrates factorial designs that include just two independent variables, and thus can be analyzed using a two-way analysis of variance (ANOVA). More specifically, this chapter deals with studies that include two predictor variables, both categorical variables with nominal modeling types, and a single numeric continuous dependent variable.

The Aggression Study

To illustrate the concept of factorial design, imagine that you are interested in conducting a study that investigates aggression in eight-year-old children. Here, aggression is defined as any verbal or behavioral act performed with the intention of harm. You want to test the following two hypotheses:

- Boys display higher levels of aggression than girls.
- The amount of sugar consumed has a positive effect on levels of aggression.

You perform a single investigation to test these two hypotheses. The hypothesis of most interest is that consuming sugar causes children to behave more aggressively. You will test this hypothesis by actually manipulating the amount of sugar that a group of school children consumes at lunch. Each day for two weeks, one group of children will receive a lunch that contains no sugar at all (this is the "0 grams of sugar" group). A second group will receive a lunch that contains a moderate amount of sugar (20 grams), and a third group will receive a lunch that contains a large amount of sugar (40 grams). Each child will then be observed after lunch, and a pair of judges will tabulate the number of aggressive acts that the child commits. The total number of aggressive acts committed by each child over the two-week period will constitute the dependent variable in the study.

You begin with a sample of 60 children, 30 boys and 30 girls. The children are randomly assigned to treatment conditions in the following way:

- 20 children are assigned to the "0 grams of sugar" treatment condition.
- 20 children are assigned to the "20 grams of sugar" treatment condition.
- 20 children are assigned to the "40 grams of sugar" treatment condition.

In making these assignments, you make sure that there are equal numbers of boys and girls in each treatment condition. For example, you verify that, of the 20 children in the 0 grams group, 10 are boys and 10 are girls.

The Factorial Design Matrix

The factorial design of this study is illustrated in Table 9.1. You can see that this design is represented by a matrix of two rows and three columns.

Table 9.1 Experimental Design Used in Aggression Study

		Predictor A: Amount of Sugar Consumed		
		Level A1: 0 Grams	Level A2: 20 Grams	Level A3: 40 Grams
Predictor B: Subject Gender	Level B1: Males	10 subjects	10 subjects	10 subjects
	Level B2: Females	10 subjects	10 subjects	10 subjects

When an experimental design is represented in a matrix such as this, it is easiest to understand if you focus on one aspect of the matrix at a time. For example, consider only the three vertical columns of Table 9.1. These columns are labeled **Predictor A: Amount of Sugar Consumed**, to represent the three levels of the sugar consumption (the independent variable). The first column represents the 20 participants in **Level A1: 0 Grams** (the participants who receive 0 grams of sugar), the second column represents the 20 participants in **Level A2: 20 Grams** (who receive 20 grams), and the last column represents the 20 participants in **Level A3: 40 Grams** (who receive 40 grams).

Now consider just the two horizontal rows of Table 9.1. These rows are labeled **Predictor B: Subject Gender**. The first row is **Level B1: Males** for the 30 male participants. The second row is **Level B2: Females** for the 30 female participants.

It is common to refer to a factorial design as an *r* by *c* design, where *r* represents the number of rows in the matrix and *c* represents the number of columns. The present study is an example of a 2 by 3 factorial design, because it has two rows and three columns. If it included four levels of sugar consumption rather than three, it would be referred to as a 2 by 4 factorial design.

You can see that this matrix consists of six different cells. A *cell* is the location in the matrix where the row for one independent variable intersects with the column for a second independent variable. For example, look at the cell where the row named B1 (males) intersects with the column headed A1 (0 grams). The entry "10 subjects" appears in this cell, which means that there were 10 participants who experienced this particular combination of "treatments" under the two independent variables. More specifically, it means that there were 10 participants who were both (a) male and (b) given 0 grams of sugar ("treatments" appears in quotation marks in the preceding sentence because subject gender is not a true independent variable that is manipulated by the researcher—it is merely a predictor variable).

Now look at the cell where the row named B2 (females) intersects with the column headed A2 (20 grams). Again, the cell contains the entry "10 subjects," which means that there was a different group of 10 children who experienced the treatments of (a) being female and (b) receiving 20 grams of sugar. You can see that there was a separate group of 10 children assigned to each of the six cells of the matrix.

Earlier, it was said that a factorial design involves two or more independent variables being manipulated so that the treatment conditions represent all possible combinations of the various levels of the independent variables, and Table 9.1 illustrates this concept. You can see that the six cells represent every possible combination of gender and amount of sugar consumed—both males and females are observed under every level of sugar consumption.

Some Possible Results from a Factorial ANOVA

Factorial designs are popular in social science research for a variety of reasons. One reason is that they allow you to test for several different types of effects in a single investigation. The nature of these effects is discussed later in this chapter.

First, however, it is important to note one drawback that is associated with factorial designs. Sometimes factorial designs produce results that can be difficult to interpret compared to the results produced in a one-way ANOVA. Fortunately, the task of interpretation is easier if you first prepare a figure that plots the results of the factorial study. This first section shows how to construct such a plot.

Figure 9.1 is a schematic that is often used to illustrate the results of a factorial study. Notice that scores on the dependent variable (or response), level of aggression displayed by the children, are plotted on the vertical axis. Remember, groups that appear higher on this vertical axis display higher mean levels of aggression.

Figure 9.1 A Significant Main Effect for Predictor A Only

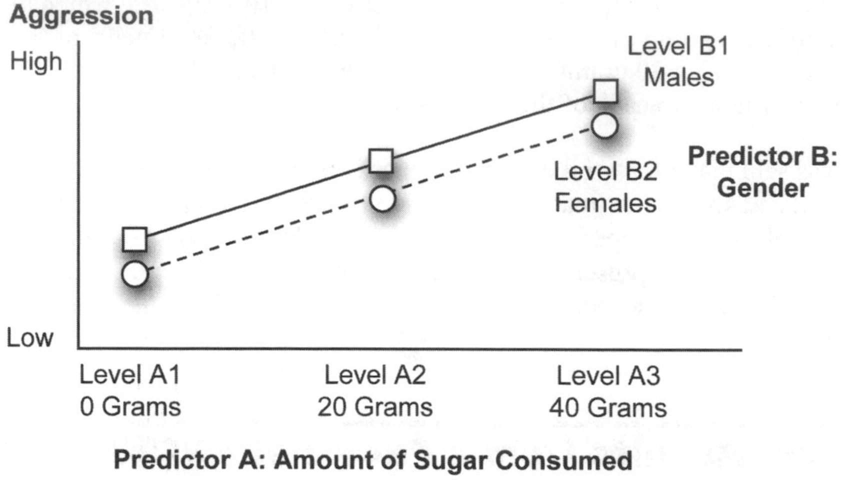

The three levels of predictor variable A (amount of sugar consumed) are plotted on the horizontal axis (the axis that runs left to right). The point at the left represents group A1 (who received 0 grams of sugar), the middle point represents group A2 (the 20-gram group), and the point at the right represents group A3 (the 40-gram group).

The two lines in the body of the figure identify the two levels of predictor variable B (subject gender). Specifically, the small squares connected by a solid line identify the mean scores on aggression displayed by the males (level B1), and the small circles connected by a dashed line show the aggression scores for the females (level B2).

In summary, the important points to remember when interpreting the figures in this chapter are as follows:

- The possible scores on the dependent variable (response) are shown on the vertical axis.

- The levels of predictor A are represented as points on the horizontal axis.

- The levels of predictor B are represented as different lines in the figure.

With this foundation, you are now ready to learn about the different types of effects that are observed in a factorial design, and how these effects appear when they are plotted in this type of figure.

Significant Main Effects

When a predictor variable (or independent variable) in a factorial design displays a significant *main effect,* this means that, in the population, there is a difference between at least two levels of that predictor variable with respect to mean scores on the dependent (response) variable. In a one-way analysis of variance there is one main effect—the main effect for the study's independent variable. In a factorial design, however, there is one main effect possible for each predictor variable examined in the study.

For example, the preceding study on aggression includes two predictor variables: amount of sugar consumed and subject gender. This means that, in analyzing data from this investigation, it is possible to obtain any of the following outcomes related to main effects.

- a significant main effect only for predictor A (amount of sugar consumed)

- a significant main effect only for predictor B (subject gender)

- a significant main effect for both predictor A and predictor B

- no significant main effects for either predictor A or predictor B.

The next section describes these effects, and shows what they look like when plotted.

A Significant Main Effect for Predictor A

Figure 9.1, shown previously, illustrates a possible main effect for predictor A (call it the sugar consumption variable). Notice that the participants in the 0-gram condition of the sugar consumption predictor variable display a relatively low mean level of aggression. When you look above the heading Level A1, you can see that both the males (represented by a square) and the females (represented by a circle) display relatively low aggression scores. Participants in the 20-gram condition show a higher level of aggression. When you look above Level A2, you can see that both boys and girls display a somewhat higher level of aggression. Finally, participants in the 40-gram condition show an even higher level of aggression. When you look above Level A3, you can see that both boys and girls in this group display fairly high levels of aggression. In short, this trend shows that there is a main effect for the sugar consumption variable.

This leads to an important point. When a figure representing the results of a factorial study displays a significant main effect for predictor variable A, it demonstrates both of the following characteristics:

- The lines for the various groups are parallel.
- At least one line segment displays a relatively steep angle.

The first of the two conditions—that the lines are parallel—ensures the two predictor variables are not involved in an interaction. This is important, because you normally do not interpret a significant main effect for a predictor variable if that predictor variable is involved in an interaction. In Figure 9.1, you can see that the lines for the two groups in the present study (the solid line for the boys and the dashed line for the girls) are parallel to one another. This suggests that there probably is not an interaction between gender and sugar consumption in the present study (the concept of interaction will be explained in greater detail later in the section titled "A Significant Interaction").

The second condition—that at least one line segment should display a relatively steep angle—can be understood by again referring to Figure 9.1. Notice that the line segment that begins at level A1 (the 0-gram condition) and extends to level A2 (the 20-gram condition) displays an upward angle (is not horizontal) because the aggression scores for the 20-gram group are higher than the aggression scores for the 0-gram group. When you obtain a significant effect for the predictor A variable, you expect to see this type of angle. Similarly, you can see that the line segment that begins at A2 and continues to A3 also displays an upward angle, consistent with a significant effect for the sugar consumption variable.

> **Note:** The lines do not have to be straight to indicate a significant main effect. The line segments that represent groups can be parallel but at different steep angles, as in the example shown to the right.

Remember that these guidelines are intended to help you understand what a main effect looks like when it is plotted in a figure such as Figure 9.1. To determine whether this main effect is statistically *significant*, it is necessary to review the results of the analysis of variance, discussed later.

A Significant Main Effect for Predictor B

You expect to see a different pattern if the main effect for the other predictor variable (predictor B) is significant. Earlier, you saw that predictor A was represented in a figure by plotting three points on the horizontal axis. In contrast, predictor B was represented by drawing different lines within the body of the figure itself—one line for each level of predictor B. In the present study, predictor B is the subject gender variable, so a solid line was used to represent mean scores for the male participants, and a dashed line was used to represent mean scores for the female participants.

When predictor B is represented in a figure by plotting separate lines for its various levels, a significant main effect for that variable is evident when the figure displays both of the following:

- The lines for the various groups are parallel, which indicates no interaction.

- At least two of the lines are relatively separated from each other.

For example, Figure 9.2 shows a main effect for predictor B (gender) in the current study. Consistent with the two preceding points, the lines in Figure 9.2 are parallel to one another and separated from one another. Notice that, in general, the males tend to score higher on the measure of aggression compared to the females. Furthermore, this tends to be true regardless of how much sugar the participants consume. Figure 9.2 shows the general trend that you expect when there is a main effect for predictor B (gender) only.

Figure 9.2 A Significant Main Effect for Predictor B Only

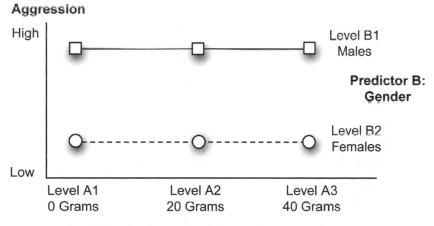

Predictor A: Amount of Sugar Consumed

A Significant Main Effect for Both Predictor Variables

It is possible to obtain significant effects for both predictor A (sugar consumed) and predictor B (gender) in the same investigation. When there is a significant effect for both predictor variables, you should see all of the following:

- The line segments for the groups are parallel, which indicates no interaction.
- At least one line segment displays a relatively steep angle (indicating a main effect for predictor A).
- At least two of the lines are relatively separated from each other (indicating a main effect for predictor B).

Figure 9.3 shows what the figure might look like under these circumstances.

Figure 9.3 Significant Main Effects for Both
Predictor A and Predictor B

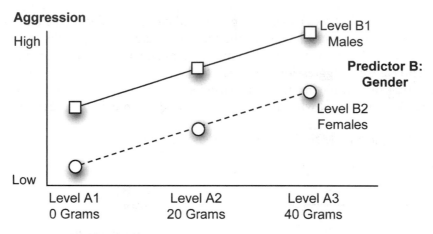

No Significant Main Effects

Figure 9.4 shows what a figure might look like if there were main effects for neither predictor A nor predictor B. Notice that the lines are parallel (indicating no interaction), none of the line segments displays a relatively steep angle (indicating no main effect for predictor A), and the lines are not separated (indicating no main effect for predictor B).

Figure 9.4 A Nonsignificant Interaction and
Nonsignificant Main Effects

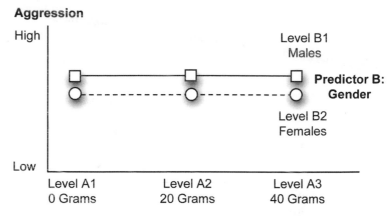

A Significant Interaction

The concept of an *interaction* can be defined in a number of ways. For example, with respect to experimental research, in which you are actually manipulating independent variables, the following definition can be used:

> An interaction is a condition in which the effect of one independent variable on the dependent variable is different at different levels of the second independent variable.

On the other hand, when conducting nonexperimental research, in which you are simply measuring naturally occurring variables rather than manipulating independent variables, it can be defined in this way:

> An interaction is a condition in which the relationship between one predictor variable and the criterion (response) is different at different levels of the second predictor variable.

These definitions might appear somewhat abstract at first glance; the concept of interaction is much easier to grasp with a visual display. For example, Figure 9.5 illustrates a significant interaction between sugar consumption and subject gender in the present study. Notice that the lines for the two groups are no longer parallel: the line for the male participants now displays a somewhat steeper angle, compared to the line for the females. This is the key characteristic of a figure that displays a significant interaction: lines that are not parallel.

Figure 9.5 Significant Interaction between Predictor A
and Predictor B

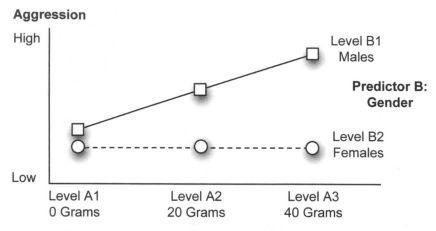

Notice how the relationships shown in Figure 9.5 are consistent with the definition of interaction—the relationship between one predictor variable (sugar consumption) and the response (aggression) is different at different levels of the second predictor variable (gender). More specifically, the figure shows that the relationship between sugar consumption and aggression is relatively strong for the male participants. Consuming larger quantities of sugar results in a dramatic increase in aggression among the male participants. Notice that the boys who consumed 40 grams of sugar displayed much higher levels of aggression than the boys who consumed 0 or 20 grams. In contrast, the relationship between sugar consumption and aggression is relatively weak among the female participants. The line for the females is fairly flat—there is little difference in aggression between the girls who consumed 0 grams versus 20 grams versus 40 grams of sugar.

Figure 9.5 shows why you do not normally interpret main effects when an interaction is significant. It is clear that sugar consumption does seem to have an effect on aggression among boys, but the figure suggests that sugar consumption probably does not have any meaningful effect on aggression among girls. To say that there is a main effect for sugar consumption might mislead readers into believing that sugar increases aggression in all children in pretty much the same way (which it apparently does not). According to Figure 9.5, whether or not sugar increases aggression depends on the child's gender.

In this situation, it makes more sense to do the following:

- Note that there was a significant interaction between sugar consumption and subject gender.
- Prepare a figure (like Figure 9.5) that illustrates the nature of the interaction.
- Test for *simple effects*.

Testing for simple effects is similar to testing for main effects, but it is done one group at a time. For example, in the preceding analysis, you would test for simple effects by first dividing your data into two groups—separate data from the male and female participants. You then perform an analysis to determine whether sugar consumption has a simple effect on aggression among just the male participants. Then, you perform a separate test to determine whether sugar consumption has a simple effect on aggression among just the female participants. The lines for gender in Figure 9.5 suggest the simple effect of sugar consumption on males is probably significant, but its effect on females is not significant. A later section shows how you how JMP can perform these tests for simple effects.

To summarize, an interaction means that the relationship between one predictor variable and the criterion is different at different levels of the second predictor variable. When an interaction is significant, you should interpret your results in terms of *simple effects*, instead of main effects.

Example with Nonsignificant Interaction

To illustrate how to do a two-way analysis of variance, let's modify the earlier study that examined the effect of rewards on commitment in romantic relationships. That study, presented in Chapter 8, "One-Way ANOVA with One Between-Groups Factor," had only one independent variable, called **Reward Group**. This chapter uses the same scenario, but with two independent variables, **Reward Group** and **Cost Group**.

In Chapter 8, participants read descriptions of a fictitious "partner 10" and rated how committed they would be to a relationship with that person. One independent variable, Reward Group, was manipulated to give three experimental conditions. Participants were assigned to either the low-, mixed-, or high-reward condition.

This following descriptions of partner 10 were provided to the participants:

- Participants in the low-reward condition were told to assume that this partner "does not enjoy the same recreational activities that you enjoy, and is not very good-looking."

- Participants in the mixed-reward condition were told to assume that "sometimes this person seems to enjoy the same recreational activities that you enjoy, and sometimes not. Sometimes partner 10 seems to be very good-looking, and sometimes not."

- Participants in the high-reward condition were told that this partner "enjoys the same recreational activities that you enjoy, and is very good-looking."

These same manipulations are used in the present study. In a sample of 30 participants, one-third is assigned to a low-reward condition, one-third to a mixed-reward condition, and one-third to a high-reward condition. This **Reward Group** factor serves as independent variable A.

However, in this study, you will also manipulate a second predictor variable (independent variable B) at the same time. The second independent variable will be called **Cost Group**, and will consist of two experimental conditions. Half of the participants (the "less" cost group) will be told to assume that they are in a relationship that does not create significant personal hardships. Specifically, when a subject reads the description of partner 10, it will say, "Partner 10 lives in the same neighborhood as you so it is easy to get together as often as you like." The other half of the participants (the "more" cost group) will be told to imagine that they are in a relationship that does create significant personal hardships. When they read the description of partner 10, it will say, "Partner 10 lives in a distant city, so it is difficult and expensive to get together."

Your study now has two independent variables. Independent variable A is type of rewards with three levels (or conditions). Independent variable B is type of costs, with two levels. The factorial design used in the study is illustrated in Table 9.2.

Table 9.2 Experimental Design Used in Investment Model Study with Two Predictors

		Predictor A: Type of Rewards		
		Level A1: Low Reward	Level A2: Mixed Reward	Level A3: High Reward
Predictor B: Type of Costs	**Level B1: Less Cost**	5 subjects ($cell_{11}$)	5 subjects ($cell_{12}$)	5 subjects ($cell_{13}$)
	Level B2: More Cost	5 subjects ($cell_{21}$)	5 subjects ($cell_{22}$)	5 subjects ($cell_{23}$)

The 2 by 3 matrix in Table 9.2 shows that there are two independent predictor variables. One independent variable has two levels and the other has three levels. There are a total of 30 participants divided into 6 cells, or subgroups. Each cell contains 5 participants. Each horizontal row in the figure (running from left to right) represents a different level of the type of costs independent variable. There are 15 participants in the less-cost condition, and 15 in the more-cost condition. Each vertical column in the figure (running from top to bottom) represents a different level of the type of rewards independent predictor variable. There are 10 participants in the low-reward condition, 10 in the mixed-reward condition, and 10 in the high-reward condition.

It is strongly advised that you prepare a similar figure whenever you conduct a factorial study, which makes it easier to enter data and reduces the likelihood of input errors.

Notice that the cell subscript numbers ($cell_{11}$, $cell_{23}$, and so forth) indicate which experimental condition a given subject experiences under each of the two independent variables. The first number in the subscript indicates the level of costs, and the second number indicates the level of rewards. This type of notation is often used to express statistical models in mathematical terms.

In subscripted notation $cell_{11}$ represents the less-cost level and low-reward level. Participants in this cell experienced the low-cost condition and the low-reward condition. $Cell_{23}$ represents more-cost level and high-reward level, which means that the five participants in this subgroup read that partner 10 lives in a distant city (more cost) and is good-looking (high reward). You interpret each design cell in this way. When working with cell notation, remember that the subscript number for the row always precedes the number for the column. That is, $cell_{12}$ is not interpreted the same as $cell_{21}$.

The JMP table uses the actual names of the levels instead of the cell subscripts. This is a useful way to enter data for a two-way ANOVA because it provides a quick way of see the treatment combination for any given subject. The JMP data table in Figure 9.6 is a listing of the data for this factorial example. The first two columns of data for the subject from line 1 provide the values "low" and "less," which tells you that this subject is in $cell_{11}$ of the factorial design matrix. Similarly, subject 16 has values "low" and "more," which tells you that this subject is in $cell_{21}$. The data do not have to be sorted by any variable to be analyzed with JMP.

Factorial Data in a JMP Table

There are two predictor variables in this study (Reward Group and Cost Group) and one response variable (Commitment). This means each subject has a score for three variables.

- First, the Reward Group variable indicates which experimental condition a subject experienced under the Type of rewards independent variable. Each subject is assigned to the "low-," "mixed-," or "high-" reward group.

- Second, the variable called Cost Group, indicates which condition the subject experienced under the Type of Costs independent variable, with values "less" and "more." Each subject is assigned to only one cost condition and one reward condition (a subject cannot be in more than one treatment group).

- Finally, the variable Commitment has the subject's rated commitment to the relationship with partner 10.

Figure 9.6 shows the data entered into a JMP table, called **commitment factorial.jmp**.

Figure 9.6 JMP Data Table with Factorial Data

		Reward Group	Cost Group	Commitment
	1	low	less	8
	2	low	less	13
	3	low	less	7
Columns (3/0)	4	low	less	14
Reward Group	5	low	less	7
Cost Group	6	mixed	less	19
Commitment	7	mixed	less	17
	8	mixed	less	20
	9	mixed	less	25
	10	mixed	less	16
	11	high	less	31
Rows	12	high	less	24
All Rows 30	13	high	less	37
Selected 0	14	high	less	30

	Reward Group	Cost Group	Commitment
15	high	less	32
16	low	more	9
17	low	more	14
18	low	more	5
19	low	more	14
20	low	more	16
21	mixed	more	22
22	mixed	more	18
23	mixed	more	20
24	mixed	more	19
25	mixed	more	21
26	high	more	24
27	high	more	30
28	high	more	24
29	high	more	26
30	high	more	31

Ordering Values in Analysis Results

In Figure 9.6, the downward arrow icon to the right of the Reward Group name in the Columns panel indicates it has the List Check column property. The List Check column property lets you specify the order in which you want variable values to appear in analysis results. In this example, you want to see the Reward Group values listed in the order "low," "mixed," and "high." Unless otherwise specified, values in reports appear in alphabetic order.

One way to order values in analysis results is to use the Column Info dialog and assign a special property called List Check to the column. To do this:

- 🖱 Right-click in the Reward Group column name area or on the column name in the Columns panel, and select **Column Info** from the menu that appears.
- 🖱 When the Column Info dialog appears, select **List Check** from the **New Property** menu, as shown to the right.

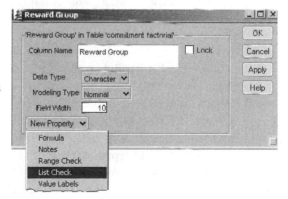

🖱 Notice that the list of values first appear in the default (alphabetic) order. Highlight values in the **List Check** list and use the **Move Up** or **Move Down** button to rearrange the values in the list, as shown in Figure 9.7.

🖱 Click **Apply**, then **OK**.

Figure 9.7 Column Info Dialog with List Check Column Property

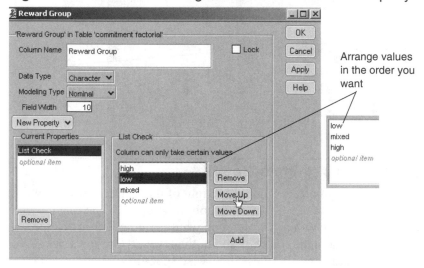

Using the Fit Model Platform

To analyze data with more than one independent predictor variable requires the Fit Model platform. The following steps lead you through a factorial analysis of the commitment data, and describe the results.

🖱 Open the commitment factorial.jmp table shown in Figure 9.6.

🖱 Choose the **Fit Model** command from the **Analyze** menu.

🖱 When the Fit Model launch dialog appears, select **Commitment** from the variable list on the left and click **Y** in the dialog to assign it as the independent response variable for the analysis.

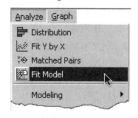

🖱 Next, select both **Reward Group** and **Cost Group** from the variable list.

✒ With these two factors selected, choose **Full Factorial** from the list of options on the **Macros** menu to enter the main effects and interaction into the model for analysis. Figure 9.8 shows the completed Fit Model dialog.

✒ Click **Run Model** in the dialog to see the factorial analysis.

Figure 9.8 Fit Model Dialog Completed for Factorial Analysis

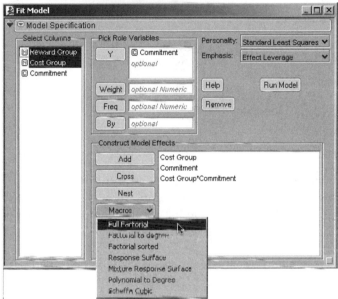

Note the items in the upper right of the dialog called **Personality** and **Emphasis**. The default commands showing are correct for this analysis. The method of analysis is denoted as the analysis *personality* in JMP. The personality you choose depends on the modeling type of the response variable. The most common method is the **Standard Least Squares**, used when there is a single continuous response variable, as in this example. The Emphasis options tailor the results to show more or fewer details. The **Effect Leverage** option is useful here because you want to examine the interaction effect, Cost Group*Reward Group, in detail. The next section describes Leverage plots.

Verifying the Results

Interpreting the results of a two-factor ANOVA is similar to interpreting the results of a one-factor ANOVA. In any analysis, you should always follow sound basic procedures. The analysis results show the analysis of the whole model on the left, and the analysis of each factor (the main effects and the interaction) on the right. Look at the portions of the whole model analysis shown in Figure 9.9 to verify that you did the analysis as intended.

Figure 9.9 Verification of the Analysis

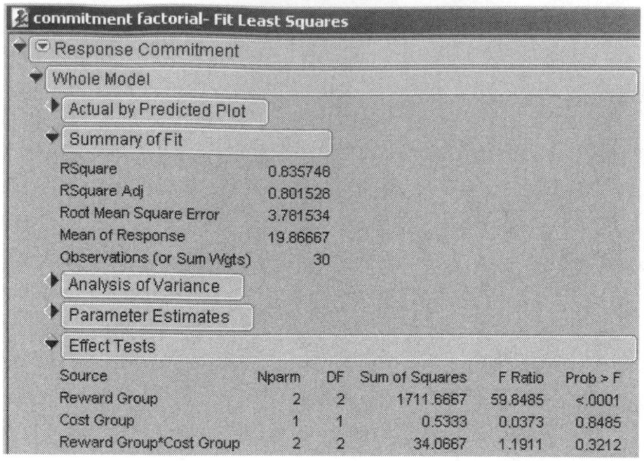

The analysis window title bar states the data table name (**commitment factorial**) and the type of analysis performed (Fit Least Squares). The Summary of Fit table shows that there were 30 observations. The Effect Tests table lists each term in the model, Reward Group, Cost Group, and the interaction term Reward Group*Cost Group.

If you want to look more closely at the analysis structure, the Parameter Estimates table (not shown) lists the levels of each effect and is discussed later.

> **Note:** All parts of a JMP analysis can be opened or closed using the icon next to each
> title. When the statistical results first appeared, all tables were open by default.
> In Figure 9.9, only the Summary of Fit and the Effect Tests tables are shown
> open—all other parts of the analysis were manually closed.

Examining the Whole Model Reports

Like most JMP analyses, the analysis of variance results begin with a graphical display. The whole model and each effect have a *leverage plot* (Sall, 1990) that graphically shows whether the whole model or an effect is significant.

Actual by Predicted Leverage Plot

The whole model Actual by Predicted leverage plot shows the data values on the Y axis and the values predicted by the factorial model on the X axis. The plot shows the line of fit with 95% confidence lines. You want to compare the line of fit with the horizontal reference line, which represents the null hypothesis. A leverage plot has the following properties, as illustrated in Figure 9.10:

- The distance from each point to the line of fit is the error (or residual) for that point. These distances represent the amount of variation not explained by fitting the model to the data.

- The distances from each point to the horizontal line (the mean of the data) is the total variation, or what the error would be if you took the effects out of the model and used the mean as the line of fit.

- In a leverage plot, the line of fit and its confidence curves quickly reveal whether the model explains enough of the variation in the data to fit significantly well. If the 95% curves cross the horizontal mean reference line, then the model fits well. If the curves do not cross the mean line—if they include or encompass the mean line—the model line fits the data no better than the mean line itself.

Figure 9.10 Leverage Plots

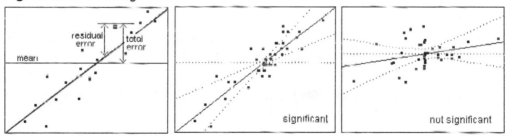

This plot is called a leverage plot because it shows the "leverage" or influence exerted by each data point in the analysis. A point has more influence on the fit if it is farther away from the middle of the plot in the horizontal direction. Further, if the points cluster in the middle of the plot, the line is not well balanced through the points and the confidence curves become wide as they extend out.

The leverage plot shown on the left in Figure 9.11 indicates that the factorial model for the commitment data appears to fit very well. This graphic is verified by the information given beneath the plot, by summary statistics and the analysis of variance statistics.

Figure 9.11 Whole Model Leverage Plot and ANOVA Report

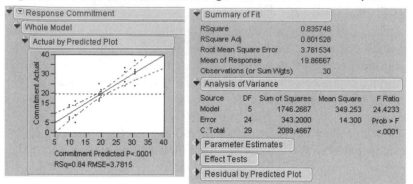

Summary of Fit Table

The Summary of Fit table for the response variable, Commitment, summarizes the basic model fit with these quantities:

- The RSquare statistic is 0.8357. This means that the factorial model explained 85.57% of the total variation in the sample. It is the ratio of the model sum of squares to the total sum of squares, found in the Analysis of Variance table. The R^2 in this example is high, and is supported by the significant F statistic in the Analysis of Variance table.

- The RSquare Adj is the R^2 statistic adjusted to make it more comparable over different models (different analyses). It is based on the ratio of mean squares instead of the ratio of sums of squares.

- The Root Mean Square Error (RMSE) is the square root of the error mean square in the Analysis of Variance table.

- The Mean of Response (the overall mean of the response variable) is 19.867, and shows as the horizontal line in the leverage plot in Figure 9.9.

- The Observations (or Sum Wgts) is the number of participants, 30, used in the analysis.

Analysis of Variance Table

You use the Analysis of Variance table to review the F statistic produced by the analysis and its associated p value. As with the one-way ANOVA, look at the F statistic to see if you can reject the null hypothesis for the whole model (Figure 9.11). There is no difference between the mean commitment score and the commitment score predicted by the statistical model that include the main effects, Reward Group and Cost Group, and their interaction.

Or, looking back at Table 9.2, you can state the null hypotheses as follows:

"In the population there is no difference in mean commitment scores between any of the groups of people identified by the cells (categories) in the factorial design."

Symbolically, you can represent the null hypothesis this way:

$$\mu_{11} = \mu_{12} = \mu_{13} = \mu_{21} = \mu_{22} = \mu_{23}$$

where μ_{11} is the mean commitment score for the population of people in the less-cost, low-reward condition, μ_{12} is the mean commitment score for the population of people in the less-cost, mixed-reward condition, and so forth.

How to Interpret a Model with No Significant Interaction

The fictitious data analyzed in this section were designed so that the interaction term would not be significant. The next sections focus on some issues that are relevant to a two-factor ANOVA.

A two-factor analysis of variance allows you to test for three types of effects:

1. The main effect of predictor A (Reward Group)
2. The main effect of predictor B (Cost Group)
3. The interaction of predictor A and predictor B (Reward Group*Cost Group).

Remember that you can interpret a main effect only if the interaction is not significant.

Verifying That the Interaction Is Not Significant

The interaction leverage plot shows to the far right of the JMP analysis of variance results. The Reward Group*Cost Group effect is summarized with a leverage plot (see Figure 9.12) that gives the same kind information for the interaction effect as for the whole model and other effects. You can quickly assess the interaction effect as not significant because the confidence curves in the leverage plot do not cross the horizontal reference line. Said another way, the reference line (which represents the null hypothesis) falls within the confidence curves of the line of fit. The p value for the significance test, $p = 0.3212$, shows beneath the plot.

Figure 9.12 Interaction Leverage Plot and Statistics

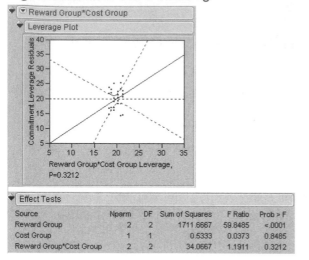

To see the details of the test for the interaction, look again at the Effect Tests table, at the bottom of Figure 9.12. The last entry, **Reward Group*Cost Group**, gives the supporting statistics. When an interaction is significant, it suggests that the relationship between one predictor and the response variable is different at different levels of the second predictor. In this case, the interaction is associated with 2 degrees of freedom, a value approximately 34.07 for the sum of squares, an F value of 1.19, and a corresponding p value of 0.3212. Remember that when you choose $\alpha = 0.05$, you view a result as being statistically significant only if the p value is less than 0.05.

This p value of 0.3212 is larger than 0.05, so you conclude that there is no interaction between the two independent predictor variables. You can therefore proceed with your review of the two main effects.

Determining Whether the Main Effects Are Statistically Significant

A two-way ANOVA tests two null hypotheses concerning main effects. The null hypothesis for the type of rewards predictor may be stated as follows:

> "In the population, there is no difference between the high-reward group, the mixed-reward group, and the low-reward group with respect to scores on the commitment variable."

The F statistic to test this null hypothesis is also found in the Effect Tests table (see Figure 9.12). To the right of the heading Reward Group, you see that this effect has 2 degrees of freedom, a sum of squares value of 1711.667, an F value of 59.85, and a p value of .0001. With such a small p value, you can reject the null hypothesis of no main effect for type of rewards, and conclude that at least one pair of reward groups are different. Later, you will review the results of the Tukey test to see which reward groups significantly differ.

The null hypothesis for the type of costs independent variable may be stated in a similar fashion:

"In the population, there is no difference between the more-cost group and the less-cost group with respect to scores on the commitment variable."

The Effect Tests table shows the F statistic for Cost Group to be 0.04, with an associated p value of 0.849. This p value is greater than 0.05, so you do not reject the null hypothesis and you conclude there is no significant main effect for type of costs.

Designing Your Own Version of the ANOVA Summary Table

Although many journals accept tables and graphs from JMP and other statistical software packages, you can extract information from the JMP analysis and create an alternate form of the analysis of variance table, similar to the one shown in Table 9.3. The information in Table 9.3 uses combined items from the Summary of Fit table, the Analysis of Variance table, and the Effect Tests table in JMP.

The first three lines of Table 9.3 are the items in the Effect Tests table. However, note that the Mean Square (MS) included in the alternative table does not show (by default) in the JMP Effect Tests table, and that the R^2 is not computed for the effects.

The next section shows how to include additional information in JMP tables and compute the quantities to complete the alternative ANOVA summary table.

Table 9.3 An Alternative ANOVA Summary Table with Some Results Missing

ANOVA Summary Table for Study Investigating the Relationship between Type of Rewards, Type of Costs, and Commitment (Nonsignificant Interaction)

Source	df	SS	MS	F	R^2
Reward Group (A)	2	1171.66	---	59.85*	---
Cost Group (B)	1	0.53	---	0.04	---
A by B Interaction	2	34.07	---	1.19	---
Within Groups	24	343.20	14.30		
Total	29	2089.47			

Note: N = 30; $p < .0001$.

Displaying Optional Information in a JMP Results Table

Many JMP results tables have optional columns that aren't shown by default. To see what is available in any table, right-click on the table and select the **Columns** command from the menu that appears. For example, when you right-click on the Effects Tests table and look at the available columns, **Mean Square** is not checked. Select **Mean Square** and add it to the Effect Tests table, as shown in Figure 9.13. The Mean Square values for the first three lines in the alternative table are now available in the JMP Effect Tests table. The Mean Square is 855.833 for Reward Group, 0.533 for Cost Group, and 17.033 for the Reward Group by Cost Group interaction.

Figure 9.13 Show Additional Information in a JMP Results Table

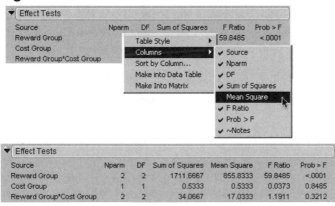

Computing Additional Statistics

In the preceding chapter, you learned that R^2 is a measure of the strength of the relationship between a predictor variable and a response variable. In a one-way ANOVA, R^2 tells you what percentage of variance in the response is accounted for (explained) by the study's predictor variable. In a two-way ANOVA, R^2 indicates what percentage of variance is accounted for by each predictor variable, as well as by their interaction term.

JMP does not compute the R^2 for each effect, but you can perform a few simple hand calculations to compute these values. To calculate R^2 for a given effect, divide the sum of squares associated with that effect by the corrected total sum of squares. For example, to calculate the R^2 value for the type of rewards main effect (Reward Group), divide the sum of squares for Reward Group found in the Effect Tests table (1711.6667) by the corrected total sum of squares (C. Total) found in the Analysis of Variance table. That is,

$$\text{Reward Group } R^2 = SS_{\text{Reward Group}} \div SS_{\text{C. Total}} = 1711.6667 \div 2089.4667 = 0.819$$

Likewise,

$$\text{Cost Group } R^2 = SS_{\text{Cost Group}} \div SS_{\text{C. Total}} = 0.5333 \div 2089.4667 = 0.0002$$

$$\text{Interaction } R^2 = SS_{\text{Interaction}} \div SS_{\text{C. Total}} = 34.0667 \div 2089.4667 = 0.0163$$

Completing the Alternative ANOVA Summary Table

Table 9.3 shows the completed alternative ANOVA summary table. Using the mean square quantities now showing in the Effect Tests table, and the computed R^2 values, you can complete the first three lines of the alternative ANOVA summary table. The Within Groups and Total quantities are the Error and the C. Total values from the Analysis of Variance table. In the table note, N is taken from the Summary of Fit table and the p value for Reward Type is taken from the Effect Tests table.

Table 9.3 Completed Alternative ANOVA Summary Table

ANOVA Summary Table for Study Investigating the Relationship between Type of Rewards, Type of Costs, and Commitment (Nonsignificant Interaction)

Source	df	SS	MS	F	R^2
Reward Type (A)	2	1171.66	855.83	59.85*	0.82
Cost Group (B)	1	0.53	0.53	0.04	0.00
A by B Interaction	2	34.07	17.03	1.19	0.02
Within Groups	24	343.20	14.30		
Total	29	2089.47			

Note: $N = 30$; $p < .0001$.

You can construct an alternative table for any type of analysis by using quantities from JMP results, or computing additional quantities from table results.

Sample Means and Multiple Comparisons

If a given main effect is statistically significant, then review the Least Squares Means plot and Tukey's HSD multiple comparison test for that main effect to see which levels are different. Tukey's test does not show automatically because it is not needed if an effect is not significant, or might not be relevant if there is a significant interaction.

Tukey's multiple comparison is available for each effect in a model using the **LSMeans Tukey HSD Test**, found on the menu on the title bar for the effect. In this example, Reward Group is the only significant effect, so look at its Tukey's multiple comparison test results in Figure 9.14.

- Select **LSMeans Plot** from the menu on the Reward Group title bar.
- Select **LSMeans Tukey's HSD** from the menu on the Reward Group title bar.

Figure 9.14 Effect Leverage and Multiple Comparison Test for Reward Group

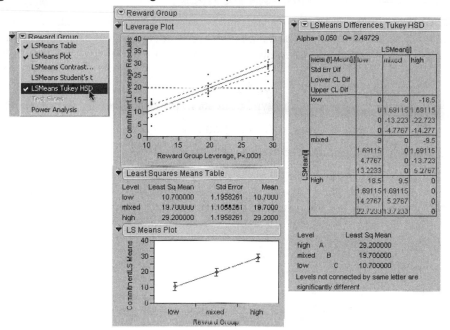

The results show a significant difference between the group means for each pair of Reward Group levels.

- The leverage plot for Reward Group shows narrow confidence curves that cross the horizontal reference line, which indicates a significant effect.

- The LSMeans Plot shows the high-reward group with the highest commitment scores, the mixed-reward group with the next highest, and the low-reward group with the lowest.

- The LSMeans Differences Tukey HSD report presents a crosstabs table with cells that list the difference between each pair of group means, the standard error of the difference, and confidence limits for the difference.

- The letter report beneath the crosstabs lists the means from high to low. Means not connected by the same letter are significantly different for the alpha level given at the top of the report ($\alpha = 0.05$). This multiple comparison statistical test is adjusted for all differences among the least squares means (Tukey, 1953; Kramer, 1956). This is an exact alpha-level test if the sample sizes are the same (as in this example) and conservative is the sample sizes are different (Hayter, 1984).

Summarizing the Analysis Results

In performing a two-way ANOVA, use the same statistical format that was used when summarizing a *t* test analysis and a one-way ANOVA. However, with two-way ANOVA, it is possible to test three null hypotheses in this study:

- the hypothesis of no interaction, stated as "In the population, there is no interaction between type of rewards and type of costs in the prediction of commitment scores"
- the hypothesis of no main effect for predictor A (type of rewards)
- the hypothesis of no main effect for predictor B (type of costs).

The list below, shown in previous chapters, is used in each case when results of a study need to be summarized and prepared for publication.

A) Statement of the problem

B) Nature of the variables

C) Statistical test

D) Null hypothesis (H_0)

E) Alternative hypothesis (H_1)

F) Obtained statistic

G) Obtained probability (*p*) value

H) Conclusion regarding the null hypothesis

I) Figure representing the results

J) Formal description of results for a paper.

Formal Description of Results for a Paper

Following is one approach that could be used to summarize the results of the preceding analysis in a scholarly paper.

Results were analyzed using two-way ANOVA, with two between-groups factors. This analysis revealed a significant main effect for type of rewards, $F(2, 24) = 59.85$; $p < 0.001$; large treatment effect, $R^2 = 0.8190$; MSE = 14.3. The sample means are displayed in Figure 9.14. Tukey's HSD test showed that participants in the high-reward condition scored significantly higher on commitment than participants in the mixed-reward condition, who, in turn, scored significantly higher than participants in the low-reward condition ($p < 0.05$). The main effect for type of costs was not significant, $F(1, 24) = 0.04$; $p = 0.849$. The interaction between type of rewards and type of costs was also not significant, $F(2, 24) = 1.19$; $p = 0.321$.

Example with a Significant Interaction

When the interaction term is statistically significant, it is necessary to follow a different procedure when interpreting the results. In most cases, this will consist of plotting the interaction in a figure, and determining which of the simple effects are significant. This section shows how to interpret results when there is a significant interaction.

For example, assume that the preceding study with 30 participants is repeated, but that this time the analysis is performed on the following data. The Reward Group and the Cost Group values are the same, but the Commitment values have been changed to produce a significant interaction (Figure 9.15). The name of the JMP table is **commitment interaction.jmp**.

Figure 9.15 Commitment Data with Significant Interaction

commitment interaction			Reward Group	Cost Group	Commitment
		1	low	less	8
Columns (3/0)		2	low	less	13
Reward Group		3	low	less	4
Cost Group		4	low	less	11
Commitment		5	low	less	5
		6	mixed	less	17
		7	mixed	less	12
Rows		8	mixed	less	20
All Rows	30	9	mixed	less	14
Selected	0	10	mixed	less	16
Excluded	0	11	high	less	31
Hidden	0	12	high	less	24
Labelled	0	13	high	less	37
		14	high	less	30
		15	high	less	32

16 low	more	18
17 low	more	10
18 low	more	26
19 low	more	15
20 low	more	14
21 mixed	more	21
22 mixed	more	13
23 mixed	more	16
24 mixed	more	12
25 mixed	more	17
26 high	more	18
27 high	more	23
28 high	more	19
29 high	more	18
30 high	more	18

To analyze the commitment interaction JMP table, use the same steps as shown previously. Complete the Fit Model dialog the same as the one shown in Figure 9.8. When you click **Run Model**, the same type of information appears. Figure 9.16 shows portions of the results. As in the previous example, the Analysis of Variance table shows that the whole model is significant. The Effect Tests table shows that Reward Group is a significant effect and Cost Group is not significant, but the interaction between Reward Group and Cost Group is significant. Before you can make a statement about the significant Reward Group effect, you must look closer at the interaction effect.

Notice that the leverage plot for the interaction has confidence curves that cross the horizontal reference line, which gives a visualization of the significant interaction effect.

Figure 9.16 Interaction Effect Shown in Effect Tests Table and Leverage Plot

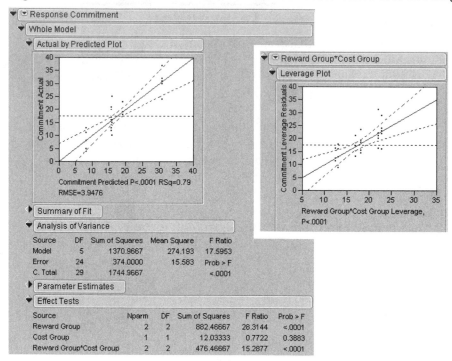

Analysis of Variance

Source	DF	Sum of Squares	Mean Square	F Ratio
Model	5	1370.9667	274.193	17.5953
Error	24	374.0000	15.583	Prob > F
C. Total	29	1744.9667		<.0001

Parameter Estimates

Effect Tests

Source	Nparm	DF	Sum of Squares	F Ratio	Prob > F
Reward Group	2	2	882.46667	28.3144	<.0001
Cost Group	1	1	12.03333	0.7722	0.3883
Reward Group*Cost Group	2	2	476.46667	15.2877	<.0001

How to Interpret a Model with Significant Interaction

This **commitment interaction.jmp** data table was constructed to produce a significant interaction between type of rewards and type of costs. The steps to follow when interpreting this interaction are different than when there is no significant interaction.

Determining If the Interaction Term Is Statistically Significant

The Effect Tests table shows the F ratio for Reward Group*Cost Group effect is 15.287 and its associated p value is less than 0.0001. Because of this significant interaction, you plot the interaction and look at the main effect levels.

> **Notes on Computation:** The numerator of the F ratio for an effect is the sum of squares for that effect divided by its corresponding degrees of freedom. For the Reward Group*Cost Group interaction effect, the numerator of the F ratio is $476.46667 \div 2 = 238.2333$. The denominator is the error sum of squares divided by its degrees of freedom (the mean square found in the Analysis of Variance table), and is $374 \div 24 = 15.5833$. The notation for an F ratio often includes its degrees of freedom, and is denoted $F(df_{numerator}, df_{denominator})$. In this case the F ratio for the interaction is $F(2, 24) = 238.2333 \div 15.5833 = 15.28773$. Compute the R^2 for the interaction as $SS_{interaction} \div SS_{total} = 476.467 \div 1744.967 = 0.27$.

Plotting the Interaction

Interactions are easiest to understand with a plot of the means for each of the cells that appear in the study's factorial design. The factorial design used in the current study involved a total of six cells (see Table 9.2). To see a plot of the mean commitment scores for these six cells choose the **Factor Profiling** command from the menu on the Whole Model title bar, and select **Interaction Plots** from its submenu (see Figure 9.17).

Figure 9.17 Plot of Interaction Effect

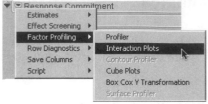

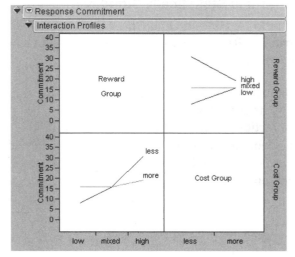

The Y axis for the plots is mean commitment score. The two interaction plots show the mean commitment score for one main effect plotted at each level of the other main effect—the lines in the body of the plots represent the conditions of one of the independent variables across the levels of the other independent variable.

When there is an interaction between two variables, it means that at least two of the lines in the interaction profile plot are not parallel to each other. Notice that none of the three lines for the Reward Group levels are parallel across cost levels, and the lines for the Cost Group across reward levels actually intersect. When there is a significant interaction, the next step is to look at simple effects.

When a two-factor interaction is significant, you can always view it from two different perspectives. In some cases you should test for simple effects from both perspectives before interpreting results. However, often the interaction is more interpretable (makes more sense) from one perspective than from the other. In the example, the reward effect is the only significant main effect, so a logical start is to look at the effect of cost type on the levels of the reward effect. In Figure 9.17, you can see that the effect of the rewards independent variable on commitment varies more for the less-cost group than for the more-cost group.

The interaction plots in this example suggests that the relationship between reward level and commitment values depends on the level of the cost group. Said another way, type of costs moderates the relationship between type of rewards and commitment.

Testing Slices (Simple Effects)

When there is a *simple effect* for independent variable A (Reward Group in this example) at a given level of independent variable B (Cost Group), it means that there is a significant relationship between independent variable A and the dependent variable (Commitment) at *that level* of independent variable B. As was stated earlier, the concept of a simple effect is easiest to understand by again looking at Figure 9.17.

First, consider the line that represents the more-cost group in Figure 9.17 (also shown to the right). This line has only a slight slope, which indicates the reward level has little effect on commitment in the more-cost group. However, the line for the less-cost group varies greatly over the reward levels, suggesting that there may be a significant relationship between cost and commitment for the participants in the high-reward group. In other 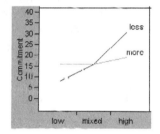 words, there may be a significant relationship between Reward Group and Commitment at the less-cost level of the Cost Group variable.

In JMP, testing for simple effects is done using the **Test Slices** command found on the menu on the interaction title bar. In fact, this command tests the simple effect of each cell in the factorial design table. You can easily look at the interaction from the perspective of either main effect. Figure 9.18 shows the Test Slices report for this example.

Figure 9.18 Testing Simple Effects with Test Slices Command

Slice Reward Group=low		Slice Reward Group=mixed		Slice Reward Group=high	
Sum of Squares	152.1	Sum of Squares	0	Sum of Squares	336.4
Numerator DF	1	Numerator DF	1	Numerator DF	1
Denominator DF	24	Denominator DF	24	Denominator DF	24
F Ratio	9.7604278075	F Ratio	0	F Ratio	21.587165775
Prob > F	0.0046118532	Prob > F	1	Prob > F	0.0001021031

Slice Cost Group=less		Slice Cost Group=more	
Sum of Squares	1322.5333333	Sum of Squares	36.4
Numerator DF	2	Numerator DF	2
Denominator DF	24	Denominator DF	24
F Ratio	42.434224599	F Ratio	1.1679144385
Prob > F	0	Prob > F	0.3280733862

The lower two tables in Figure 9.18 show the effect of reward on each of the cost groups. The computed *F* ratio for the less-cost group is 42.43, with a corresponding *p* value close to zero. The reward level has a significant effect on the less-cost group. However, the *F* ratio for the more-cost group is only 1.16 with the insignificant *p* value of 0.328. Therefore, you fail to reject the null hypothesis for the more-cost group, and conclude that there is not a significant simple effect for the reward-level factor at the more-cost group.

> **Note:** The computation of the *F* ratio for testing slices uses the error mean square generated by the two-factor analysis of variance with interaction (15.583 in Figure 9.16), with 24 degrees of freedom in the denominator. When testing slices, this whole model can be referred to as the *omnibus* model because the analysis includes all the participants. The numerator for each slice is obtained by analyzing each factor separately in a one-way analysis of variance, not shown in this example, but computed automatically by the Test Slices analysis.

You can verify the *F* statistic given by the Test Slices report by taking a subset of each cost-factor level and using the Fit Y by X platform to see a one-way ANOVA. For example, if you analyze the subset of the participants in the less-cost group, the one-way Analysis of Variance table gives a Mean Square for Reward Group of 661.267. Using this Mean Square as the numerator for the simple effect (test slices) test, and the error Mean Square, 15.583, from the factorial (omnibus) analysis shown in this example as the denominator, gives

$$F = (661.27) \div (15.58) = 42.44$$

This is the significant *F* value for testing the reward effect in the less-cost group.

A one-way ANOVA on the more-cost group gives an error mean square of 18.20, and the resulting simple effect test is

$$F = (18.20) \div (15.58) = 1.17$$

with nonsignificant *F* statistic, as shown in the **Slice Cost Group=more** table in Figure 9.18.

The three tables at the top of Figure 9.18 let you investigate the interaction from the perspective of possible simple effects for the type of costs at three different levels of the type of rewards factor. The interaction plot for that perspective is shown to the right. The horizontal axis represents the type of costs factor and has a midpoint for the less-cost group and one for the more-cost group. Within the body of the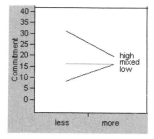
figure itself, there are lines representing the low-, mixed-, and high-reward groups. The plot and slice tests indicate that cost has a significant simple effect for the high- and low-reward groups, but no effect for the mixed-reward group.

When a two-factor interaction is significant, you can always view it from two different perspectives. Furthermore, the interaction is often more interpretable (makes more sense) from one perspective than from the other.

Formal Description of Results for a Paper

Results from this analysis could be summarized in the following way for a published paper:

Results were analyzed using two-way analysis of variance, with two between-group factors. This revealed a significant type of rewards by type of costs interaction, $F(2, 24) = 15.29$, $p < 0.0001$, MSE = 15/563, and $R^2 = 0.273$. The nature of this interaction is displayed in Figure 9.17.

Subsequent analyses demonstrated that there was a simple effect for type of rewards at the less-cost level of the type of costs factor, $F(2, 24) = 42.44$, $p < 0.00001$. As Figure 9.17 shows, the high-reward group displayed higher commitment scores than the mixed-reward group, which, in turn, demonstrated higher commitment scores than the low-reward group. The simple effect for type of rewards at the more-cost level of the type of costs factor was nonsignificant, $F(2, 24) = 1.17$, $p > 0.05$.

Summary

The factorial design introduced in this chapter has a single dependent response variable and two independent predictor (between-group) variables. The two predictor variables are manipulated so that treatment conditions include all combinations of levels of the predictor variables.

Analysis of data from a factorial design lets you test several hypotheses.

- Tests for main effects tell whether manipulating either or both of the predictor variables has a significant effect on the response.

- Tests for interaction determine if the effect of one predictor variable on the response is different at different levels of the other predictor variable.

- When there is significant interaction, testing for simple effects (testing slices) determines which levels of the predictor variables significantly affect the response.

A more complex design can have multiple response variables as well as one or more predictor variables. The next chapter introduces multivariate analysis of variance (MANOVA) for a situation with two responses and a single between-subjects predictor variable.

Assumptions for Factorial ANOVA with Two Between-Subjects Factors

Modeling type
> The response variable should be assessed on an interval or ratio level of measurement. Both predictor variables should be nominal-level variables (categorical variables).

Independent observations
> A given observation should not be dependent on any other observation in another cell (for a more detailed explanation of this assumption, see Chapter 7: "*t* Tests: Independent Samples and Paired Samples").

Random sampling
> Scores on the response variable should represent a random sample drawn from the populations of interest.

Normal distributions

Each cell should be drawn from a normally distributed population. If each cell contains more than 30 participants, the test is robust against moderate departures from normality.

Homogeneity of variance

The populations represented by the various cells should have equal variances on the response. If the number of participants in the largest cell is no more than 1.5 times greater than the number of participants in the smallest cell, the test is robust against violations of the homogeneity assumption.

References

Hayter, A. J. 1984. "A Proof of the Conjecture That the Tukey-Kramer Multiple Comparisons Procedure Is Conservative." *Annals of Mathematical Statistics* 12:61–75.

Kramer, C. Y. 1956. "Extension of Multiple Range Tests to Group Means with Unequal Numbers of Replications." *Biometrics* 12:309–310.

Sall, J. P. 1990. "Leverage Plots for General Linear Hypotheses." *American Statistician* 44 (4): 303–315.

SAS Institute Inc. 2002. *JMP Statistics and Graphics Guide.* Cary, NC: SAS Institute Inc.

Tukey, J. 1953. "Problems of Multiple Comparison." Manuscript, Princeton University.

Multivariate Analysis of Variance (MANOVA) with One Between-Subjects Factor

> **Overview.** This chapter shows how to use the Fit Model platform in JMP to perform a one-way multivariate analysis of variance (MANOVA). You can think of MANOVA as an extension of ANOVA that allows for the inclusion of multiple response variables in a single test. This chapter focuses on the *between-groups* design, in which each subject is exposed to only one condition under the independent predictor variable. Examples show how to summarize both significant and nonsignificant MANOVA results.

Introduction: Multivariate Analysis of Variance (MANOVA)

Multivariate analysis of variance (MANOVA) with one between-groups factor is appropriate when the analysis involves

- a single nominal or ordinal predictor variable that defines groups
- multiple numeric continuous response variables.

MANOVA is similar to ANOVA in that it tests for significant differences between two or more groups of participants. The important difference between MANOVA and ANOVA is that ANOVA is appropriate when the study involves just one response variable, but MANOVA is needed when there is more than one response variable. With MANOVA, you can perform a single test that determines whether there is a significant difference between treatment groups when compared simultaneously on all response variables.

To illustrate one possible use of MANOVA, consider the hypothetical study on aggression described in Chapter 8, "One-Way ANOVA with One Between-Subjects Factor." In that study, each of 60 children is assigned to one of three treatment groups based on the amount of sugar consumed at lunch:

- 0 grams
- 20 grams
- 40 grams.

A pair of observers watched each child after lunch, and recorded the number of aggressive acts displayed by that child. The total number of aggressive acts displayed by a given child over a two-week period served as that child's score on the aggression dependent variable. It is appropriate to analyze data from the aggression study with ANOVA because there is a single dependent variable—level of aggression.

Now consider how the study could be modified so that it is appropriate to analyze the data using a multivariate procedure, MANOVA. Imagine that, as the researcher, you want to have more than one measure of aggression. After reviewing the literature, you believe that there are at least four different types of aggression that children display:

- aggressive acts toward children of the same sex
- aggressive acts toward children of the opposite sex
- aggressive acts toward teachers
- aggressive acts toward parents.

Assume that you want to replicate the earlier study, but this time you want the observers to note the number of times each child displays an aggressive act in each of the four preceding categories. At the end of the two-week period, you have scores for each child on each of the categories. These four scores constitute four dependent variables.

You now have a number of options as to how you can analyze your data. One option is to perform four ANOVAs, as described in Chapter 8. In each ANOVA the independent variable is "amount of sugar consumed." In the first ANOVA the dependent variable is "number of aggressive acts directed at children of the same sex," in the second ANOVA the dependent variable is "number of aggressive acts directed at children of the opposite sex," and so forth.

However, a better alternative is to perform a single test that assesses the effect of the independent variable on all four of the dependent variables simultaneously. This is what MANOVA can do. Performing a MANOVA allows you to test the following null hypothesis:

> "In the population, there is no difference between the various treatment groups when they are compared simultaneously on the dependent variables."

Here is another way of stating this null hypothesis:

> "In the population, all treatment groups are equal on all dependent variables."

MANOVA produces a single *F* statistic that tests this null hypothesis. If the null hypothesis is rejected, it means that at least two of the treatment groups are significantly different with respect to at least one of the dependent variables. You can then perform follow-up tests to identify the pairs of groups that are significantly different and the specific dependent variables on which they are different. In doing these follow-up tests, you might find that the groups differ on some dependent variables (such as "aggressive acts directed toward children of the same sex") but not on another dependent variables (such as "aggressive acts directed toward teachers").

A Multivariate Measure of Association

Chapter 8 introduced the R^2 statistic, a measure of association often computed by an analysis of variance. Values of R^2 range from 0 to 1, where higher values indicate a stronger relationship between the predictor variable and the response variable in the study. If the study is a designed experiment, you can view R^2 as an index of the magnitude of the treatment effect.

This chapter introduces a *multivariate* measure of association that can be used when there are multiple response variables (as in MANOVA). This multivariate measure of association is called *Wilks' lambda*. Values of Wilks' lambda range from 0 to 1, but the way you interpret lambda is the opposite of the way you interpret R^2. With lambda, small values (near 0) indicate a relatively strong relationship between the predictor variable and the multiple response variables taken as a group, while larger values (near 1) indicate a relatively weak relationship. The *F* statistic that tests the significance of the relationship between the predictor and the multiple response variables is based on Wilks' lambda.

The Commitment Study

To illustrate multivariate analysis of variance (MANOVA), assume you replicate the study that examined the effect of rewards on commitment in romantic relationships. However, this time modify the study to obtain scores on three dependent variables instead of just one.

It was hypothesized that the rewards people experience in romantic relationships have a causal effect on their commitment to those relationships. The one-way ANOVA analysis in Chapter 8 tested this hypothesis by conducting a type of role-playing experiment. All 18 participants in the experiment were asked to engage in similar tasks in which they read the descriptions of 10 potential romantic "partners." For each partner, the participants imagined what it would be like to date this person, and rated how committed they would be to a relationship with that person. For the first nine partners, every subject saw exactly the same description. However, the different experimental groups saw a slightly different description for partner 10.

For example, the six participants in the high-reward condition saw the following final sentence describing partner 10:

"This person enjoys the same recreational activities that you enjoy, and is said to be very good-looking."

For the six participants in the mixed-reward condition, the last part of the description of partner 10 was worded as follows:

"Sometimes this person seems to enjoy the same recreational activities that you enjoy, and sometimes not. Some people think partner 10 is very good-looking, but others think not."

For the six participants in the low-reward condition, the last sentence in the description of partner 10 read this way:

"This person does not enjoy the same recreational activities that you enjoy, and is not considered by most people to be very good-looking."

In your current replication of the study from Chapter 8, you manipulate this "type of rewards" independent variable in exactly the same way. The only difference between the present study and the one described earlier involves the number of dependent variables obtained. In the earlier study, there was only one response variable, called Commitment. Commitment was defined as the participants' rating of their commitment to remain in the relationship. Scores on this variable were created by summing participant responses to four questionnaire items.

However, in the present study you intend to obtain two additional dependent variables:

- participants' ratings of their *satisfaction* with their relationship with partner 10
- participants' rating of *how long they intend to stay* in the relationship with partner 10.

Here, satisfaction is defined as the participants' emotional reaction to partner 10, and the intention to stay is defined as the participants' rating of how long they intend to maintain the relationship with partner 10. Assume that satisfaction and intention to stay in the relationship are measured with multiple items from a questionnaire, similar to those used previously to assess commitment.

You can see that there will be four values for each subject in the study. One value is for the classification variable that indicates whether the subject is assigned to the high-reward group, the mixed-reward group, or the low-reward group. The remaining three scores are the subject's scores on the measures of commitment, satisfaction, and intention to stay in the relationship. In conducting the MANOVA, you want to determine whether there is a significant relationship between the type of rewards predictor variable and the three response variables, taken as a group.

Performing a MANOVA with the Fit Model Platform

To do a multivariate analysis of variance in JMP, use the **Fit Model** command on the **Analyze** menu. The Fit Model dialog accepts multiple response variables when you specify the model. When there are multiple Y (response) variables, the *personality* types available are restricted to **Standard Least Squares** and **Manova**. The personality type specified for an analysis tells the Fit Model platform what kind of analysis to do.

When there are multiple Y variables, the default personality is **Standard Least Squares**, which does a *univariate* ANOVA for each response (four separate ANOVAs). In this context, univariate means *one response variable*. Each univariate ANOVA is the same analysis described in Chapter 8. The **Standard Least Squares** personality produces a univariate analysis of variance for the Commitment response, a second ANOVA for the Satisfaction response, and a third analysis for the variable called Stay in Relationship. This chapter introduces the **Manova** personality.

Figure 10.1 shows an example of the Fit Model dialog with multiple Y variables, and the available personality types. The **Manova** personality does the multivariate analysis of variance described in the next sections of this chapter. In the example you complete the dialog and select the **Manova** personality, as shown in Figure 10.1.

In the context of a MANOVA, *multivariate* means *multiple response variables*. For your purposes, these multivariate results consist of Wilks' lambda and the *F* statistic derived from Wilks' lambda.

Figure 10.1 Fit Model Platform with Multiple Responses

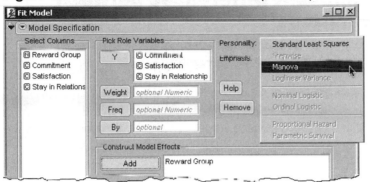

There is a specific sequence of steps to follow when interpreting these results.

1. First, review the multivariate F statistic derived from Wilks' lambda. If this multivariate F statistic is significant, you can reject the null hypothesis of no overall effect for the predictor variable. In other words, you reject the null hypothesis that, in the population, all groups are equal on all response variables. At that point, you perform and interpret a univariate ANOVA for each response.

2. To interpret the ANOVAs, first identify those response variables for which the univariate F statistic is significant. If the F statistic is significant for a given response variable, then go on to interpret the results of the Tukey's HSD (Honestly Significant Difference) multiple comparison test to determine which pairs of groups are significantly different from one another.

3. However, if the multivariate F statistic that is computed in the MANOVA is not significant, you cannot reject the null hypothesis that all groups have equal means on the response variables in the population. In most cases, your analysis should terminate at that time. You don't usually proceed with univariate ANOVAs.

4. Similarly, even if the multivariate F statistic is significant, you should not interpret the results for any specific response variable that did not display a significant univariate F statistic. This is consistent with the general guidelines for univariate ANOVA presented in Chapter 8, "Oneway ANOVA with One Between-Subjects Factor."

Example with Significant Differences between Experimental Conditions

Assume the study is complete and the data are entered into a JMP table. The data table, **multivariate commitment.jmp**, is shown in Figure 10.2.

There is one predictor variable in this study called Reward Group. This variable can assume the values "low" to represent participants in the low-reward group, "mixed" for participants assigned to the mixed-reward group, or "high" for participants in the high-reward group.

There are three response variables in this study. The first response variable, Commitment, is a measure of the subject's commitment to the relationship. The second response, Satisfaction, measures the subject's satisfaction with the relationship, and the third response, called Stay in Relationship, measures the subject's intention to stay in the relationship. Each variable has a possible score that ranges from a low of 4 to a high of 36.

Figure 10.2 Multivariate Data for Commitment Study

	Reward Group	Commitment	Satisfaction	Stay in Relationship
1	low	13	10	14
2	low	8	10	8
3	low	12	12	15
4	low	4	10	13
5	low	9	5	7
6	low	8	5	12
7	mixed	33	30	36
8	mixed	29	25	29
9	mixed	35	30	30
10	mixed	34	28	26
11	mixed	28	30	26
12	mixed	27	26	25
13	high	36	30	28
14	high	32	32	29
15	high	31	31	27
16	high	32	36	36
17	high	35	30	33
18	high	34	30	32

Columns panel: multivariate commitment — Columns (4/0): Reward Group, Commitment, Satisfaction, Stay in Relationship — Rows: All Rows 18, Selected 0, Excluded 0, Hidden 0, Labelled 0

Note: The downward arrow icon to the right of the Reward Group name in the Columns panel of the data table indicates it has the List Check column property, which lets you specify the order in which you want variable values to appear in analysis results. In this example, you want to see the Reward Group values listed in the order "low," "mixed," and "high." Unless otherwise specified, values in reports appear in alphabetic order. See the section "Ordering Values in Analysis Results," in Chapter 9, "Factorial ANOVA with Two Between-Subjects Factors," for details about using the List Check property.

Testing for Significant Effects with the Fit Model Platform

The following example shows a step-by-step example of how to use the Fit Model platform in JMP to perform a multivariate analysis of variance, and explains results.

- Open the table shown in Figure 10.2, called **multivariate commitment.jmp**, with one predictor variable, Reward Group, and three response variables, Commitment, Satisfaction, and Stay in Relationship.
- Choose **Fit Model** from the **Analyze** menu.
- When the Fit Model dialog appears, complete the dialog as shown previously in Figure 10.1. Be sure you select **Manova** from the **Personality** menu.
- Click **Run Model** to see the initial results, shown in Figure 10.3.

When you do a MANOVA, the results unfold as you provide additional information to the platform. To begin, you see the Response Specification panel on the left in Figure 10.3, which lets you specify the next step in the analysis. The initial results also include the overall means, group means, and means plots.

Initial Means and Plots

As with most JMP analyses, initial results show summary statistics, as well as supporting plots or graphs whenever possible. For example, a table of overall means for the response variables and a plot of those means are displayed below the Response Specification panel. You can quickly see that the overall means don't appear to vary across the responses. The next table and corresponding plots are the means for each response in the three levels of the predictor variable, Reward Group. The plot for these means indicates that responses for the high-reward group and the mixed-reward group are similar, but the levels for the low-reward group are much lower for each response.

Note that the Response Specification panel also shows the total observations in the sample, N, and the error degrees of freedom, DFE, for the analysis. These items give you a verification that data are what you intended—that the number of observations (N) in the data table is correct, and the number of predictor groups is N minus the error degrees of freedom ($18 - 15 = 3$ groups).

Figure 10.3 Initial Results of MANOVA Analysis

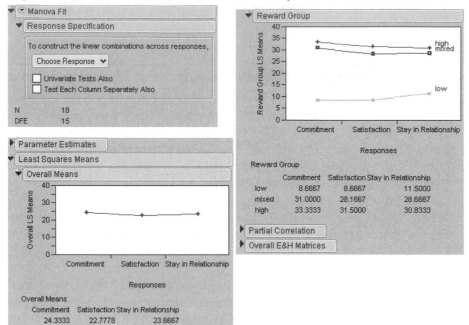

The Response Specification Panel

You use the Response Specification panel to specify the response design for the kind of multivariate tests you want to see. The menu on the left in Figure 10.4 shows the choices available for multivariate comparisons. The choice you make defines the *M matrix* used in the analysis. The columns in the M matrix define a set of transformation variables used to compute the multivariate tests. It is easiest to understand the M matrix and corresponding multivariate tests by looking at different M matrices. To do this, select different options from the **Choose Response** menu and look at the M matrix appended to the Response Specification panel by the choice you make.

One of the most commonly used response designs is Identity, which is used for this example. It is the standard design used to test the multivariate responses as a group. Descriptions and details for all the choices are in the *JMP Statistics and Graphics Guide* (2003).

- 🖰 Choose **Identity** from the **Choose Response** menu on the Response Specification panel.
- 🖰 Click **Run** to see the MANOVA results.

Figure 10.4 Response Specification Panel

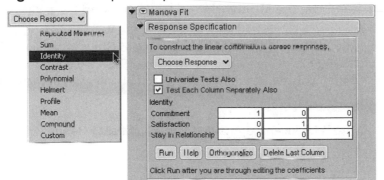

Steps to Interpret the Results

You can state the multivariate null hypothesis for the present study as follows:

> "In the population, there is no difference between the high-reward group, the mixed-reward group, and the low-reward group when they are compared simultaneously on commitment, satisfaction, and intention to stay in the relationship."

In other words, in the population, the three groups are equal on all response variables. Symbolically, you can represent the null hypothesis this way:

$$\begin{bmatrix} M_{11} \\ M_{21} \\ M_{31} \end{bmatrix} - \begin{bmatrix} M_{12} \\ M_{22} \\ M_{32} \end{bmatrix} = \begin{bmatrix} M_{13} \\ M_{23} \\ M_{33} \end{bmatrix}$$

Each M represents a population mean for one of the treatment conditions on one of the response variables. The first number in each subscript identifies the response variable, and the second number identifies the experimental group. In this example, the List Check option orders the Reward Group values as "low," "mixed," and "high," so M_{32} refers to the population mean for the third response variable (Stay in Relationship) in the second experimental group (the mixed-reward group), while M_{13} refers to the population mean for the first response variable (Commitment) in the third experimental group (the high-reward group).

Reviewing Wilks' Lambda, the Multivariate *F* Statistic, and Its *p* Value

The *F* statistic appropriate to test this null hypothesis is in the Whole Model table of the Manova Fit results. Note in Figure 10.5 that, because Reward Group is the only effect, the results for the effect and the whole model are the same. The Whole Model table provides the results from four different multivariate tests, but you want to focus only on Wilks' lambda.

Figure 10.5 Results from Multivariate Analysis with Significant Results

Manova Fit

Identity

Whole Model

Test	Value	Approx. F	NumDF	DenDF	Prob>F
Wilks' Lambda	0.0297553	20.7878	6	26	<.0001
Pillai's Trace	0.9962929	4.6322	6	28	0.0022
Hotelling-Lawley	31.732004	63.4640	6	24	<.0001
Roy's Max Root	31.704393	147.9538	3	14	<.0001

Intercept

Test	Value	Exact F	NumDF	DenDF	Prob>F
F Test	154.99086	671.6270	3	13	<.0001

Reward Group

Test	Value	Approx. F	NumDF	DenDF	Prob>F
Wilks' Lambda	0.0297553	20.7878	6	26	<.0001
Pillai's Trace	0.9962929	4.6322	6	28	0.0022
Hotelling-Lawley	31.732004	63.4640	6	24	<.0001
Roy's Max Root	31.704393	147.9538	3	14	<.0001

Information for Wilks' lambda shows in the first line of the table. Read the information from this row only. Where the row headed Wilks' Lambda intersects with the column headed Value, you find the computed value of the Wilks' lambda statistic. For the current analysis, lambda is 0.0297553. Remember that small values (closer to 0) indicate a relatively strong relationship between the predictor variable and the multiple response variables, and so this obtained value (rounded to 0.03) indicates that there is a strong relationship between Reward Group and the response variables. This is an exceptionally low value for lambda, much lower than you are likely to encounter in actual research.

This value for Wilks' lambda indicates that the relationship between the predictor and the response is strong, but to see if it is statistically significant, look at the multivariate *F* statistic, shown under the heading Approx. F. You can see that the multivariate *F* for this analysis is approximately 20.79. There are 6 numerator degrees of freedom and 26 denominator degrees of freedom associated with this test, as seen under the headings Num DF and Den DF. The probability value for this *F* appears under the heading Pr > F, and you can see that this *p* value is very small—less than 0.0001. Because this *p* value is less than the standard cutoff of 0.05, you can reject the null hypothesis of no differences between groups in the population. In other words, you conclude that there is a significant multivariate effect for type of rewards.

Univariate ANOVAs and Multiple Comparisons

When the MANOVA F test is significant, you might want to look at the univariate ANOVA and Tukey's HSD test for each individual response. An easy way to do this is to use the Fit Model dialog as shown previously for the multivariate analysis, but change the personality to Standard Least Squares.

- ⃝ Click the Fit Model dialog to make it the active window. Or, if you previously closed this dialog, select **Fit Model** from the **Analyze** menu again and enter the same responses and factors as before (see Figure 10.1).
- ⃝ Use the default Personality, **Standard Least Squares**, and change the Emphasis to **Minimal Report**, as shown in Figure 10.6.

Figure 10.6 Personality and Emphasis in the Fit Model Dialog

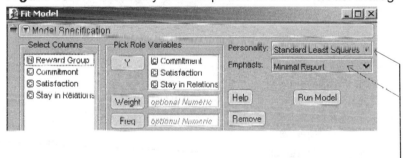

The **Standard Least Squares** personality (as opposed to **Manova**) performs an analysis of variance on each of the specified Y responses. The **Minimal Report** selection from the **Emphasis** menu suppresses some of the results, which is appropriate in this example because you are only interested in the univariate ANOVA Tukey's HSD tests.

- ⃝ Click **Run Model** to see the results. Because the only results of interest are the analysis of variance and Tukey HSD tests, first close all open reports, which gives the outline nodes shown on the left in Figure 10.7.
- ⃝ Open the Analysis of Variance reports for each response. The analysis of variance for each response variable shows on the right in Figure 10.7. All three responses have significant F tests with probabilities less than 0.0001, which means you want to look at Tukey's HSD for each response.

Figure 10.7 Univariate ANOVA for the Three Responses

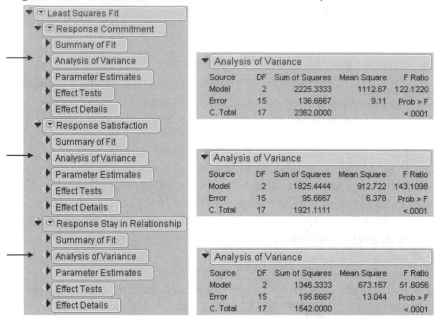

- Least Squares Fit
  - Response Commitment
    - Summary of Fit
    - → Analysis of Variance
    - Parameter Estimates
    - Effect Tests
    - Effect Details

Analysis of Variance

Source	DF	Sum of Squares	Mean Square	F Ratio
Model	2	2225.3333	1112.67	122.1220
Error	15	136.6667	9.11	Prob > F
C. Total	17	2362.0000		<.0001

  - Response Satisfaction
    - Summary of Fit
    - → Analysis of Variance
    - Parameter Estimates
    - Effect Tests
    - Effect Details

Analysis of Variance

Source	DF	Sum of Squares	Mean Square	F Ratio
Model	2	1825.4444	912.722	143.1098
Error	15	95.6667	6.378	Prob > F
C. Total	17	1921.1111		<.0001

  - Response Stay in Relationship
    - Summary of Fit
    - → Analysis of Variance
    - Parameter Estimates
    - Effect Tests
    - Effect Details

Analysis of Variance

Source	DF	Sum of Squares	Mean Square	F Ratio
Model	2	1346.3333	673.167	51.6056
Error	15	195.6667	13.044	Prob > F
C. Total	17	1542.0000		<.0001

To see Tukey's HSD multiple comparison tests, first open the Effect Details report for each of the three response variables. This report begins with the Least Squares Means table, as shown here. The Effect Details table shows the mean and standard error for each group level.

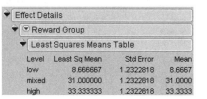

Effect Details
- Reward Group
  - Least Squares Means Table

Level	Least Sq Mean	Std Error	Mean
low	8.666667	1.2322818	8.6667
mixed	31.000000	1.2322818	31.0000
high	33.333333	1.2322818	33.3333

Note: The LSMeans (Least Squares Means) and the Mean are the same when there are no interaction terms in the model. However, they can be different for a main effect when the design is unbalanced (group sample sizes are different).

Each response shows the Reward Group effect. Select **LSMeans Tukey HSD** from the menu on the Reward Group title bar for each response.

Figure 10.8 shows the analysis results for the Commitment response.

Figure 10.8 Tukey's HSD Multiple Comparison Tests for Commitment Response

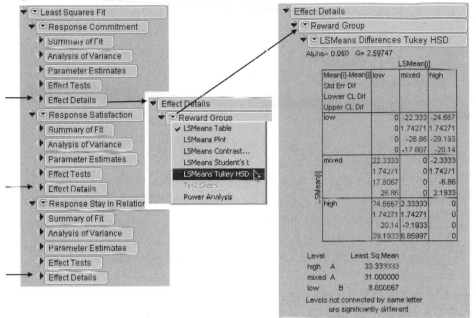

The alpha level of 0.05 is at the top of the report. The multiple comparison statistical results in Figure 10.8 show as a crosstabs table followed by the letter report.

- The crosstabs report is a two-way table that gives the difference in the group means, standard error of the difference, and confidence levels. Significant differences are highlighted (in red on screen).

- The letter report is a quick way to see differences. Groups (levels) connected by the same letter are not significantly different. The Tukey HSD tests confirm the graphical results seen previously in Figure 10.3. For each of the responses, the high- and mixed-reward groups are not statistically different from each other, but both are different from the low-reward group.

Summarizing the Analysis Results

This section shows how to summarize the results of the multivariate test discussed in the previous section. When the multivariate test is significant, proceed to the univariate ANOVAs and summarize them using the format presented in Chapter 8. To see multiple comparisons between group levels, do a univariate ANOVA for each response and look at Tukey's HSD, as described in the previous section.

You can use the following statistical interpretation format, similar to previously shown formats, to summarize the multivariate results from a MANOVA:

A) Statement of the problem
B) Nature of the variables
C) Statistical test
D) Null hypothesis (H_0)
E) Alternative hypothesis (H_1)
F) Obtained statistic
G) Obtained probability (p) value
H) Conclusion regarding the multivariate null hypothesis
I) If the MANOVA p value is significant, perform ANOVA on each response
J) For responses with significant ANOVA results, look at Tukey's HSD multiple comparisons
K) Conclusion regarding group differences for individual responses.

You could summarize the preceding MANOVA analysis in this way:

A) Statement of the problem:

The purpose of this study was to determine whether there was a difference between people in a low-reward relationship, people in a mixed-reward relationship, and people in a high-reward relationship with respect to their commitment to the relationship, their satisfaction with the relationship, and their intention to stay in the relationship.

B) Nature of the variables:

This analysis involved one predictor variable and three response variables. The predictor variable was type of rewards, a nominal variable that could assume three values—low-reward, mixed-reward, and high-reward. The three continuous numeric response variables were commitment, satisfaction, and intention to stay in the relationship.

C) Statistical test:

Wilks' lambda, derived through a one-way MANOVA, between-subjects design.

D) Null hypothesis (H_0):

$$\begin{bmatrix} M_{11} \\ M_{21} \\ M_{31} \end{bmatrix} = \begin{bmatrix} M_{12} \\ M_{22} \\ M_{32} \end{bmatrix} = \begin{bmatrix} M_{13} \\ M_{23} \\ M_{33} \end{bmatrix}$$

"In the population, there is no difference between people in a low-reward relationship, people in a mixed-reward relationship, and people in a high-reward relationship when they are compared simultaneously on commitment, satisfaction, and intention to stay in the relationship."

E) Alternative hypothesis (H_1):

In the population, there is a difference between people in at least two of the conditions (low-, mixed-, and high-reward conditions), when they are compared simultaneously on commitment, satisfaction, and intention to stay.

F) Obtained statistic:

Wilks' lambda = 0.03, with corresponding $F(6, 26) = 20.79$.

G) Obtained probability (p) value:

$p < 0.0001$.

H) Conclusion regarding the null hypothesis.

Reject the multivariate null hypothesis.

I) Perform ANOVA on each response variable:

Difference between groups was significant for all response variables; $p < 0.0001$.

J) Multiple comparison results:

For each response, Tukey's HSD multiple comparisons showed that the high- and mixed-reward groups were not statistically different from each other but both were different from the low-reward group.

Formal Description of Results for a Paper

You could summarize this analysis in the following way for a professional journal:

> Results were analyzed using a one-way MANOVA, between-groups design. This analysis revealed a significant multivariate effect for type of rewards, Wilks' lambda = .03, $F(6, 26) = 20.79$; $p < 0.0001$. Subsequent univariate ANOVA of the response variables showed significant differences between groups for each response. For each response, Tukey's HSD multiple comparisons showed that the high- and mixed-reward groups were not statistically different from each other but both were different from the low-reward group.

Example with Nonsignificant Differences between Experimental Conditions

Assume that you perform MANOVA a second time, but this time the analysis is performed on the data table called **multivariate nocommitment.jmp**. This table is constructed to provide nonsignificant results. Proceed with a multivariate analysis as shown in the previous section. Use **Fit Model** on the **Analyze** menu. In the Fit Model dialog:

   🖱 Select **Commitment**, **Satisfaction**, and **Stay in Relationship** from the Select Column lists and click **Y**.
   🖱 Select **Reward Group** and click **Add.**
   🖱 Click **Run** to see the results.

The results are shown below the data table in Figure 10.9.

Figure 10.9 Data and Multivariate Analysis with No Significant Results

		Reward Group	Commitment	Satisfaction	Stay in Relationship
	1	low	16	19	18
	2	low	10	15	17
	3	low	18	14	14
	4	low	16	20	10
	5	low	15	13	17
	6	low	12	15	11
	7	mixed	16	20	13
	8	mixed	18	14	16
	9	mixed	13	10	14
	10	mixed	17	13	19
	11	mixed	14	10	15
	12	mixed	19	16	18
	13	high	20	18	16
	14	high	18	15	19
	15	high	13	14	17
	16	high	12	18	15
	17	high	16	17	18
	18	high	14	19	15

Columns (4/0)
- Reward Group
- Commitment
- Satisfaction
- Stay in Relationship

Rows
All Rows	18
Selected	0
Excluded	0
Hidden	0
Labelled	0

Manova Fit
Identity
Whole Model

Test	Value	Approx F	NumDF	DenDF	Prob>F
Wilks' Lambda	0.726563	0.7504	6	26	0.6147
Pillai's Trace	0.283410	0.7795	6	28	0.5995
Hotelling-Lawley	0.362606	0.7252	6	24	0.6336
Roy's Max Root	0.3196269	1.4916	3	14	0.2599

You can see that this analysis resulted in a value for Wilks' lambda of 0.726563. Because this is a relatively large number (closer to 1), it indicates that the relationship between type of rewards and the three response variables is weaker with these data than with the previous data.

This analysis produced a multivariate F of only 0.75, which, with 6 and 26 degrees of freedom, was nonsignificant ($p = 0.615$). Therefore, this analysis fails to reject the multivariate null hypothesis of no group differences in the population. Because the multivariate F is not significant, your analysis usually terminates at this point—you would not go on to interpret univariate ANOVAs for the responses.

Summarizing the Results of the Analysis

Because this analysis tested the same null hypothesis that you tested earlier, you would prepare items A through E in the same manner described before. Therefore, this section presents only items F through H of the statistical interpretation format:

F) **Obtained statistic:**

Wilks' lambda = 0.73, corresponding $F = 0.75$.

G) **Obtained probability (*p* value):**

$p = 0.615$.

H) **Conclusion regarding the multivariate null hypothesis:**

Fail to reject the null hypothesis.

Formal Description of Results for a Paper

Results were analyzed using a one-way MANOVA, between-groups design. This analysis failed to reveal a significant multivariate effect for type of rewards, Wilks' lambda = 0.73, $F(6, 26) = 0.75; p = 0.615$.

Summary

This chapter examined the situation where groups of subjects are measured for two criteria under levels of a single nominal (grouping) predictor variable. Each group was assigned to one level of the predictor variable.

There are different methods of analyzing data collected this way.

- Perform simple ANOVAs on each of the response variables. This approach does not let you address the multiple responses simultaneously.

- Perform a multivariate analysis of variance (MANOVA), which addresses the null hypothesis, "In the population, all treatment groups are equal on all dependent variables." The MANOVA produces a single F statistic to test this hypothesis.

You learned about Wilks' lambda, a multivariate measure of association. Small values of lambda (near 0) indicate a relatively strong relationship between the predictor variable and the multiple response variables (taken together). Larger lambda values (near 1) indicate a relatively weak relationship. The multivariate F statistic is derived from Wilks' lambda.

In this chapter (and previous chapters) different independent subject groups are assigned to the predictor variable treatment levels. In the next chapter, the same group of respondents is exposed to all treatment conditions of the predictor variable. This is called a *repeated-measures* design.

Assumptions Underlying MANOVA with One Between-Subjects Factor

Level of measurement, data type, and modeling type

Each response variable should be a continuous numeric variable (an interval or ratio level of measurement). The predictor variable should be a categorical (nominal-level) variable (a categorical variable).

Independent observations

Across participants, a given observation should not be dependent on any other observation in any group. It is acceptable for the various response variables to be correlated with one another. However, a given subject's score on any response variable should not be affected by any other subject's score on any response variable. For a more detailed explanation of this assumption, see Chapter 7, "*t* Tests: Independent Samples and Paired Samples."

Random sampling

Scores on the response variables should represent a random sample drawn from the populations of interest.

Multivariate normality

In each group, scores on the various response variables should follow a multivariate normal distribution. Under conditions normally encountered in social science research, violations of this assumption have only a very small effect on the Type I error rate (the probability of incorrectly rejecting a true null hypothesis). On the other hand, when the data are platykurtic (form a relatively flat distribution), the power of the test may be significantly attenuated. The power of the test is the probability of correctly rejecting a false null hypothesis. Platykurtic distributions may be transformed to better approximate normality (see Stevens, 2002, or Tabachnick and Fidell, 2001).

Homogeneity of covariance matrices

In the population, the response-variable covariance matrix for a given group should be equal to the covariance matrix for each of the remaining groups. This is the multivariate extension of the *homogeneity of variance* assumption in univariate ANOVA.

To illustrate, consider a simple example with two groups and three response variables (V1, V2, and V3). To satisfy the homogeneity assumptions, the variance of V1 in group 1 must equal (in the population) the variance of V1 in group 2. The same must be true for the variances of V2 and V3. In addition, the covariance between V1 and V2 in group 1 must equal (in the population) the covariance between V1 and V2 in group 2. The same must be true for the remaining covariances (between V1 and V3 and between V2 and V3).

It becomes clear that the number of corresponding elements that must be equal increases dramatically as the number of groups increases and/or as the number of response variables increases. For this reason, the homogeneity of covariance assumption is rarely satisfied in real-world research. Fortunately, the Type I error rate associated with MANOVA is relatively robust against typical violations of this assumption so long as the sample sizes are equal. However, the power of the test tends to be attenuated when the homogeneity assumption is violated.

References

SAS Institute Inc. 2003. *JMP Statistics and Graphics Guide.* Cary, NC: SAS Institute Inc.

Stevens, J. 2002. *Applied Multivariate Statistics for the Social Sciences.* 4th ed. Mahwah, NJ: Lawrence Erlbaum Associates.

Tabachnick, B. G., and L. S. Fidell. 2001. *Using Multivariate Statistics.* 4th ed. New York: HarperCollins.

One-Way ANOVA with One Repeated-Measures Factor

> **Overview.** This chapter shows how to perform one-way repeated-measures ANOVA using the Fit Model platform in JMP. The focus is on repeated-measures designs, in which each participant is exposed to every condition under the independent variable. This chapter describes the necessary conditions for performing a valid repeated-measures ANOVA, discusses alternative analyses to use when the validity conditions are not met, and reviews strategies for minimizing sequence effects.

Introduction: What Is a Repeated-Measures Design?

A one-way repeated-measures ANOVA is appropriate when

- the analysis involves a single nominal predictor variable (independent variable) and a single numeric continuous response variable (dependent variable)
- each participant is exposed to each condition under the independent variable.

The *repeated-measures design* derives its name from the fact that each participant provides *repeated scores* on the response variable. That is, each participant is exposed to every treatment condition under the study's independent variable and contributes a response score for each of these conditions. Perhaps the easiest way to understand the repeated-measures design is to contrast it with the *between-subjects* design, in which each participant takes part in only one treatment condition.

For example, Chapter 8, "One-Way ANOVA with One Between-Subjects Factor," presents a simple experiment that uses a between-subjects design. In that fictitious study, participants were randomly assigned to one of three experimental conditions. In each condition, they read descriptions of a number of romantic partners and rated their likely commitment to each partner. The purpose of that study was to determine whether the level of rewards associated with a given partner affected the participant's rated commitment to that partner. The level of rewards was manipulated by varying the description of one specific partner (partner 10) that was presented to the three groups. Level of rewards was manipulated in this way:

- Participants in the low-reward condition read that partner 10 provides few rewards in the relationship.
- Participants in the mixed-reward condition read that partner 10 provides mixed rewards in the relationship.

- Participants in the high-reward condition read that partner 10 provides many rewards in the relationship.

After reading one of these descriptions, each participant rated how committed he or she would probably be to partner 10.

This study is called a between-subjects study because the participants were divided into different treatment groups (reward groups) and the independent variable was manipulated *between* these groups. In a between-subjects design, each participant is exposed to *only one* level of the independent variable. In this case, a given participant read either the low-reward description, the mixed-reward description, or the high-reward description. No participant read more than one description of partner 10.

On the other hand, each participant in a repeated-measures design is exposed to every level of the independent variable and provides scores on the dependent variable under each of these levels. For example, you could modify the preceding study so that it becomes a one-factor repeated-measures design. Imagine that you conduct a study with a single group of 20 participants instead of three treatment groups. You ask each participant to go through a stack of descriptions of potential romantic partners and give a commitment rating to each partner. Imagine further that all three versions of partner 10 appear somewhere in this stack, and that a given participant responds to each of these versions.

For example, a given participant might find the third potential partner to be the low-reward version of partner 10 (assume that you renamed this fictitious partner "partner 3"). This participant gives a commitment rating to partner 3 and moves on to the next partner. Later, the 11th partner happens to be the mixed-reward version of original partner 10 (now renamed as "partner 11"). The participant rates partner 11. Finally, the 19th partner happens to be the high-reward version of partner 10 (now renamed as "partner 19"). The participant rates partner 10 and finishes the questionnaire.

This study is now a repeated-measures design because each participant is exposed to all three levels of the independent variable. One way to analyze the data is to create one variable for each of the commitment ratings made under the three different conditions:

- One variable (call it Low) contains the commitment ratings for the low-reward version of the fictitious partner.

- One variable (Mixed) contains the commitment ratings for the mixed-reward version of the fictitious partner.

- One variable (High) contains the commitment ratings for the high-reward version of the fictitious partner.

To analyze repeated-measures data in this form, you compare the mean scores for these three variables. Perhaps you hypothesize that the commitment score for High is significantly higher than the commitment scores for Low or Mixed.

Make note of the following two cautions before moving on.

1. Remember that you need a special type of statistical procedure to analyze data from a repeated-measures study. You should not analyze the data using the Fit Y by X platform to do a one-way anova, as illustrated in Chapter 9.
2. Second, the fictitious study described here was used to illustrate the nature of a repeated-measures research design—do not view it as an example of a good repeated-measures research design. In fact, the preceding study suffers from several serious problems. Repeated-measures studies are vulnerable to problems not encountered with between-subjects designs. Some of the problems associated with this design are discussed later, in the "Sequence Effects" section.

Example: Significant Differences in Investment Size Across Time

To demonstrate the use of the repeated measures ANOVA, this chapter presents a new fictitious experiment that examines a different aspect of the investment model (Rusbult, 1980). Recall from earlier chapters that the investment model is a theory of interpersonal attraction. The study describes variables that determine commitment to romantic relationships and other interpersonal associations.

Designing the Study

Some of the earlier chapters described investigations of the investment model that involved the use of fictitious partners. They included written descriptions of potential romantic partners to which the participants responded as if they were real people. Assume that critics of the previous studies are very skeptical about the use of this analogue methodology and contend that investigations using fictitious partners do not generalize to how individuals actually behave in the real world. To address these criticisms, this chapter presents a different study that could be used to evaluate aspects of the investment model using actual couples.

This study focuses on the *investment size* construct from the investment model. Investment size refers to the amount of time, effort, and personal resources that an individual puts into a relationship with a romantic partner. People report heavy investments in a relationship when they have spent a good deal of time with their romantic partner, when they have a lot of shared activities or friends that they would lose if the relationship were to end, and so forth.

Assume that it is desirable for couples to believe they have invested time and effort in their relationships. Further, this perception is desirable because research has shown that couples are more likely to stay together when they feel they have invested more in their relationships.

The Marriage Encounter Intervention

Given that high levels of perceived investment size are a good thing, assume that you are interested in finding interventions that are likely to increase perceived investments in a marriage. Specifically, you read research indicating that a program called the *marriage encounter* is likely to increase perceived investments. In a marriage encounter program, couples spend a weekend together under the guidance of counselors, sharing their feelings, learning to communicate, and engaging in exercises intended to strengthen their relationship.

Based on what you have read, you hypothesize that couples' perceived investment in their relationships increases immediately after participation in a marriage encounter program. In other words, you hypothesize that, if couples are asked to rate how much they have invested in a relationship both before and immediately after the marriage encounter weekend, the *post* (after) ratings will be significantly higher than the *pre* (before) ratings. This is the primary hypothesis for your study.

However, being something of a skeptic, you assume further that you do not expect these increased investment perceptions to endure. Specifically, you believe that if couples rate their perceived investment at a follow-up point three weeks after the marriage encounter weekend, these ratings will have declined to their initial pre-level ratings, observed just before the weekend. In other words, you hypothesize that there will not be a significant difference between investment ratings obtained before the weekend and those obtained three weeks after. This is the secondary hypothesis for your study.

To test these hypotheses, you conduct a study that uses a single-group experimental design with repeated measures. The design of this study is illustrated in Figure 11.1.

Figure 11.1 Single-Group Experimental Design with Repeated Measures

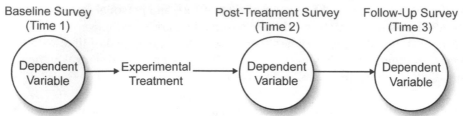

Suppose you recruit 20 couples who are about to go through a marriage encounter program. The response variable in this study is perceived investment measured with a multiple-item questionnaire. Higher scores on the scale reflect higher levels of perceived investment.

Each couple gives investment ratings at the three points in time illustrated by the three circles in Figure 11.1. Each circle represents one dependent variable. Specifically,

- a *baseline survey* obtains investment scores at Time 1, just before the marriage encounter weekend

- a *post-treatment* survey obtains investment scores at Time 2, immediately after the encounter weekend

- a *follow-up* survey obtains investment scores at Time 3, three weeks after the encounter weekend.

Notice that an experimental treatment, the marriage encounter program, appears between Time 1 and Time 2.

Problems with Single-Group Studies

Keep in mind that the study described here uses a weak research design. To understand why, remember the main hypothesis you want to test is that investment scores increase significantly from Time 1 to Time 2 because of the marriage encounter program. Imagine for a moment that the results obtained when you analyze the study data show that Time 2 investment scores are significantly higher than Time 1 scores.

These results do not provide strong evidence that the marriage encounter manipulation caused the increase in investment scores. There are obvious alternative explanations for the increase. For example, perhaps investment scores naturally increase over time among married couples regardless of whether they participate in a marriage encounter program.

Perhaps an increase in score is due to simply having a restful weekend and not due to the marriage encounter program at all. There is a long list of possible alternative explanations.

The point is that you must carefully design repeated-measures studies to avoid confounding the response and other problems. This study is for illustration only, and is not designed very carefully. A later section in this chapter, "Sequence Effects," discusses some of the problems associated with repeated-measures studies and reviews strategies for dealing with them. In addition, Chapter 12, "Factorial ANOVA with Repeated-Measures Factors and Between-Subjects Factors," shows how you can make the present single-group design stronger by including a control group.

Predicted Results

The primary hypothesis is that couples' ratings of investment in their relationships increases immediately following the encounter weekend. The secondary hypothesis is that this increase is a transient or temporary effect rather than a permanent change. These hypothesized results appear graphically in Figure 11.2. Notice that mean investment scores increase from Time 1 to Time 2, and then decrease again at Time 3.

Figure 11.2 Hypothesized Results for the Investment Model Study

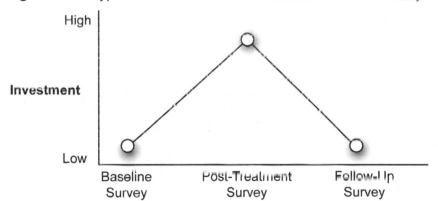

You test the primary hypothesis by comparing post-treatment (Time 2) investment scores to the baseline scores (Time 1). Comparing the follow-up scores (Time 3) to the baseline scores (Time 1) shows whether any changes observed at post-treatment are maintained, and you use this comparison to test the secondary hypothesis.

Assume this study is exploratory because studies of this type have not previously been undertaken. Therefore, you are uncertain what methodological problems might be encountered or the magnitude of possible changes. This example could serve as a pilot study to assist in the design of a more definitive study and the determination of an appropriate sample size. Chapter 12, "Factorial ANOVA with Repeated-Measures Factors and Between-Subjects Factors," presents a follow-up study based on the pilot data from this project.

A Univariate Approach for Repeated-Measures Analysis Using the Fit Y by X Platform

Twenty couples participated in this study. Suppose that only the investment ratings made by the wife in each couple are analyzed. The sample consists of data from the 20 married women who participated in the marriage encounter program.

The response variable in this study is the size of the investment the participants (the wives) believe they have made in their relationships. Investment size is measured by a scale that gives a continuous numeric value.

Reviewing the JMP Data Table

The JMP data table called **marriage univar.jmp** lists the investment scores obtained at the three different points in time for 20 participating couples. Each couple (each wife) contributes three rows to the data table.

- The nominal variable, ID, identifies each couple with values "1" through "20." Each couple contributes three lines of data corresponding to the three time periods.

- The nominal variable, time, identifies the time period with values "Pre," "Post," and "Followup."

- The continuous numeric investment score variable, investment, lists the investment scores recorded at each time period.

Figure 11.3 shows a partial listing of the 60 observations in the data table.

Figure 11.3 Partial Listing of the **marriage univar** Data Table

		ID	time	investment
marriage univar	1	1	Pre	8
	2	1	Post	10
Columns (3/0)	3	1	Followup	10
ID	4	2	Pre	10
time	5	2	Post	13
investment	6	2	Followup	12
	7	3	Pre	7
Rows	8	3	Post	10
All Rows 60	9	3	Followup	12
Selected 0	10	4	Pre	6
Excluded 0	11	4	Post	9
Hidden 0	12	4	Followup	10
Labelled 0	13	5	Pre	7
	14	5	Post	8
	15	5	Followup	9
	16	6	Pre	11

Plotting the Mean Investment Scores

You can quickly compare the mean investment scores for the three time periods by using the Chart platform, as follows:

- Choose **Chart** from the **Graph** menu.
- In the Chart launch dialog, select **time** in the Select Columns list and click **X, Level**.
- Select **investment** in the Select Columns list and choose **Mean** from the **Statistics** menu in the dialog.
- Click **OK** to see a bar chart of means.
- To label the bars, shift-click to highlight the bars on the chart and all rows in the data table. Select **Label/Unlabel** from the **Rows** menu.

Figure 11.4 shows the completed Chart launch dialog and a bar chart of the group means with labels. The Mean investment score for Pre is 8.6, Post is 10.75, and Followup is 11.0.

Figure 11.4 Chart Dialog and Bar Chart of Group Means

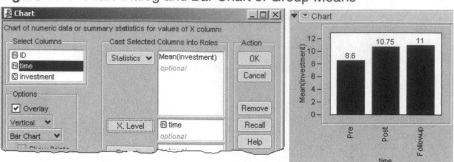

Looking at Descriptive Statistics

First, look at descriptive statistics for the investment. This serves two purposes.

- Scanning the sample size, minimum value, and maximum value for each variable lets you check for obvious data entry errors.

- You sometimes need the means and standard deviations for the response variables to interpret significant differences found in the analysis. Also, the means for within-participants variables are not routinely included in the output of the Fit Model platform repeated measures analysis, discussed later in this chapter.

One way to see a table of descriptive statistics is to use the **Summary** command on the **Tables** menu. Here are the steps to create a table of summary statistics.

- 🖑 With the **marriage univar** table active, choose **Summary** from the **Tables** menu.
- 🖑 When the Summary dialog appears, select **time** in the variable selection list on the left of the dialog and click **Group**.
- 🖑 Then, select **time** in the Group list and click the **a...Z/Z...a** button below the list. This causes the values of time to be displayed in descending order in the summary table (instead of the default ascending order).
- 🖑 Select **investment** in the variable selection list.
- 🖑 Click the **Statistics** button in the launch dialog and choose **Mean** to see Mean(investment) in the statistics list. Continue in this manner and choose **Std Dev**, **Min**, and **Max** in the statistics list. The completed dialog should look like the example at the top in Figure 11.5.
- 🖑 Click **OK** to generate the new data table with summary statistics shown at the bottom in Figure 11.5.

Figure 11.5 Summary Dialog and Table of Summary Statistics

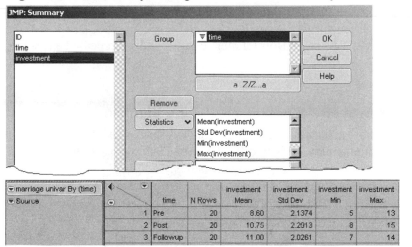

The value in the N Rows column is 20 for each variable—there are no missing values. The investment Min and investment Max values indicate no obvious errors in data. The mean investment score for Pre is 8.6, Post is 10.75, and Followup is 11.0.

A review of the group means shown in Figure 11.4 and Figure 11.5 suggests support for your study's primary hypothesis, but does not appear to support the study's secondary hypothesis. Notice that mean investment scores increase from the baseline survey at Time 1 (Pre) to the post-treatment survey at Time 2 (Post). If this increase is statistically significant, it would be consistent with your primary hypothesis that perceived investment increases immediately following the marriage encounter weekend.

However, notice that the mean investment scores remain at a relatively high level at the Time 3 follow-up survey, three weeks after the program. This trend is not consistent with your secondary hypothesis that the increase in perceived investment is short-lived.

However, at this point, you are only "eyeballing" the means. It is not clear whether any of the differences are statistically significant. To determine significance, you must analyze the data using a repeated-measures analysis. The next section shows a simple way to do this with using the Fit Y by X platform. Later sections show a repeated-measures ANOVA using the Fit Model platform.

Using the Fit Y by X Platform with a Matching Variable

One approach to analyze data from a simple repeated-measures design with no between-subject effects (grouping effects) is to do a one-way analysis of variance and use a matching variable to identify the repeating measures.

The Fit Y by X platform offers this feature.

- Choose **Fit Y by X** from the **Analyze** menu.
- Select **investment** from the Select Columns list and click **Y, Response** in the launch dialog. Select **time** from the Select Columns list and click **X, Factor** in the launch dialog.
- Click **OK** to see the scatterplot of investment by time, which is the initial Fit Y by X platform result.
- Choose the **Matching Columns** command from the menu on the Oneway Analysis title bar, as illustrated on the left in Figure 11.6. This command displays a dialog prompting you to select a matching variable.
- When the Matching Columns dialog appears, choose **ID** as the matching variable. Click **OK** to see the analysis shown on the right in Figure 11.6.

When you choose a matching variable in the Fit Y by X one-way ANOVA platform, the analysis uses that variable to compute the within-subjects variation. With the within-subjects variation removed from the residual, the remaining error is the correct denominator to compute for the correct F statistic to test the repeated-measures between-subjects effect.

Figure 11.6 Fit Y by X Results of Matching Fit Analysis

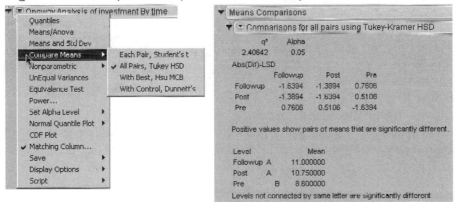

Because the *F* statistic for the time variable is significant ($p < 0.0001$), use the **Compare Means** option found on the menu on the Oneway Analysis title bar to see which of the means differ.

   🖱 Return to the Oneway Analysis of Investment by Time results and select **Compare Means → All Pairs, Tukey's HSD** from the menu on the title bar. The results of the multiple comparison tests are appended to the analysis and shown in Figure 11.7.

Note: This multiple comparison test does not consider the advantage of using the matching variable effect, and therefore gives a conservative result.

Figure 11.7 Compare Group Means with Tukey's HSD

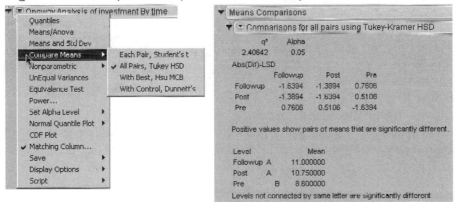

Keep in mind that the Fit Y by X platform can only be used for repeated-measures analysis when there are no between-subject (grouping) effects, as in this example.

Steps to Interpret the Results

1. **Make sure the results look reasonable.** Review all the results. First, check the number of observations given in the Summary table (N Rows in Figure 11.5). Next, check the number of levels for the response variable—the three levels of Time are shown in the bar chart with the mean values above the bars (Figure 11.4). Also, the Matching Fit report gives two degrees of freedom for the time response variable, which is the number of levels (Pre, Post, and Followup) used in the analysis, minus one (3 − 1 = 2).

2. **Review the appropriate *F* statistic and its associated probability value.** The first step to interpret the results of the analysis is to review the *F* value of interest. The relevant *F* value is shown in the Matching Fit report for the time variable (the within-subjects variable). This *F* value tells whether to reject the null hypothesis that states:

   "In the population, there is no difference in investment scores obtained at the three points in time."

   This hypothesis can be represented symbolically in this way:

   H_0: T1 = T2 = T3

   where T1 is the baseline pre-weekend mean investment score, T2 is the mean score following a marriage encounter weekend, and T3 is the mean score three weeks after the marriage encounter weekend.

   The Matching Fit report (shown again here) for the repeated-measures time variable gives a univariate *F* ratio that tests this null hypothesis. The *F* statistic for the time effect is 30.28, with $p < 0.0001$. Because the p value

Matching Fit					
Source	SS	DF	MS	F Ratio	Prob>F
Whole Model	290.4833	21	13.83254	12.028	<.0001
time	69.6333	2	34.81667	30.275	<.0001
ID	220.8500	19	11.62368	10.108	<.0001
Error1	43.7000	38	1.15000		

is less than 0.05, you can reject the null hypothesis of no difference in mean levels of commitment in the population. In other words you conclude that there is an effect for the repeated-measures variable, time.

Computational note: This numerator for this statistic is the mean square for the Time effect (with 2 degrees of freedom). The denominator is the mean square for the error, labeled Error1 in the Matching Fit report (with 38 degrees of freedom). This is the appropriate denominator for the repeated-measures F statistic because it does not include the between-subjects error. The F statistic is computed as 34.817 ÷ 1.105 = 30.275.

3. **Prepare your own version of an ANOVA table.** The next step is to formulate the ANOVA summary table shown in Table 11.1. Note that the quantities needed for this table are either in the Matching Fit report or easily computed:

   - The ID variable in the report is the between-subjects effect, with 19 DF.

   - The within-subjects sum of squares (SS) and Mean Squares (MS) information is composed of the treatment (time) effect and the within-subjects residual (Error1).

   - The total within-subjects quantities are the sum of the treatment and residual terms.

   - The total DF is the sum of the total within-subjects DF and the between-subjects DF.

Table 11.1 ANOVA Table for Study of Investment Using a Repeated-Measures Design

Source	DF	SS	MS	F	p
Between Participants	19				
Within Subjects	40	113.33			
Treatment	2	69.6	34.82	30.28	$p < 0.0001$
Residual	38	43.7	1.15		
Total	59				

N = 20.

4. **Review the results of contrasts.** The significant effect for time tells you that investment scores obtained at one point in time are significantly different from scores obtained at some other point in time. When you select **Compare Means → All Pairs, Tukey HSD**, the means comparison results are displayed at the bottom of the analysis report and comparison circles are shown to the right of the scatterplot. See the section "Using JMP for Analysis of Variance" in Chapter 8, "One-Way ANOVA with One Between-Subjects Factor," for an explanation of comparison circles.

Options on the Means Comparisons report display the following results:

- The alpha value of 0.05 appears at the top of the report (see Figure 11.7).
- Next, a table of the actual absolute difference in the means minus the LSD. The Least Significant Difference (LSD) is the difference that would be significant. Pairs with a positive value are significantly different.

The Tukey letter grouping shows the mean scores for the three groups on the response variable (investment in this study) and whether or not these group means are different at the 0.05 alpha level. You can see that the mean follow-up score is 11.0, the mean post-study score is 10.75, and the mean pre-study score is 8.6. The letter A identifies the follow-up score and the post-study score, and the letter B identifies the pre-study score. Therefore, you conclude that the pre-study mean score is significantly different from the post-study and follow-up mean scores but the post-study and follow-up scores are not different.

Summarizing the Results of the Analysis

The standard statistical interpretation format used for between-subjects ANOVA (as previously described in Chapter 8) is also appropriate for the repeated-measures design. The outline of this format appears again here.

A) Statement of the problem

B) Nature of the variables

C) Statistical test

D) Null hypothesis (H_0)

E) Alternative hypothesis (H_1)

F) Obtained statistic

G) Obtained probability (p value)

H) Conclusion regarding the null hypothesis

I) ANOVA summary table

J) Figure representing the results.

Because most sections of this format appear in previous chapters, they are not repeated here. Instead, the formal description of the results follows.

Formal Description of Results for a Paper

Mean investment size scores across the three trials are displayed in Figure 11.5. Results analyzed using one-way analysis of variance (ANOVA) repeated-measures design revealed a significant effect for the treatment (time), with $F(2,38) = 30.28$; $p < 0.0001$. Contrasts showed that the baseline measure was significantly lower than the post-treatment and follow-up mean scores. The post-treatment and follow-up mean scores were not significantly different.

Repeated-Measures Design versus Between-Subjects Design

An alternative to the repeated-measures design is a between-subjects design, as described in Chapter 8, "One-Way ANOVA with One Between-Subjects Factor." For example, you could follow a between-subjects design and measure two groups immediately following the weekend marriage encounter program (at Time 2). In this between-subjects study, one group of couples attends the weekend program and the other group does not attend. If you conduct the study well, you can then attribute any differences in the group means to the weekend experience.

In both the repeated-measures design and the between-subjects design, the sums of squares and mean squares are computed in the same way. However, an advantage of the repeated-measures design is that each participant also serves as control. Because each participant serves in each treatment condition, variability in scores due to individual differences between participants is not a factor in determining the size of the treatment effect. The between-subjects variability is removed from the error term in the F test computation (see Table 11.1). This computation usually allows for a more sensitive test of treatment effects because the between-subjects variance can be much larger than the within-subject variance. This within-subject variance is smaller because multiple observations from the same participant tend to be positively correlated, especially when measured across time.

A repeated-measures design also has the advantage of increased efficiency because this design requires only half the number of participants that are needed in a between-subjects design. This can be an important consideration when the targeted study population is limited. These statistical and practical differences illustrate the importance of careful planning of the statistical analysis when designing an experiment.

Weaknesses of the One-Way Repeated-Measures Design

The primary limitation of the present design is the lack of a control group. Because all participants receive the treatment in this design, there is no comparison that evaluates whether observed changes are the result of the experimental manipulation. For example, in the study just described, increases in investment scores might occur because of time spent together and so might have nothing to do with the specific program activities during the weekend (the treatment). Chapter 12, "Factorial ANOVA with Repeated-Measures Factors and Between-Subjects Factors," shows how to remedy this weakness with the addition of an appropriate control group.

Another potential problem with this type of design is that participants might be affected by a treatment in a way that changes responses to subsequent measures. This problem is called a *sequence effect* and is discussed in the next section.

Sequence Effects

Another important consideration in a repeated measures design is the potential for certain experimental confounds. In particular, the experimenter must control for order effects and carryover effects.

Order Effects

Order effects result when the ordinal position of the treatments biases participant responses. For example, suppose an experimenter studying perception requires participants to perform a reaction-time task in each of three conditions. Participants must press a button on a response pad while waiting for a signal, and after receiving the signal they must press a different button. The dependent variable is reaction time and the independent variable is the type of signal. The independent variable has three levels:

- Condition 1—flash of light
- Condition 2—an audio tone
- Condition 3—both the light and tone simultaneously.

Each test session consists of 50 trials. A mean reaction time is computed for each session. Assume that these conditions are presented in the same order for all participants (morning, before lunch, and after lunch).

The problem with this research design is that reaction-time scores can be adversely affected after lunch by fatigue. Responses to Condition 3 might be more a measure of fatigue than a true treatment effect. Suppose the experiment yields mean scores for 10 participants as shown in Table 11.2. It appears that presentation of both signals (tone and light) causes a delayed reaction time compared to the other two treatments (tone alone or light alone).

Table 11.2 Mean Reaction Time (msecs) for All Treatment Conditions

Type of Signal		
Tone	**Light**	**Light and Tone**
650	650	1125

An alternative explanation for the preceding results is that a fatigue effect in the early afternoon causes the longer reaction times. According to this interpretation, you expect each of the three treatments to have longer reaction times if presented during the early afternoon period. If you collect the data as described, there is no way to determine which explanation is correct.

To control for this problem, you must vary the treatment order. This technique, called *counterbalancing*, presents the conditions in a different order to different participants. Table 11.3 shows a research design that use counterbalancing.

Table 11.3 Counterbalanced Treatment Conditions to Control for Sequence Effects

	Treatment Order		
	Early AM	**Late AM**	**Early PM**
Participant 1	Tone	Light	Both
Participant 2	Both	Tone	Light
Participant 3	Light	Both	Tone
Participant 4	Tone	Both	Light
Participant 5	Light	Tone	Both
Participant 6	Both	Light	Tone

Note in Table 11.3 that each treatment occurs an equal number of times at each point of measurement. To achieve complete counterbalancing, you must use each combination of treatment sequences with an equal number of participants.

Complete counterbalancing becomes impractical as the number of treatment conditions increases. For example, there are only 6 possible sequences of 3 treatments, but this increases to 24 sequences with 4 treatments, 120 sequences with 5 treatments, and so forth. Usually, counterbalancing is possible only if the independent variable assumes a relatively small number of values.

Carryover Effects

Carryover effects occur when an effect from one treatment changes (carries over to) the participants' responses in the following treatment condition. For example, suppose you investigate the sleep-inducing effect of three different drugs. Drug 1 is given on Night 1 and sleep onset latency is measured by electroencephalogram. The same measure is collected on Night 2 and Night 3, when Drug 2 and Drug 3 are administered. If Drug 1 has a long half-life, then it can still exert an effect on sleep latency at Night 2. This carryover effect makes it impossible to accurately assess the effect of Drug 2.

To avoid potential carryover effects, the experimenter can separate the experimental conditions by some period of time, such as one week. Counterbalancing also provides some control over carryover effects. If all treatment combinations can be given, then each treatment will be followed (and preceded) by each other treatment with equal frequency. However, counterbalancing is more likely to correct order effects than to correct carryover effects. Counterbalancing is advantageous because it enables the experimenter to measure the extent of carryover effects and to make appropriate adjustments to the analysis.

Ideally, careful consideration of experimental design allows the experimenter to avoid significant carryover effects in a study. This is another consideration in choosing between a repeated-measures and a between-subjects design. In the study of drug effects described above, the investigator can avoid any possible carryover effects by using a between-subjects design. In that design, each participant only receives one of the drug treatments.

Univariate or Multivariate ANOVA for Repeated-Measures Analysis

The analysis described previously in the section "A Univariate Approach for Repeated-Measures Analysis Using the Fit Y by X Platform" was conducted as a conventional univariate repeated-measures ANOVA. However, using the univariate approach requires that the data fulfill two assumptions:

- The scores from experimental treatments have a multivariate normal distribution in the population. Normality is impossible to prove but becomes more likely as sample size increases.

- The common covariance matrix has a specific type of pattern called *sphericity* (homogeneity of covariance).

Homogeneity of covariance refers to the covariance between participants for any two treatments. One way to conceptualize this is that participants should have the same rankings in scores for all pairs of levels of the independent variable. For example, if there are three treatment conditions (T1, T2, and T3), the covariance between T1 and T2 should be comparable to the covariance between T2 and T3, and to the covariance between T1 and T3. Homogeneity of covariance is sufficient for test validity, but a less specific type of covariance pattern, called sphericity, is a necessary condition for the validity of a univariate analysis of a repeated-measures design (Huynh and Mandeville, 1979; Rouanet and Lepine, 1970).

The violation of homogeneity of covariance (sphericity) is particularly problematic for repeated-measures designs. In the case of between-subjects designs, the analysis still produces a robust F test even when this assumption is not met (provided sample sizes are equal). On the other hand, a violation of this assumption with a repeated-measures design leads to an increased probability of a Type I error (rejection of a true null hypothesis). Therefore, you must take greater care when analyzing repeated-measures designs either to prove that the assumptions of the test have been met or to alter the analysis to account for the effects of the violation.

For a more detailed discussion of the validity conditions for the univariate ANOVA and alternative approaches to the analysis of repeated measures data, see Barcikowski and Robey (1984) or LaTour and Miniard (1983).

The next section approaches repeated-measures analysis as a multivariate problem, and tests sphericity for this marriage encounter data example.

Reviewing the Data Table for a Multivariate Analysis

As previously described, the sample consists of data from the 20 married women who participated in the marriage encounter program. The response variable is the size of the investment the participants (the wives) believe they have made in their relationships. Investment size is measured by a scale that gives a continuous numeric value.

The JMP data table called **marriage.jmp** lists the investment scores obtained at the three different points in time arranged as three columns. The nominal variable called ID identifies each couple. The investment score variable Pre lists the investment scores recorded at Time 1, Post records the scores at Time 2, and Followup records the scores at Time 3 (Figure 11.8).

Figure 11.8 Listing of Data for Multivariate Repeated-Measures Analysis

	ID	Pre	Post	Followup
1	1	8	10	10
2	2	10	13	12
3	3	7	10	12
4	4	6	9	10
5	5	7	8	9
6	6	11	15	14
7	7	8	10	9
8	8	5	8	8
9	9	12	11	12
10	10	9	12	12
11	11	10	14	13
12	12	7	12	11
13	13	8	8	9
14	14	13	14	14
15	15	11	11	12
16	16	7	8	7
17	17	9	8	10
18	18	8	13	14
19	19	10	12	12
20	20	6	9	10

Columns (4/0): ID, Pre, Post, Followup

Rows — All Rows 20, Selected 0, Excluded 0, Hidden 0, Labelled 0

Using the Fit Model Platform for Repeated-Measures Analysis

The Fit Model platform can analyze repeated-measures data using multivariate methods (MANOVA). The three commitment measurements (Pre, Post, and Followup) are three response variables and the fitting Personality is **Manova**. In a between-subjects study, the predictor variable (or model effect) appears in the Construct Model Effects area of the Fit Model dialog. The Fit Model dialog in Figure 11.9 is similar to the one shown in Chapter 10, "Multivariate Analysis of Variance (MANOVA) with One Between-Subjects Factor," except there is no model effect.

Figure 11.9 Fit Model Dialog for Multivariate Repeated-Measures Analysis

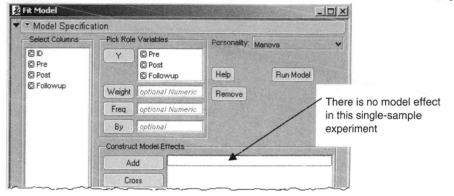

There is no model effect in this single-sample experiment

When you click **Run Model** in the Fit Model dialog, the initial multivariate analysis appears, as shown on the left in Figure 11.10. This initial result plots the means of the three response variables, which again indicates the difference between the Pre and Post scores, but shows little difference between the Post and Followup scores.

To continue with the repeated-measures analysis:

- Click the **Univariate Tests Also** check box. This option gives the sphericity test and univariate analysis of the data.
- Click the **Test Each Column Separately Also** check box. This option sets up contrasts between the Pre and Post mean scores, and between the Pre and Followup mean scores.
- Choose **Repeated Measures** from the **Choose Response** menu, as shown above.
- Because the measures are repeated, an additional dialog prompts for a name to call the repeated measure. The default name is Time, which is a typical type of repeated measure, and appropriate for this example.
- Click **OK** in the Specification of Repeated Measures dialog to see the completed analysis.

Figure 11.10 Initial MANOVA Results for Repeated-Measures Analysis

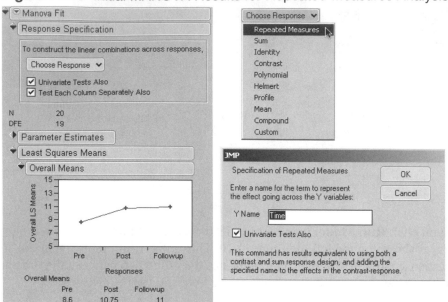

Note: Most JMP results include tables, charts, statistics, and other statistical tools geared for any possible request or model you might specify. This is especially true for the MANOVA reports because they are equipped to handle complex situations. However, the example in this chapter is very simple, and only requires that you look at a single line of results. The figures in this chapter show unneeded outline nodes as closed in the analysis and show as open only what you need to see.

Examining the Sphericity Text

The MANOVA results include both univariate and multivariate F tests for the within-subjects time effect. To decide which test to use, look at the Sphericity Test table, which is the first table in the Within Subjects section, as shown here. The sphericity test, called Mauchly's criterion, has a significant F test with p value (prob > Chisq) of 0.0323. This marginally significant test suggests the data display a significant departure from sphericity. In other words you reject the null hypothesis of the homogeneity of covariance. However, this test is very sensitive and any

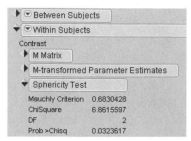

deviation from sphericity results in a significant F test. You can compensate for small to moderate deviations from sphericity by using a modified F test, or if there is a severe departure from sphericity ($p < 0.0001$) then multivariate tests should be used.

Examining the Multivariate Analysis of Variance

Figure 11.11 shows the results of the repeated-measures analysis for the time effect. The line in the Time report labeled **Univar unadj Epsilon** gives the Exact F as 30.2754 with 2 degrees of freedom in the numerator and 38 degrees of freedom in the denominator. This is the same univariate analysis given by the Fit Y by X platform shown previously. However, to account for the deviation in sphericity, you can refer to the modified F tests or to the multivariate F test, also given in the Time report.

The primary concern when the sphericity pattern is not present is that the F test will be too liberal and will lead to inappropriate rejection of the null hypothesis (Type I error). Modifications to compensate for nonsphericity make the test more conservative.

The currently accepted method to modify the F test to account for deviations from sphericity is to reduce the degrees of freedom in the numerator and the denominator associated with the F value by some fraction epsilon (ε). Greenhouse and Geisser (1959) offer a technique for estimating the epsilon degrees-of-freedom adjustment. The degrees of freedom are multiplied by the correction factor to yield a number that is either lower or unchanged. Therefore, a given test has fewer degrees of freedom and requires a greater F value to achieve a given p level. With epsilon at a value of 1 (meaning the assumption has been met), the degrees of freedom are unchanged. To the extent that sphericity is not present, epsilon is reduced and this further decreases the degrees of freedom to produce a more conservative test.

The computations for the Greenhouse-Geisser (G-G) epsilon are performed by the MANOVA when you check **Univariate Tests Also** in the initial dialog (see Figure 11.10). An even more conservative adjustment, called the Huynh-Feldt or H-F epsilon (Huynh and Feldt, 1970), is also included.

Although the exact procedure to follow depends somewhat on characteristics specific to a given data set, there is some consensus for use of the following general guidelines:

- You should use the adjusted univariate test when the G-G epsilon value is greater than or equal to 0.75.

- If the G-G epsilon is less than 0.75, a multivariate analysis (MANOVA) is a more powerful test.

A G-G epsilon value of 0.759, shown in Figure 11.11, suggests using the adjusted univariate F test ($F = 30.28$, $p < 0.0001$) or the multivariate F test, discussed next.

Figure 11.11 Repeated-Measures Analysis of Time

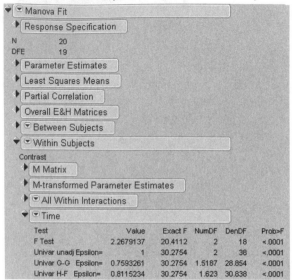

Note: Don't forget that the Time variable in the multivariate report is the repeated-measures variable constructed from the Pre, Post, and Followup variables in the data.

The Multivariate *F* Test

As a repeated-measures design consists of within-subjects observations across treatment conditions, the individual treatment measures can be viewed as separate, correlated dependent variables. The data are easily conceptualized as multivariate even though the design is univariate, and the data can be analyzed with multivariate statistics. In this type of analysis, each level of the repeated factor is treated as a separate variable. The first line in the Time report shown in Figure 11.11 gives the multivariate F value of 20.4112, with $p < 0.0001$.

The MANOVA test has an advantage over the univariate test in that it requires no assumption of sphericity. Some statisticians recommend that the MANOVA be used frequently, if not routinely, with repeated-measures designs (Davidson, 1972). This argument is made for several reasons.

In selecting a test statistic, it is always desirable to choose the most powerful test. A test is said to have power when it is able to correctly reject the null hypothesis. In many situations, the univariate ANOVA is more powerful than the MANOVA and is therefore the better choice if the sphericity assumptions are met. However, the test for sphericity is not very powerful with a small sample. Therefore, only when n is large (20 greater than the number of treatment levels, or $n > k + 20$) does the test for sphericity have sufficient power. Remember that the multivariate test becomes just as powerful as the univariate test as n grows larger. However, keep in mind that with a small n there is no certainty that the assumptions underlying the univariate ANOVA have been met, but with a large n the MANOVA is equally if not more powerful than the univariate test.

Others argue that the univariate approach offers a more powerful test for many types of data and should not be so readily abandoned. Tabachnick and Fidell (2001) feel that the MANOVA should be reserved for those situations that cannot be analyzed with the univariate ANOVA.

With the current example, the sample is composed of 20 participants and falls below the sensitivity threshold for sphericity tests described above ($n < k + 20$). In other words, the decision to reject the null hypothesis of significant departure from the homogeneity of variance assumption (Mauchly's criterion of 0.683, $p < 0.05$) might not be correct. Fortunately, the standard F, adjusted F values, and multivariate statistics suggest rejection of the study's null hypothesis of no difference in investment scores across the three points in time.

To interpret the multivariate results, follow the same steps outlined previously.

Steps to Review the Results

1. **Make sure that everything looks reasonable.** First, check the number of observations listed with the initial analysis in the Response Specification dialog to make certain data from all participants are included in the analysis. If any data are missing, then all data for that participant are automatically dropped from the analysis. Next, check the number of levels assigned to the response variable—there are three levels of Time in the means plot with the means listed below it (Figure 11.10).

2. **Review the appropriate *F* statistic and its associated probability value.** The first step in interpreting the results of the analysis is to review the *F* value and associated statistics. The relevant *F* value is shown in the Time report for the Within Subjects section of the analysis. This *F* value tells you whether to reject the null hypothesis that states,

> "In the population, there is no difference in investment scores obtained at the three points in time."

This hypothesis can be represented symbolically in this way:

H_0: T1 = T2 = T3

where T1 is the baseline mean investment score before a marriage encounter weekend, T2 is the mean score following the marriage encounter weekend, and T3 is the mean score three weeks after the marriage encounter weekend.

The report for the repeated-measures Time variable, shown to the right, gives the multivariate *F* ratio that tests this null hypothesis. The exact *F* of 20.4112 with 2 and 18 degrees of freedom has a

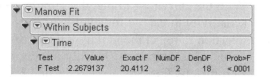

significant probability of less than 0.0001. This *F* statistic is significant because its *p* value is less than 0.05, so you can reject the null hypothesis of no differences in mean levels of commitment in the population. In other words, you conclude that there is an effect for the repeated-measures variable, Time.

3. **Review the contrast results.** The significant effect of Time tells you that investment scores obtained at one point in time are significantly different from scores obtained at some other point in time. You still do not know which scores significantly differ. To determine this, consult the planned comparisons given when you check **Test Each Column Separately Also** in the initial multivariate results dialog (see Figure 11.10). By default, this option sets up contrasts between the first and second time scores (Pre and Post mean scores), and between the first and third time scores (Pre and Followup mean scores). Figure 11.12 shows these results.

Figure 11.12 Results of Contrasts Given by the Fit Model MANOVA

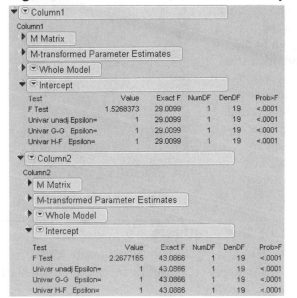

Test	Value	Exact F	NumDF	DenDF	Prob>F
F Test	1.5268373	29.0099	1	19	<.0001
Univar unadj Epsilon=	1	29.0099	1	19	<.0001
Univar G-G Epsilon=	1	29.0099	1	19	<.0001
Univar H-F Epsilon=	1	29.0099	1	19	<.0001

Test	Value	Exact F	NumDF	DenDF	Prob>F
F Test	2.2677165	43.0866	1	19	<.0001
Univar unadj Epsilon=	1	43.0866	1	19	<.0001
Univar G-G Epsilon=	1	43.0866	1	19	<.0001
Univar H-F Epsilon=	1	43.0866	1	19	<.0001

You can use the contrast between the Pre and Post mean scores to evaluate the primary hypothesis that investment increases significantly after the marriage intervention at post-treatment (Time 2) compared to baseline (Time 1).

The multivariate F ratio for this contrast is found in the table labeled Intercept, in the Column1 report. You can see that the obtained F value is 29.0099. It has 1 and 19 degrees of freedom and is significant at $p < 0.0001$. This F test supports the primary hypothesis that investment scores display a significant increase from Time 1 to Time 2 as illustrated by the plot in the initial MANOVA results shown in Figure 11.10.

The contrast between the baseline scores (Pre) and the scores taken three weeks after the program (Followup) is shown in the Intercept table of the Column2 report. The F value for this contrast is also statistically significant ($p < 0.0001$), which indicates that these two treatment means are different. Inspection of the means in Figure 11.10 shows that the follow-up mean is greater than the baseline mean. The second hypothesis stated that there would *not* be an increase in investment scores Time 1 (Pre) and Time 3 (Followup). The significant contrast does not support this hypothesis of no change. The increase in perceived investment size is maintained three weeks after the marriage encounter treatment.

The contrast reports have generic labels, Column1 and Column2, denoting the first contrast and the second contrast. The contrast variables can be seen by looking at the M matrix for each contrast, as shown to the right. In each case, the contrasts are identified by the set of weights (−1 and 1) that sum to zero. In a repeated-measures analysis, the first response variable listed in the Fit Model dialog always acts as the baseline measure. The contrast result tables are labeled Intercept because the estimate of a group mean in a simple means comparison is its Y intercept.

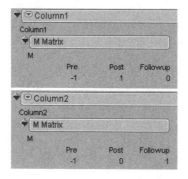

Summarizing the Results of the Analysis

The standard statistical interpretation format used for between-subjects ANOVA (as previously described in Chapter 8) is also appropriate for the repeated-measures design. The outline of this format appears again here:

A)   Statement of the problem

B)   Nature of the variables

C)   Statistical test

D)   Null hypothesis (H_0)

E)   Alternative hypothesis (H_1)

F)   Obtained statistic

G)   Obtained probability (*p* value)

H)   Conclusion regarding the null hypothesis

I)   ANOVA summary table

J)   Figure representing the results.

Because most sections of this format appear in previous chapters, they are not repeated here. Instead, the formal description of the results follows.

Formal Description of Results for a Paper

Mean investment size scores across the three trials are displayed in Figure 11.4. Results analyzed using one-way analysis of variance (ANOVA) repeated-measures design revealed a significant effect for the treatment (Time), with $F(2,18) = 20.4112$; $p < 0.0001$. Contrasts showed that the baseline measure was significantly lower than the post-treatment trial, $F(1,19) = 29.0099$, $p < 0.0001$, and the follow-up trial, $F(1,19) = 43.0866$; $p < 0.0001$.

Summary

This chapter described the one-way repeated-measures ANOVA. This kind of design provides data in which each participant in a single group is exposed to every level of a single independent variable. Often the repeated measurements of the study group are taken at different points in time. This design is compared to the between-subjects design described in Chapter 8, "One-Way ANOVA with One Between-Subjects Factor."

Single-group repeated-measures designs can suffer from a number of problems.

- Lack of a control group means there is no comparison that evaluates whether observed changes are actually the result of experimental manipulation.

- This kind of single-group experiment does not provide information to detect possible order effects. Order effects result when the ordinal position of the treatments biases participant responses.

- Carryover effects occur when an effect from one treatment changes (carries over to) the participants' responses to the next treatment condition.

This chapter also introduced both the univariate approach and the multivariate approach to analyzing repeated-measures designs, and discussed the homogeneity of variance necessary for a valid univariate analysis. Lack of sphericity (homogeneity of variance) suggests the need for a modified F test or a multivariate analysis.

Adding a second group improves the repeated-measures design. The next chapter shows how to design and analyze a repeated-measures study with more than one group.

Assumptions of the Multivariate Analysis of Design with One Repeated-Measures Factor

Level of measurement

Repeated-measures designs are so named because they normally involve obtaining repeated measures on some response variable from a single sample of participants. This response variable should be assessed on an interval or ratio level of measurement. The predictor variable should be a nominal-level variable (a categorical variable), which typically codes "time," "trial," "treatment," or some similar construct.

Independent observations

A given participant's score in any one condition should not be affected by any other participant's score in any of the study's conditions. However, it is acceptable for a given participant's score in one condition to be dependent upon his or her own score in a different condition. This is another way of saying, for example, that it is acceptable for participants' scores in Condition 1 to be correlated with their scores in Condition 2 or Condition 3.

Random sampling

Scores on the response variable should represent a random sample drawn from the populations of interest.

Multivariate normality

The measurements obtained from participants should follow a multivariate normal distribution. Under conditions normally encountered in social science research, violations of this assumption have only a very small effect on the Type I error rate (the probability of incorrectly rejecting a true null hypothesis).

References

Barcikowski, R., and R. Robey. 1984. "Decisions in Single Group Repeated Measures Analysis: Statistical Tests and Three Computer Packages." *American Statistician* 38:148–150.

Davidson, M. 1972. "Univariate versus Multivariate Tests in Repeated-Measures Experiments." *Psychological Bulletin* 77:446–452.

Greenhouse, S., and S. Geisser. 1959. "On Methods in the Analysis of Profile Data." *Psychometrika* 24:95–112.

Huynh, H., and L. S. Feldt. 1970. "Conditions under Which Mean Square Ratios in Repeated Measurements Designs Have Exact F-Distributions." *Journal of the American Statistical Association* 65:1582–1589.

Huynh, H., and G. Mandeville. 1979. "Validity Conditions in Repeated Measures Designs." *Psychological Bulletin* 86:964–973.

LaTour, S., and P. Miniard. 1983. "The Misuse of Repeated Measures Analysis in Marketing Research." *Journal of Marketing Research* 20:45–57.

Rouanet, H., and D. Lepine. 1970. "Comparison between Treatments in a Repeated-Measurement Design: ANOVA and Multivariate Methods." *British Journal of Mathematical and Statistical Psychology* 23:147–163.

Rusbult, C. E. 1980. "Commitment and Satisfaction in Romantic Associations: A Test of the Investment Model." *Journal of Experimental Social Psychology* 16:172–186.

Tabachnick, B. G., and L. S. Fidell. 2001. *Using Multivariate Statistics*. 4th ed. New York: HarperCollins.

Factorial ANOVA with Repeated-Measures Factors and Between-Subjects Factors

> **Overview.** This chapter shows how to use the JMP Fit Model platform to analyze a two-way ANOVA with repeated measures, and includes guidelines for the hierarchical interpretation of the analysis. This chapter shows how to do the following:
> - interpret main effects in the absence of an interaction
> - interpret a significant interaction between the repeated-measures factor and the between-subjects factor
> - look at post-hoc multiple-comparison tests.

Introduction: Mixed-Design ANOVA

This chapter discusses analysis of variance procedures appropriate for studies that have both repeated-measures factors and between-subjects factors. These research designs are sometimes called *mixed-model designs*.

A mixed-model ANOVA is similar to the factorial ANOVA discussed in Chapter 9, "Factorial ANOVA with Two Between-Group Factors." The procedure discussed in Chapter 9 assumes there is a continuous numeric response variable and categorical (nominal or ordinal) predictor variables. However, the predictor variables in a repeated-measures ANOVA differ from those in a factorial ANOVA. Specifically, a mixed-design ANOVA assumes that the analysis includes

- at least one nominal-scale predictor variable that is a *between-subjects* factor, as in Chapter 8, "One-Way ANOVA with One Between-Subjects Factor"

- at least one nominal-scale predictor variable that is a *repeated-measures* factor, as in Chapter 11, "One-Way ANOVA with One Repeated-Measures Factor."

These designs are called mixed designs because they include a mix of between-subjects fixed effects, repeated-measures factors, and subjects. The repeatedly sampled subjects in a repeated-measures design are a random sample from the population of all possible subjects. This type of variable (subject) is called a random effect because it represents a random sample of all possible subjects.

Extension of the Single-Group Design

It is useful to think of a factorial mixed design as an extension of the single-group repeated-measures design presented in Chapter 11. A one-factor repeated-measures design takes repeated measurements on the response variable from only one group of participants. The occasions at which measurements are taken are often called *times* (or *trials*), and there is typically some type of experimental manipulation that occurs between two or more of these occasions. It is assumed that, if there is a change in scores on the response from one time to another, it is due to experimental manipulation. However, this assumption can often be challenged, given the many possible confounds associated with this single-group design.

Chapter 11 shows how to use a one-factor repeated-measures research design to test the effectiveness of a marriage encounter program to increase participants' perceptions of how much they have invested in their romantic relationships. The concept of *investment size* was drawn from Rusbult's investment model (Rusbult, 1980). Investment size refers to the amount of time and personal resources that an individual has put into the relationship with a romantic partner. In the preceding chapter, it was hypothesized that the participants' perceived investment in their relationships would increase immediately following the marriage encounter weekend.

Figure 12.1 illustrates the one-factor repeated-measures design used to test this hypothesis in Chapter 11. The circles represent the occasions when the dependent variable (response or criterion variable) is measured. The response variable in the study is Investment Size. You can see that the survey measures the response variable at three points in time:

- A baseline survey (Time 1) was administered immediately before the marriage encounter weekend.

- A post-treatment survey (Time 2) was administered immediately after the marriage encounter weekend.

- A follow-up survey (Time 3) was administered three weeks after the marriage encounter weekend.

Figure 12.1 Single-Group Experimental Design with Repeated Measures

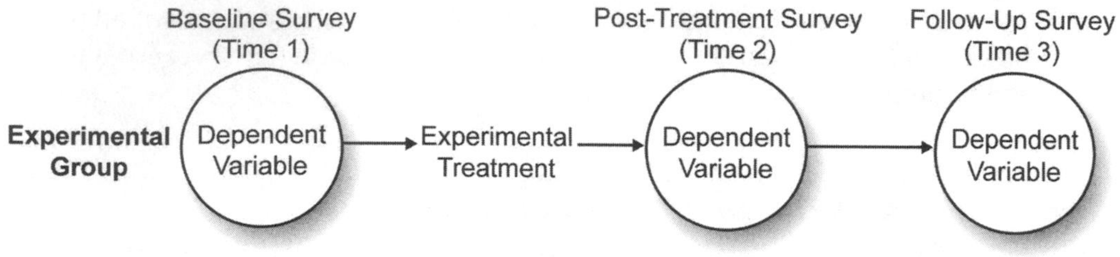

In the previous chapter the JMP data table **marriage univar.jmp** shows the predictor variable in this study, Time, with three levels called Pre (Time 1), Post (Time 2), and Followup (Time 3). Each subject in the study provides data for each of these three levels of the predictor variable, which means that Time is a repeated-measures factor.

Figure 12.1 shows an experimental treatment (the marriage encounter program) positioned between Times 1 and 2. You therefore hypothesize that the investment scores observed immediately after this treatment (Post) will be significantly higher than scores observed just before this treatment (Pre).

However, Chapter 11 points out a weakness in this single-group repeated-measures design, if used inappropriately. The fictitious single-group study that assessed the effectiveness of the marriage encounter program is an example of a weak design. The statistical evidence that supported the hypothesis that Post (Time 2) scores are significantly higher than Pre (Time 1) scores was weak because there could be many alternative explanations for your findings that have nothing to do with the effectiveness of the marriage encounter program.

For example, if perceived investment increased from Time 1 to Time 2, you could argue that this change was due to the passage of time rather than the marriage encounter program. This argument asserts that the perception of investments naturally increases over time in all couples, regardless of whether they participate in a marriage encounter program. Because your study includes only one group, you have no evidence to show that this alternative explanation is wrong.

As the preceding discussion suggests, data obtained from a single-group experimental design with repeated measures generally provide weak evidence of cause and effect. To minimize the weaknesses associated with this design, it is often necessary to expand it into a two-group experimental design with repeated measures (a mixed design).

Figure 12.2 illustrates how to expand the current investment model study into a two-group design with repeated measures. This study begins with a single group of participants and randomly assigns each to one of the two experimental conditions. The experimental group goes through the marriage encounter program and the control group does not go through the program. You administer the measure of perceived investment to both groups at three points in time:

- a baseline survey at Time 1 (**Pre**), immediately before the experimental group goes through the marriage encounter program

- a post-treatment survey at Time 2 (**Post**), immediately after the experimental group went through the program

- a follow-up survey at Time 3 (**Followup**), three weeks after the experimental group went through the program.

Figure 12.2 Two-Group Experimental Design with Repeated Measures

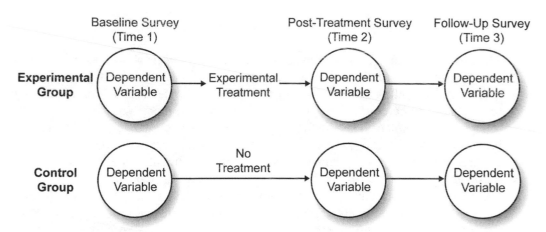

Advantages of the Two-Group Repeated-Measures Design

Including the control group makes this a more rigorous study compared to the single-group design presented earlier. The control group lets you test the plausibility of alternative explanations that can account for the study results.

For example, consider a best-case scenario, in which you obtain the results that appear in Figure 12.2. These fictitious results show the mean investment scores displayed by the two groups at three points in time. The scores for the experimental group (the group that experienced the marriage encounter training) displayed a relatively low level of perceived investment at Time 1—the mean investment score is approximately 7 on a 12-point scale. The score for this group visibly increased following the marriage encounter program at Time 2 and Time 3—mean investment scores are approximately 11.5. These findings support your hypothesis that the marriage encounter program positively affects perceptions of investment.

In contrast, consider the line in Figure 12.3 that represents the mean scores for the control group (the group that did not go through the marriage encounter program). Notice that the mean investment scores for this group begin relatively low at Time 1—about 6.9 on the 12-point scale—and do not appear to increase at Time 2 or Time 3 (the line remains relatively flat). This finding is also consistent with your hypothesis. Because the control group did not receive the marriage encounter training, you did not expect them to demonstrate large improvements in perceived investments over time.

The results in Figure 12.3 are consistent with the hypothesis about the effects of the program, and the presence of the control group lets you discount many alternative explanations for the results. For example, it is no longer tenable for someone to argue that the increases in perceived investment are due simply to the passage of time (sometimes referred to as *maturation effects*). If that were the case you would expect to see similar increases in perceived investments in the control group, yet these increases are not evident. Similarly, it is not tenable to say the post-treatment scores for the experimental group are significantly higher than those for the control group because of some type of bias in assigning participants to conditions. Figure 12.3 shows that the two groups demonstrated similar levels of perceived investment at the baseline survey of Time 1 (before the experimental treatment). Because this two-group design protects you from these alternative explanations (as well as some others), it is often preferable to the single-group procedure discussed in Chapter 11.

Figure 12.3 A Significant Interaction between Time and Group

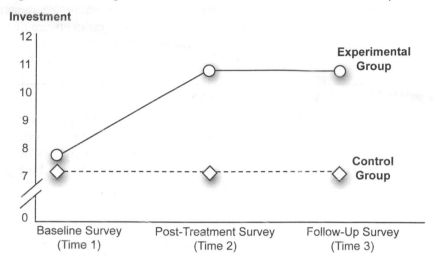

Random Assignment to Groups

It is useful at this point to review the important role played by randomization in a mixed-design study. In the experiment just described, it is essential that volunteer couples be randomly assigned to either the treatment or control group. In other words, the couples must be *assigned* to conditions by you (the researcher) and these assignments must be made in a completely *random* (unsystematic) manner, such as by using a coin toss or a table of random numbers.

The importance of random assignment becomes clear when you consider what might happen if participants are not randomly assigned to conditions. For example, imagine that the experimental group in this study consisted only of people who voluntarily chose to participate in the marriage encounter program, and that the control group consisted only of couples who had chosen not to participate in the program. Under these conditions, it would be reasonable to argue that a *selection bias* could affect the outcome of the study. Selection biases can occur any time participants are assigned to groups in a nonrandom fashion. When this happens, the results of the experiment might reflect pre-existing differences instead of treatment effects. In the study just described, couples that volunteered to participate in a marriage encounter weekend are probably different in some way from couples that do not volunteer. The volunteering couples could have different concerns about their relationships, different goals, or different views about themselves that influence their scores on the measure of investment. Such pre-existing differences could render any results from the study meaningless.

One way to rectify this problem in the present study is to ensure that you recruit all couples from a list of couples that volunteer to participate in the marriage encounter weekend. Then, randomly assign the couples to treatment conditions to ensure that the couples in both groups are as similar as possible at the start of the study.

Studies with More Than Two Groups

To keep things simple, this chapter focuses only on two-group repeated measures designs. However, it is possible for these studies to include more than two groups under the between-subjects factors. For example, the preceding study included only two groups—a control group and an experimental group. A similar study could three groups of participants, such as

- the control group
- experimental group 1, which experiences the marriage encounter program
- experimental group 2, which experiences traditional marriage counseling.

The addition of the third group lets you test additional hypotheses. The procedures for analyzing the data from such a design are a logical extension of the two-group procedures described in this chapter.

Possible Results from a Two-Way Mixed-Design ANOVA

Before analyzing data from a mixed design, it is instructive to review some of the results possible with this design. This will help illustrate the power of this design and will lay a foundation for concepts to follow.

A Significant Interaction

In Chapter 9, "Factorial ANOVA with Two Between-Subjects Factors," you learned that a significant interaction means that the relationship between one predictor variable and the response is different at different levels of the second predictor variable. In experimental research, the corresponding definition is as follows:

> "The effect of one independent variable on the response variable is different at different levels of the second independent variable."

To better understand this definition, refer back to Figure 12.3, which illustrates the interaction between Time and Group.

In this study, Time consists of three variables that code the repeated-measures factor—Pre (Time 1) scores, Post (Time 2) scores, and Followup (Time 3) scores. Group is the variable that codes the between-subjects factor (experimental group or control group). When there is a significant interaction between a repeated-measures factor and a between-groups factor, it means that the relationship between the repeated-measures factor and the response is different for the different groups coded under the between-subjects factor.

The two lines in Figure 12.3 illustrate this interaction. To understand this interaction, focus first on the line in the figure that illustrates the relationship between time and investment scores for the experimental group. Notice that the mean for the experimental group is relatively low at Time 1 but significantly higher at Time 2 and Time 3. This shows that there is a significant relationship between time and perceived investment for the experimental group.

Next, focus on the line that illustrates the relationship between time and perceived investment for just the control group. Notice that this line is flat—there is little change from Time 1 to Time 2 to Time 3, which shows there is no relationship between time and investment size for the control group.

Combined, these results illustrate the definition of an *interaction*. The relationship between one predictor variable (Time) and the response variable (Investment Size) is different at levels of the second predictor variable—the relationship between Time and Investment Size is significant for the experimental group, but nonsignificant for the control group.

To determine whether an interaction is significant, consult the appropriate reports in the JMP analysis results. However, it is also sometimes possible to identify a likely interaction by reviewing a figure that illustrates group means. For example, consider each line that goes from Time 1 to Time 2 in Figure 12.3. Notice that the line for the experimental group is not parallel to the line for the control group. The line for the experimental group has a steeper angle. Nonparallel lines are the hallmark of an interaction. Whenever a plot contains a line segment for one group that is not parallel to the corresponding line segment for a different group, it can mean that there is an interaction between the repeated-measures factor and the between-subjects factor.

> **Note:** In some (but not all) studies that employ a mixed design, the central hypothesis can require that there be a significant interaction between the repeated-measures variable and the between-subjects variable. For example, in the present study that assessed the effectiveness of the marriage encounter program, a significant interaction shows that the experimental group displays a greater increase in investment size compared to the control group.

Significant Main Effects

If a given predictor variable is not involved in any significant interactions, you then determine whether that variable displays a significant main effect. When a predictor variable displays a significant main effect, there is a difference between at least two levels of that predictor variable with respect to scores on the response variable.

The number of possible main effects is equal to the number of predictor variables. The present investment model study has two predictor variables giving main effects—the repeated-measures factor (time) and the between-subjects factor (group).

A Significant Main Effect for Time

For example, Figure 12.4 shows a possible main effect for the time variable that is an increasing linear trend. Notice that the investment scores at Time 2 are somewhat higher than the scores at Time 1, and that the scores at Time 3 are somewhat higher than the scores at Time 2. Whenever the values of a predictor variable are plotted on the horizontal axis of a figure, a significant main effect for that variable is indicated when the line segments display a relatively steep slope.

Figure 12.4 A Significant Effect for Time Only

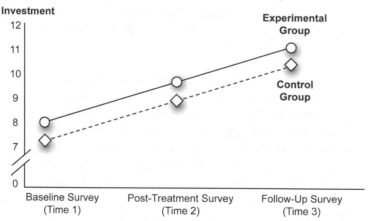

Here, the horizontal axis identifies the values of time as "Baseline Survey (Time 1)," "Post-Treatment Survey (Time 2)," and "Follow-Up Survey (Time 3)." There is a relatively steep slope in the line that goes from Time 1 to Time 2, and from Time 2 to Time 3. These results typically indicate a significant main effect. However, always check the appropriate F test for the main effect in the JMP results to verify that it is significant.

The statistical analysis averages the main effect for time over the groups in the study. In Figure 12.4, there is an overall main effect for time after collapsing across the experimental group and control group.

A Significant Main Effect for Group

The previous figure illustrates a significant time effect. If the group effect is significant but time is not, you expect to see a different pattern. The values of the predictor variable for time are shown as three separate points on the horizontal axis of the figure. In contrast, the values of the predictor variable for group are coded by drawing separate lines for the two groups under this variable. The line for the experimental group shows circles at each point and the line for the control group shows diamonds.

When a predictor variable, such as Group, is represented in a figure by plotting separate lines for its various levels, a significant main effect for that variable is evident when at least two of the lines are separate from each other, as in Figure 12.5.

Figure 12.5 A Significant Main Effect for Group Only

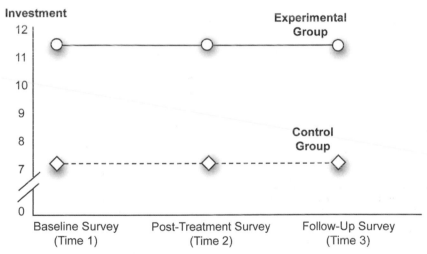

To interpret Figure 12.5, first determine which effects are probably *not* significant. You can see that all line segments for the experimental group are parallel to their corresponding segments for the control group. This tells you that there is probably not a significant interaction between time and group. Next, you can see that none of the line segments display a relatively steep angle, which tells you that there is probably not a significant main effect for time.

However, the line that represents the experimental group appears to be separated from the line that represents the control group. Now look at the individual data points. At Time 1, the experimental group displays a mean investment score that appears to be much higher than the one displayed by the control group. This same pattern of differences appears at Time 2 and Time 3. Combined, these are the results that you expect to see if there is a significant main effect for Group. Figure 12.5 suggests that the experimental group consistently demonstrated higher investment scores than the control group.

A Significant Main Effect for Both Predictor Variables

It is possible to obtain significant main effects for both predictor variables. Such an outcome appears in Figure 12.6. Note that the line segments display a relatively steep slope (indicative of a main effect for time), and the lines for the two groups are also relatively separated (indicative of a main effect for group).

Figure 12.6 Significant Main Effect for Both Time and Group

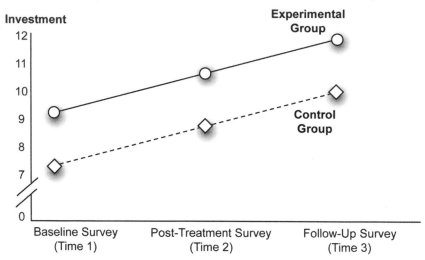

A Nonsignificant Interaction, Nonsignificant Main Effects

Of course, there is no law that says that any of your effects have to be significant (as every researcher knows all too well!). This is evident in Figure 12.7. In this figure the lines for the two groups are parallel, which suggests a probable nonsignificant interaction. The lines have a slope of zero, which indicates that the main effect for time is not significant. Finally, the line for the experimental group is not separated from the line for the control group, which suggests that the main effect for Group is also not significant.

Figure 12.7 A Nonsignificant Interaction, and Nonsignificant Main Effects

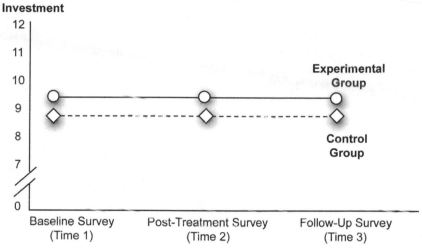

Problems with the Mixed-Design ANOVA

The proper use of a control group in a mixed-design investigation can help remedy some of the weaknesses associated with the single-group repeated-measures design. However, there are other problems that can affect a repeated measures design even when it includes a control group.

For example, Chapter 11, "One-Way ANOVA with One Repeated-Measures Factor," described a number of sequence effects that can confound a study with a repeated-measures factor. Specifically, repeated-measures investigations often suffer from

- *order effects* (effects that occur when the ordinal position of the treatments introduces response biases)

- *carryover effects* (effects that occur when the effect from one treatment changes, or carries over, to the participants' responses in the following treatment conditions).

You should always be sensitive to the possibility of sequence effects with any type of repeated-measures design. In some cases, it is possible to successfully deal with these problems through the proper use of counterbalancing, spacing of trials, additional control groups, or other strategies. Some of these approaches were discussed previously in the section "Sequence Effects" in Chapter 11.

Example with a Nonsignificant Interaction

The fictitious example presented here is a follow-up to the pilot study described in Chapter 11. The results of that pilot study suggest that scores on a measure of perceived investment significantly increase following a marriage encounter weekend. However, you can argue that those investment scores increased as a function of time and not because of the experimental manipulation. In other words, the investment scores could have increased simply because the couples spent time together, and this would have occurred with any activity, not just a marriage encounter experience.

To address concerns about this possible confound (as well as some others inherent in the one-group design), replicate the study as a two-group design. An experimental group again takes part in the marriage encounter program and a control group does not. You obtain repeated measures on perceived investment from both groups, as illustrated earlier in Figure 12.2.

Remember that this two-group design is generally considered to be superior to the design described in Chapter 11. You should use it in place of an uncontrolled design whenever possible.

The primary hypothesis is that the experimental group will show a greater increase in investment scores at post-treatment and follow-up than the control group. To confirm this hypothesis, you want to see a significant group by time interaction in the ANOVA. You also hypothesize (based on the results obtained from the pilot project described in Chapter 11) that the increased investment scores in the treatment group will still be found at follow-up. To determine whether the results are consistent with these hypotheses, it is necessary to check group means, review the results from the test of the interaction, and consult a number of post-hoc tests, to be described later.

Using the JMP Distribution Platform

The response variable in this study is perceived investment size: the amount of time and effort that participants believe they have invested in their relationships (their marriages). Perceived investment is assessed with a survey constructed such that higher scores indicate greater levels of investment. Assume that these scores are assessed on a continuous numeric scale and that responses to the scale have been shown to demonstrate acceptable psychometric properties (validity and reliability).

The JMP data table has a column for each of the three points in time. The data table, named **mixed nointeraction.jmp** (Figure 12.8), has these variables:

- Pre lists investment scores obtained at Time 1 (the baseline survey).

- Post has investment scores obtained at Time 2 (the post-treatment survey).

- Followup contains investment scores obtained at Time 3 (the follow-up survey).

- Participant denotes each participant's identification number (values range from 1 through *n*, where *n* = the number of participants).

- Group designates membership in either the "experimental" or "control" group.

Begin by looking at descriptive statistics for the data. Use the Distribution platform to obtain descriptive statistics for all variables. Simple statistics for each variable provide an opportunity to check for obvious data entry errors.

Using the Distribution platform is important when analyzing data from a mixed-design study because the output of the Fit Model platform does not include a table of simple statistics for within-participants variables in a mixed-design ANOVA.

You want to look at means and other descriptive statistics for the investment score variables—Pre, Post, and Followup. You need to have the overall means (based on the complete sample) as well as the means by Group. That is, you need means on the three variables for both the control group and the experimental group. The Distribution platform gives this information.

Figure 12.8 Data Table for Repeated-Measures Results with No Significant Effects

	Participant	Group	Pre	Post	Follow up
1	1	control	8	10	10
2	2	control	10	13	12
3	3	control	7	10	12
4	4	control	6	9	10
5	5	control	7	8	9
6	6	control	11	15	14
7	7	control	8	10	9
8	8	control	5	8	8
9	9	control	12	11	12
10	10	control	9	12	12
11	11	experimental	10	14	13
12	12	experimental	7	12	11
13	13	experimental	8	8	9
14	14	experimental	13	14	14
15	15	experimental	11	11	12
16	16	experimental	7	8	7
17	17	experimental	9	8	10
18	18	experimental	8	13	14
19	19	experimental	10	12	12
20	20	experimental	6	9	10

The next section shows you how JMP Preferences can be used to specify only the portions of analysis results you want to see. The Distribution analysis that follows uses these preferences.

Using Preferences (Optional)

The Distribution platform in JMP automatically (by default) shows histograms with an outlier box plot, a Quantiles table, and a Moments table, for all numeric variables specified in its launch dialog.

In this example you only want to see means and other simple statistics. One way to simplify or tailor the results from any JMP platform is with platform preferences. To change preferences for the Distribution platform, do the following:

- Choose the **Preferences** command from the **File** menu (**Edit** menu on the Macintosh).
- When the initial Preferences panel appears, click the **Platforms** tab.
- Select **Distribution** from the list of JMP platforms.
- Deselect all the default results except for the Moments table, as shown in Figure 12.9.

With these preferences in effect, when you request results from the Distribution platform, only the Moments table appears.

> **Note:** Don't forget to reset preferences to the ones you want (usually the defaults noted in the list of platform features).

Figure 12.9 Use Platform Preferences to Tailor JMP Results

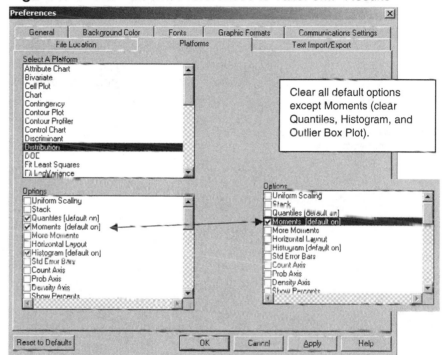

Distribution Platform Results

You want to see summary statistics for the three commitment measures over the whole sample and for each group separately. To do this, you need to launch the Distribution platform twice.

- First, launch the Distribution platform and specify the three response measures, Pre, Post, and Followup, in the launch dialog.
- Click **OK** to see the overall statistics.
- Next, again launch the Distribution platform dialog. Specify the three responses, and also select **Group** as the **By** variable. Click **OK** to see results for each group.

These two Distribution analyses produce the tables in Figure 12.10. By using preferences, only the Moments tables show. If you want to see more, the menu on the title bars of each table accesses all plots and tables given by the Distribution platform.

The results for the combined sample appear at the top of Figure 12.10. Review these overall means to get a sense for any general trends in the data. Remember that the variables Pre, Post, and Followup are investment scores obtained at Time 1, Time 2, and Time 3. The Pre variable displays a mean score of 8.60—the average investment score was 8.60 just before the marriage encounter weekend. The mean score on Post shows that the average investment score increased to 10.75 immediately after the marriage encounter weekend, and the mean score on Followup shows that investment scores averaged 11.00 three weeks following the program. These means seem to display a fairly large increase in investment scores from Time 1 to Time 2, which suggests a significant effect for time when you review ANOVA results.

Figure 12.10 Summary Statistics for Whole Sample and for Groups

Distributions

Pre		Post		Followup	
Moments		Moments		Moments	
Mean	8.6	Mean	10.75	Mean	11
Std Dev	2.1373865	Std Dev	2.2912878	Std Dev	2.0261449
Std Err Mean	0.4779342	Std Err Mean	0.5123475	Std Err Mean	0.4530598
upper 95% Mean	9.6003277	upper 95% Mean	11.822356	upper 95% Mean	11.948265
lower 95% Mean	7.5996723	lower 95% Mean	9.6776443	lower 95% Mean	10.051735
N	20	N	20	N	

Group=control

Distributions

Pre		Post		Followup	
Moments		Moments		Moments	
Mean	8.3	Mean	10.6	Mean	10.8
Std Dev	2.2135944	Std Dev	2.2211108	Std Dev	1.8737959
Std Err Mean	0.7	Std Err Mean	0.7023769	Std Err Mean	0.5925463
upper 95% Mean	9.88351	upper 95% Mean	12.188887	upper 95% Mean	12.140433
lower 95% Mean	6.71649	lower 95% Mean	9.011113	lower 95% Mean	9.4595672
N	10	N	10	N	10

Group=experimental

Distributions

Pre		Post		Followup	
Moments		Moments		Moments	
Mean	8.9	Mean	10.9	Mean	11.2
Std Dev	2.1317703	Std Dev	2.4698178	Std Dev	2.2509257
Std Err Mean	0.6741249	Std Err Mean	0.781025	Std Err Mean	0.7118052
upper 95% Mean	10.424977	upper 95% Mean	12.666801	upper 95% Mean	12.810215
lower 95% Mean	7.3750234	lower 95% Mean	9.1331988	lower 95% Mean	9.5897847
N	10	N	10	N	10

It is important to remember that, in order for the hypotheses to be supported, it is not adequate to merely observe a significant effect for time. Instead, it is necessary to obtain a significant time by group interaction. Specifically, the increase in investment scores over time must be greater among participants in the experimental group than it is among those in the control group. To see whether such an interaction has occurred, you can first look at a figure that plots data for the two groups separately and then consult the F test for the interaction in the statistical analyses. Figure 12.11 shows a plot of the means from the Distribution results in Figure 12.10.

This JMP overlay plot requires that the data be restructured. The next section shows how to manipulate the raw data and produce the overlay plot shown here. A similar plot is automatically produced by the Fit Model platform when you analyze the data later.

Figure 12.11 Profile Plot of Mean Investment Scores

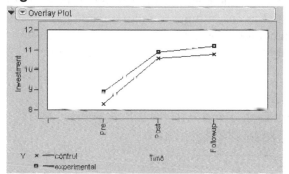

Data Manipulation in JMP

The overlay plot in Figure 12.11 is sometimes called a profile plot. A profile plot shows the response of one factor under different settings of another factor, and is automatically produced by the Fit Model platform when you analyze the data. To produce this plot from the raw data table requires the use of **Tables** menu commands. The data must be summarized and manipulated to arrange results suitable for the **Overlay Plot** command. These kinds of table manipulations are often necessary but not always explained in statistical texts. To produce the overlay plot above using the **mixed nointeraction.jmp** data table, do the following:

- Select **Tables** → **Summary** to produce a table of mean investment scores for each group at each time.

- Select **Tables** → **Transpose** to rearrange (transpose) the summarized table.

- Change the name of the first column in the transposed table from Label to Time.

- Use the Column Info dialog for Time to assign it a **List Check** property, and change the default alphabetic order of the values so they are chronological ("Pre," "Post," then "Followup"). By default, most platforms arrange values in alphabetic order. See the section "Ordering Values in Analysis Results" in Chapter 9, "Factorial ANOVA with Two Between-Subjects Factors," for details about using the List Check property.

The remainder of this section shows the mouse steps for these actions, with more detailed explanation.

    🖰   With the **mixed nointeraction.jmp** table active, choose **Summary** from the **Tables** menu.

    🖰   When the Summary dialog appears, select **Pre**, **Post**, and **Followup** from the variable list on the left of the dialog, and select **Mean** from the **Statistics** menu in the Summary dialog.

    🖰   Select **Group** from the variable list and click **Group** in the Summary dialog.

    🖰   Select **Column** from the **statistics column name format** menu in the Summary dialog, to assign the original column names to columns in the new table.

    🖰   Click **OK** to see the Summary table shown with the completed Summary dialog in Figure 12.12.

Figure 12.12 Completed Summary Dialog and Resulting JMP Table

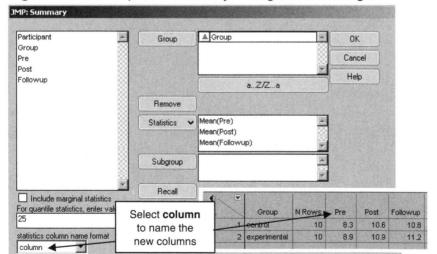

However, this summarized table is not in the form needed to produce an overlay plot of mean investment scores for the groups shown previously in Figure 12.11. To plot the investment means at each time period for each group requires that the time periods be rows in the table, and the groups be columns. The Transpose feature in JMP can rearrange the table:

- With the new summary table active, choose the **Transpose** command from the **Tables** menu.
- When the Transpose dialog first appears, all columns show in the **Columns to be transposed** list. Highlight the N rows variable and click the **Remove** button in the dialog.
- Select **Pre**, **Post**, and **Followup** from the Columns list on the left of the dialog and click **Add**.
- Select **Group** from the Columns list and click the **Label** button in the Transpose dialog. The completed dialog should look the one shown in Figure 12.13.
- Click the **Transpose** button in the dialog to see the transposed table. Note that the values of Group, "control" and "experimental," are column names in the new table.
- In the new transposed table, change the name of the column called Label to Time, to make the overlay plot more easily readable.

Figure 12.13 Transpose Dialog and Transposed Table

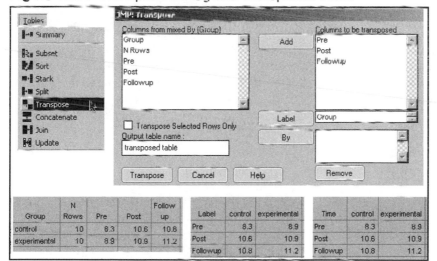

The transposed table of summarized data now lends itself to the Overlay Plot platform.

- 🖱 With the transposed table active, choose **Overlay Plot** from the **Graph** menu, as shown here.
- 🖱 Complete the dialog as shown in Figure 12.14. Be sure to clear the Sort X check box so that the X axis is arranged as found in the data table.
- 🖱 Click **OK** to see the overlay profile plot shown in Figure 12.11, and again here.
- 🖱 On the overlay plot, double-click on the Y axis name to edit it. Change the name from "Y" to "Investment."
- 🖱 If the points are not connected, use the **Y Options → Connect Points** command found on the menu on the Overlay Plot title bar.

On the plot, the line with the **x** markers illustrates mean scores for the control group. These mean scores are the Pre, Post, and Followup means that appear in the summary statistics shown in Figure 12.10 and in the summary table. The line with the square markers illustrates mean scores for the experimental group.

The general pattern of means does not suggest an interaction between time (Pre, Post, and Followup) and Group. When two variables are involved in an interaction, the lines that represent the various groups tend not to be parallel. So far, the lines for the control group and experimental group do appear to be parallel. This might mean that the interaction is nonsignificant. However, the only way to be sure is to analyze the data and compute the appropriate statistical test. The next section shows how to do this.

Figure 12.14 Completed Overlay Plot Dialog and Results

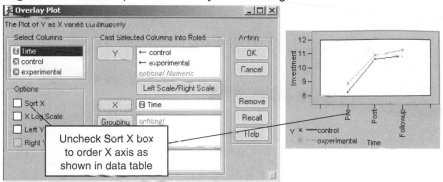

Testing for Significant Effects with the Fit Model Platform

The Fit Model platform can test for interaction between time and group in the **mixed nointeraction.jmp** data table. To do this, choose **Fit Model** from the **Analyze** menu and complete the dialog as shown in Figure 12.15. The three Time variables, Pre, Post, and Followup, record the investment scores and are the response (**Y**) variables. Group is the model effect.

Figure 12.15 Fit Model Dialog for Repeated Measures Analysis

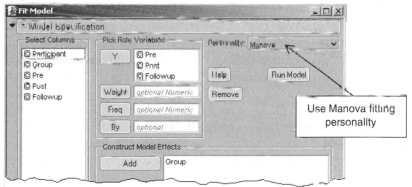

When you click **Run Model** in the Fit Model dialog, the initial Manova Fit panel in Figure 12.16 appears.

Figure 12.16 Initial MANOVA Results for the No Interaction Data

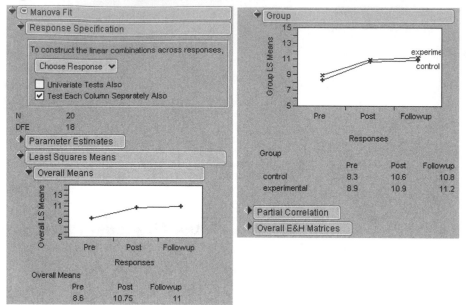

Notes on the Response Specification Dialog

The Response Specification dialog, at the top of the Manova Fit platform, lets you interact with the MANOVA analysis. Each time you select a response design (other than Repeated Measures) from the **Choose Response** menu, the structure of the analysis is displayed in the Response Specification dialog. When you click **Run**, JMP performs that analysis and appends it to the bottom of the report, following any previous analysis you requested.

> **Note:** The structure of the Repeated Measures design is a combination of the Contrast and Sum response designs. The Repeated Measures design uses the Sum design to compute the between-subjects statistics and the Contrast design to compute the within-subjects statistics.

Computations for the MANOVA are a function of a linear combination of responses. The response design selection from the **Choose Response** menu defines this linear combination. The response structure is called the **M** matrix and is included with each requested multivariate analysis. You can see this structure in the Response

Specification dialog when you select a response model or open the **M** matrix in the analysis report. The columns of the **M** matrix define the set of transformation variables for the multivariate analysis. Looking at the **M** matrix tells you which kind of multivariate analysis has been done.

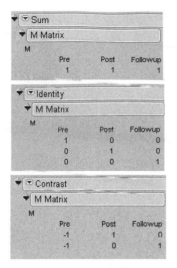

The following list describes three commonly used response designs with associated **M** matrices.

- The **Sum** design is the sum of the responses, giving a single response value.

- The **Identity** design uses each separate response, as shown by the identity matrix.

- The **Contrast** design compares each response to the first response. You can see in the Contrast **M** Matrix that this kind of analysis compares the pre-weekend response to both the post-weekend response and the followup response.

Testing Each Column Separately

When you select the **Test Each Column Separately Also** check box in the initial MANOVA dialog, the analysis results include a univariate ANOVA for each column in the design specification. Note that each column of the design specification appears as a row of the **M** matrix (the **M** matrix is a transpose of the design specification).

The example in Figure 12.17 shows a Contrast design and its **M** matrix. This example requests an ANOVA for each column in the design, which produces a contrast of Pre and Post responses, and a second contrast of Pre and Followup responses.

Figure 12.17 Response Design and M Matrix

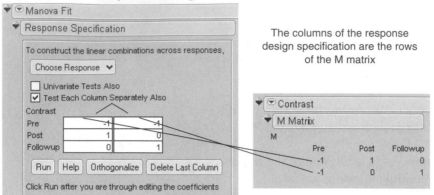

The Repeated Measures Response Design

To continue with the analysis, choose **Repeated Measures** from the **Choose Response** menu in the Response dialog, as shown in Figure 12.18. This is called selecting a response model. When you choose the **Repeated Measures** response, the Specification of Repeated Measures dialog shown in Figure 12.18 appears, prompting you for a name that identifies the Y (repeated) variables. By default this name is Time, which is often the nature of the repeated measures and is appropriate for this example.

Selecting **Repeated Measures** does not display a single design, as shown previously, because the repeated-measures analysis uses the Sum response design to compute between-subjects statistics and a Contrast response design for the within-subjects analysis. The **Test Each Column Separately Also** box is checked so that the post-weekend and follow-up responses are compared to the pre-weekend baseline response.

There are initial parameter estimates, least squares means with plots (shown previously in Figure 12.16), correlations among the response columns, and **E** and **H** matrices. The **E** and **H** matrices are the error and hypothesis cross-product matrices. More details about these results can be found in the *JMP Statistics and Graphics Guide* (2003).

Figure 12.18 Initial Results from MANOVA Analysis

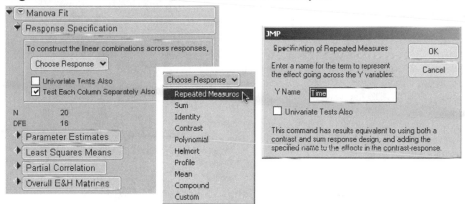

Results of the Repeated-Measures MANOVA Analysis

Click **OK** in the Specification of Repeated Measures dialog to see the repeated measures analysis appended to the initial analysis (Figure 12.19).

> **Note:** The MANOVA reports contain more information than is needed in this example so unneeded outline nodes are closed. Figure 12.19 shows everything you need for this example.

You see both between-subjects results and within-subjects results. In a repeated measures design, selecting **Test Each Column Separately Also** causes contrasts between time periods to be shown. These contrasts are appended to the report, and are shown on the right in Figure 12.19. The next section, "Steps to Review the Results," includes explanation of the results.

Figure 12.19 MANOVA Analysis for Investment Study

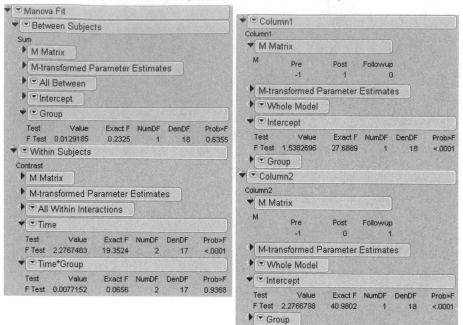

Steps to Review the Results

1. **Make sure that everything looks reasonable.** As always, before interpreting the results it is best to review the results for possible signs of problems. Most of these steps with a mixed-design ANOVA are similar to those used with between-subjects designa. That is, check the number of observations listed in the initial MANOVA results, labeled N, to make certain the analysis included all the observations. The initial MANOVA results show that N is 20. Also note that the degrees of freedom for error (DFE) is 18, which tells you that there are two group levels (N – DFE = number of group levels—"control" and "experimental").

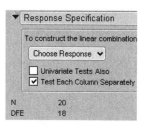

2. **Determine whether the interaction term is statistically significant.** First check the interaction effect. If the interaction effect is not statistically significant, then you can proceed with interpretation of main effects. This chapter follows the same general procedure recommended in Chapter 10, "Multivariate Analysis of Variance (MANOVA) with One Between-Subjects Factor," in which you first check for interactions and then proceed to test for main effects and examine post-hoc analyses.

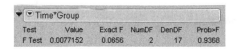

The results of the between-subjects multivariate significance test for the time by group interaction shows an *F* value of 0.0656 with associated probability value of 0.9368. You conclude that there is no interaction between time and group. The control and experimental groups respond the same over time.

The primary hypothesis for this study required a significant interaction but you now know that the data do not support such an interaction. If you actually conducted this research project, your analyses might terminate at this point. However, in order to illustrate additional steps in the analysis of data from a mixed-design, this section looks at the tests for main effects.

3. **Determine if the group effect is statistically significant.** In this chapter, the term group effect is used to refer to the effect of the between-subjects factor. In the present study, the variable named Group represented this effect.

The group effect is of no real interest in the present investment-model investigation because support for the study's central hypothesis required a significant interaction. Nonetheless, it is still useful to plot group means and review the statistic for the group effect in order to validate the methodology used to assign participants to groups. For example, if the effect for Group proved to be significant and if the scores for one of the treatment groups were consistently higher than the corresponding scores for the other (particularly at Time 1), it could indicate that the two groups were not equivalent at the beginning of the study. This might suggest that there was some type of bias in the selection and assignment process. Such a finding could invalidate other results from the study.

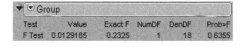

The test for the Group effect shows that the obtained *F* value for the group effect is only 0.2325 with 1 and 18 degrees of freedom and is not significant. The *p* value for this *F* is 0.6355, much greater than 0.05. This indicates that there is not an overall difference between experimental and control groups with respect to their mean investment scores. Figure 12.11 previously illustrated this finding, which shows that there is very little separation between the line for the experimental group and the line for the control group.

4. **Determine if the time effect is statistically significant.** In this chapter, the term *time effect* refers to the effect of the repeated-measures factor. Remember that the word time was designated in the initial platform dialog to code the repeated-measures factor, which consists of the variables Pre, Post, and Followup. A main effect for time suggests that there is a significant difference between investment scores obtained at one time and the investment scores obtained at least one other time during study.

The multivariate test for the time effect appears just above the time by group interaction effect. You can see that the

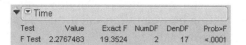

multivariate F for the time effect is 19.3524. This computed F value with 2 and 17 degrees of freedom is significant at $p < 0.0001$. That is, the overall within-subjects effect across the three time variables is highly significant. Time is a significant effect for both groups.

5. **Prepare your own version of the ANOVA summary table.** Table 12.1 summarizes the preceding analysis of variance.

Table 12.1 MANOVA Summary Table for Study Investigating Changes in Investments Following an Experimental Manipulation (Nonsignificant Interaction)

Source	Numerator df	Denominator df	F	p
Between Subjects				
Group (A)	1	18	0.23	0.6355
Within Subjects				
Time (B)	2	17	19.35**	< 0.0001
A by B Interaction	2	17	0.07	0.9368

Note: N = 20.

6. **Review the results of the contrasts.** The specification of the multivariate analysis included the **Test Each Column Separately Also** option, checked in the initial multivariate results (see Figure 12.18). For a repeated-measures analysis, this option performs the contrast described by the M matrix of the Within Subjects analysis, shown here. The first row of the M matrix compares the Pre scores to the Post scores. The second row compares the Pre scores to the Followup scores.

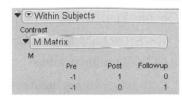

The contrast analyses show at the bottom of the Within Subjects analysis. The contrast analyses are labeled Column 1 and Column 2 in the report. The Column 1 report compares the Time 1 (Pre) investment scores with Time 2 (Post) scores. Column 2 compares Time 1 (Pre) with Time 3 (Followup). The comparison of Time 1 to Time 2 has an F ratio of 27.6889, which is significant at $p < 0.0001$. You can reject the null hypothesis that there is no difference between Time 1 (Pre) investment scores and Time 2 (Post) investment scores in the population.

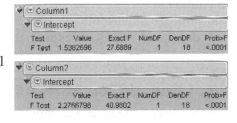

The analysis that compares Time 1 to Time 3 has an F ratio of 40.9802 with associated $p < 0.0001$. It is clear that investment scores at Time 3 are also significantly higher than Time 1 scores. Recall that the plot of means also suggested Time 2 scores were significantly higher than Time 1 scores for both experimental and control groups.

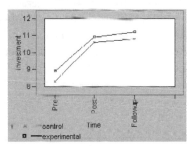

Summarizing the Analysis Results

The results of this mixed-design analysis can be summarized using the standard statistical interpretation format presented earlier in this text:

A) Statement of the problem

B) Nature of the variables

C) Statistical test

D) Null hypothesis (H_0)

E) Alternative hypothesis (H$_1$)

F) Obtained statistic

G) Obtained probability (p) value

H) Conclusion regarding the null hypothesis

I) ANOVA summary table

J) Figure representing the results.

This summary could be performed three times: once for the null hypothesis of no interaction in the population; once for the null hypothesis of no main effect for group; and once for the null hypothesis of no main effect for time. As this format has been already described, it does not appear again here.

Formal Description of Results for a Paper

Here is one way that you could summarize the present results for a published paper:

> Results were analyzed using a two-way analysis of variance with repeated measures on one factor. The group by time interaction was not significant, $F(2,17) = 0.07, p = 0.9368$. The main effect of group also was not significant, $F(1,18) = 0.23, p = 0.63$. However, this analysis did reveal a significant effect for time, $F(2,17) = 19.75, p < 0.0001$. Post-hoc contrasts found that investment scores at the post-treatment and follow-up times were significantly higher than scores observed at baseline (Time 1), with $p < 0.0001$.

Example with a Significant Interaction

The initial hypothesis stated that the investment scores would increase more for the experimental group than for the control group. Support for this hypothesis requires a significant time by group interaction. This section analyzes a different set of fictitious data, where participant responses provide the desired interaction. You will see that it is necessary to perform a number of different follow-up analyses when there is a significant interaction in a mixed-design anova. Figure 12.20 shows the JMP data table, **mixed interaction.jmp**.

Figure 12.20 Investment Data with Interaction

	Participant	Group	Pre	Post	Followup
1	1	control	10	10	10
2	2	control	12	13	12
3	3	control	10	10	10
4	4	control	9	9	9
5	5	control	8	8	9
6	6	control	12	13	12
7	7	control	10	9	10
8	8	control	8	9	10
9	9	control	11	12	11
10	10	control	12	11	12
11	11	experimental	10	14	13
12	12	experimental	7	12	11
13	13	experimental	8	9	11
14	14	experimental	13	14	14
15	15	experimental	11	11	12
16	16	experimental	7	8	8
17	17	experimental	9	8	10
18	18	experimental	8	13	14
19	19	experimental	11	12	12
20	20	experimental	6	9	10

Use the Fit Model platform to run the same analysis as described previously in the section "Testing for Significant Effects with the Fit Model Platform." Figure 12.21 shows results for the data with a significant interaction. Before you select the Repeated Measures response design, note that the interaction shows clearly in the Group profile plot, which is part of the initial analysis, The profile plot shows that the means of each group over time are not parallel, denoting interaction between the variables Time and Group.

Figure 12.21 Repeated-Measures Analysis with Significant Interaction

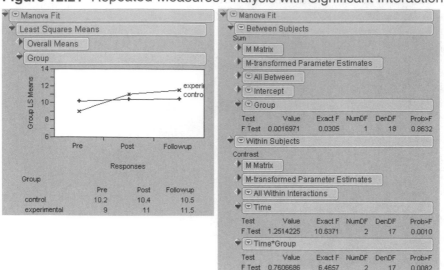

Testing Simple Effects

Testing *simple effects* (called *testing slices* in JMP) determines whether there is a significant relationship between one predictor variable and the response for participants at one level of the second predictor variable. The concept of testing slices to look at simple effects is covered in Chapter 9, "Factorial ANOVA with Two Between-Subjects Factors."

In the present study you might want to determine whether there is a significant change in investment over time among those participants in the experimental group. If there is, you can state that there is a simple effect for time at the experimental level of group.

To better understand the meaning of this, consider the line in the interaction plot that represents the experimental group. This line plots investment scores for the experimental group at three points in time. If there is a simple effect for time in the experimental group, it means that investment scores obtained at one point in time are significantly different from investment scores at one other point in time. If the marriage encounter program is effective, you expect to see a simple effect for time in the experimental group—investment scores should improve over time in this group. You do not expect to see a simple effect for time in the control group because subjects in that group did not experience the program.

To test for the simple effects of the repeated-measures factor (time, in this case), it is necessary to do the following:

- Consider participants separately based on their classification under the group variable, Group.

- Perform a one-way ANOVA with one repeated-measures factor on just those participants in the second treatment group ("experimental").

- Perform a one-way ANOVA with one repeated-measures factor on just those participants in the first treatment group ("control").

To do this:

- Select the **mixed interaction.jmp** table so that it is active.
- Highlight all rows where the group membership is "control."
- With those rows highlighted, choose **Exclude/Include** from the **Rows** menu. This is a row state command that assigns the excluded state to the highlighted rows. JMP now excludes these rows from all further analyses until you highlight them and again choose **Rows** → **Exclude/Include**. Figure 12.22 shows the "control" rows excluded from further analysis.

Figure 12.22 Exclude/Include Command to Exclude Rows from Analysis

		Participant	Group	Pre	Post	Followup
	1	1	control	10	10	10
	2	2	control	12	13	12
	3	3	control	10	10	10
	4	4	control	9	9	8
	5	5	control	8	8	9
	6	6	control	12	10	12
	7	7	control	10	9	10
	8	8	control	8	9	10
	9	9	control	11	12	11
	10	10	control	12	11	12

(Rows menu shown: Exclude/Unexclude Ctrl+E, Hide/Unhide, Label/Unlabel, Colors, Markers, Next Selected F3, Previous Selected F2, Row Selection)

- Use **Fit Model** to do a MANOVA (see Figure 12.15), but do not include the Group variable as an effect because all rows are in the experimental group. When the initial dialog appears, select the **Test Each Column Separately Also** check box and choose **Repeated Measures** from the **Choose Response** menu (see Figure 12.18).

Repeat this process for the "control" group:

- With the excluded "control" group rows highlighted, again select **Exclude/Include** from the **Rows** menu to include them in subsequent analyses.
- Remove the highlighting from the "control" group rows and highlight all rows where the group membership is "experimental."
- With those rows highlighted, choose **Exclude/Include** from the **Rows** menu to exclude the "experimental" rows from the next analysis.
- As before, use Fit Model to do a MANOVA. Do not include the Group variable as an effect because all rows in the analysis are in the "control" group. When the initial dialog appears, select the **Test Each Column Separately Also** check box, and choose **Repeated Measures** from the **Choose Response** menu.

The results of the simple effects analysis for each level of the Group variable are shown in Figure 12.23.

- The simple effect test of Time at the "experimental" level of Group appears on the left in Figure 12.23. The F value of 9.8997 with 2 and 8 degrees of freedom gives a significant simple effect for Time in the experimental group ($p = 0.0069$).

- The simple effect test of Time at the "control" level of Group appears on the right in Figure 12.23. The F value of 0.9524 with 2 and 8 degrees of freedom gives a nonsignificant simple effect for Time in the control group ($p = 0.4256$).

Figure 12.23 Results of Simple Effects Analysis of Time in Experimental Group (left) and Control Group (right)

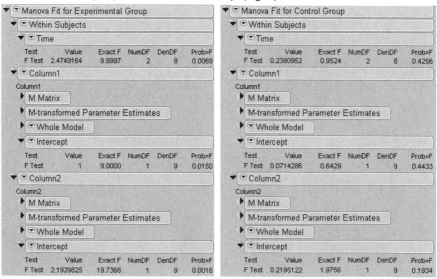

Because the simple effect for Time is significant in the experimental group, you can interpret the planned contrasts that appear beneath the test for Time. These contrasts show

- there is a significant difference between mean scores obtained at Time 1 (Pre) versus those obtained at Time 2 (Post), labeled Column1 in the results. The *F* value with 1 and 9 degrees of freedom is 9.00, with $p - 0.0150$.

- there is also a significant difference between mean scores obtained at Time 1 (Pre) and those obtained at Time 3 (Followup) with $F(1, 9) = 19.7368$ and $p = 0.0016$.

Because the simple effect for the control group is not significant, you don't need to look at contrasts for that group.

Although these results seem straightforward, there are two complications in the interpretation. First, you performed multiple tests (one test for each level under the Group factor), which means that the significance level initially chosen needs to be adjusted to control for the experiment-wise probability of making a Type I error (rejection of a true null hypothesis or finding a significant difference when there is no difference in the population). Looney and Stanley (1989) recommend dividing the initial alpha (0.05 in the present study) by the number of tests conducted, and using the result as the required significance level for each test. This adjustment is also referred to as a Bonferroni correction. For example, assume the alpha is initially set at 0.05 (per convention) in this investigation. You perform two tests for simple effects, so divide the initial alpha by 2, giving an actual alpha of 0.025. When you conduct tests for simple effects, you conclude that an effect is significant only if the *p* value that appears in the output is less than 0.025. This approach to adjusting alpha is viewed as a rigorous adjustment. There are other approaches to protect the experiment-wise error rate, not included in this book.

The second complication involves the error term used in the analyses. With the preceding approach, the error term used in computing the *F* ratio for a simple effect is the mean square error from the one-way ANOVA based on data from one group. An alternative approach is to use the mean square error from the test that includes the between-subjects factor, the repeated-measures factor, and the interaction term, found in the original analysis (the "omnibus" model) shown in Figure 12.21. This second approach has the advantage of using the same yardstick (the same mean square error) in the computation of both simple effects. Some authors recommend this procedure when the mean squares for the various groups are homogeneous. On the other hand, the first approach, in which different error terms are used in the different tests, has the advantage (or disadvantage, depending upon your perspective) of providing a slightly more conservative *F* test because it involves fewer degrees of freedom for the denominator. For a discussion of the alternative approaches for testing simple effects, see Keppel (1982, pp. 428–431) and Winer (1971, pp. 527–528).

This section discussed simple effects only for the Time factor. It is also possible to perform tests for three simple effects for the Group factor. Specifically, you can determine whether there is a significant difference between the experimental group and the control group with respect to the following:

- investment scores obtained at Time 1 (pre-weekend)

- investment scores obtained at Time 2 (post-weekend)

- investment scores obtained at Time 3 (follow-up).

Because there are only two levels of Group, you can use the Fit Y by X platform as described in Chapter 8, "One-Way ANOVA with One Between-Subjects Factor" to test simple effects for Group. Analyzing these simple effects is left as an exercise for the reader as these analyses do not yield any significant results.

When you look at the big picture of the Group by Time interaction discovered in this analysis, it becomes clear that the interaction is due to a simple effect for Time at the "experimental" group level of Group. On the other hand, there is no evidence of a simple effect for Group at any level of Time.

Steps to Review the Output

1. Determine whether the interaction term is statistically significant. Once you scan the results in the usual manner to verify that there are no obvious errors in the program or data, check the appropriate statistics to see whether the interaction between the between-subjects factor and the repeated-measures factor is significant. The interaction results are part of the Within Subjects analysis, and appear as the Time*Group report. The exact F value for this interaction term is 6.4657. With 2 and 17 degrees of freedom, the p value for this F is 0.0082, which indicates a significant interaction. This finding is consistent with the hypothesis that the relationship between the Time variable and investment scores is different for the two experimental groups.

2. Examine the interaction. The initial MANOVA results show the interaction plot shown here to the right and in Figure 12.21. Notice that the line for the experimental group is not parallel to the line for the control, which is what you expect with a significant Time by Group interaction. You can can see that the line for the 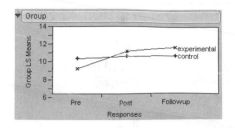 control group is relatively flat. There does not appear to be an increase in perceived investment from Pre to Post to Followup times. In contrast, notice the angle displayed by the line for the experimental group. There is an apparent increase in investment size from Pre to Post, and another slight increase from Post to Followup.

These findings are consistent with the hypothesis that there is a larger increase in investment scores in the experimental group than in the control group. However, to have more confidence in this conclusion, you want to test for simple effects.

3. Look at simple effects. The previous section described the steps to look at simple effects, which showed that the group by time interaction in this analysis was due to a simple effect for Time at the "experimental" group level of Group. There was no evidence of a simple effect for Group at any level of Time.

4. Prepare your own version of the ANOVA summary table. Table 12.2 presents the ANOVA summary table from the analysis (see Figure 12.21).

Table 12.2 MANOVA Summary Table for Study Investigating Changes in Investments Following an Experimental Manipulation (Significant Interaction)

Source	Numerator df	Denominator df	F	p
Between Subjects				
Group (A)	1	18	0.03	0.8632
Within Subjects				
Time (B)	2	17	10.64	0.0010
A by B Interaction	2	17	6.47	0.0082

Note: N = 20.

Formal Description of Results for a Paper

Results were analyzed using a two-way ANOVA with repeated measures on one factor. The Group by Time interaction was significant, $F(2,17) = 6.466$, $p = 0.0082$. Tests for simple effects showed that the mean investment scores for the control group displayed no significant differences across time, $F(2,8) = 0.9524$, $p = 0.4256$. However, the experimental group did display significant increases in perceived investment across time, $F(2,8) = 9.8997$, $p < 0.0096$. Post-hoc contrasts showed that the experimental group has significantly higher scores at post-test [$F(1,9) = 9.00$, $p < 0.05$] and follow-up [$F(1,9) = 19.74$, $p < 0.01$] compared to baseline.

An Alternative Approach to a Univariate Repeated-Measures Analysis

The examples in this chapter used the Fit Model platform to perform a multivariate repeated-measures analysis. The examples in Chapter 11, "One-Way ANOVA with One Repeated-Measures Factor," also used the Fit Model multivariate platform, but included univariate F tests, which are an option on the multivariate control panel (see Figure 12.16). Univariate F tests are appropriate if conditions of sphericity are met.

Sphericity is a characteristic of a difference-variable covariance matrix obtained when performing a repeated-measures ANOVA. The concept of sphericity was discussed in greater detail in Chapter 11.

If sphericity is not a concern and you want to do a univariate analysis of repeated-measures data, there is an alternate way to specify the model in the Fit Model dialog. This section shows that method and compares the results to those given by the multivariate platform.

First, to perform a univariate repeated-measures analysis, the data need to be restructured. Figure 12.24 shows partial listings of the data table called mixed nointeraction.jmp, and the restructure of that data, called stacked nointeraction.jmp. You can find an example that shows how to use the Stack command to restructure data in Chapter 7, "*t* Tests: Independent Samples and Paired Samples."

In the stacked table, the repeated measures are stacked into a single column and identified by the variable called Time with values "Pre," "Post," and "Followup." Notice that Participant, Group, and Time are character (nominal) variables.

Figure 12.24 Repeated-Measures Data (left) and Stacked Data (right)

Participant	Group	Pre	Post	Follow up
9 9	control	12	11	12
10 10	control	9	12	12
11 11	experimental	10	14	13
12 12	experimental	7	12	11

	Participant	Group	Time	Investment
25	9	control	Pre	12
26	9	control	Post	11
27	9	control	Followup	12
28	10	control	Pre	9
29	10	control	Post	12
30	10	control	Followup	12
31	11	experimental	Pre	10
32	11	experimental	Post	14
33	11	experimental	Followup	13
34	12	experimental	Pre	7
35	12	experimental	Post	12
36	12	experimental	Followup	11

Use this stacked table to perform the univariate repeated measures analysis. Follow these steps to specify the model in the Fit Model dialog:

- With the **stacked interaction.jmp** table active, choose **Fit Model** from the **Analyze** menu.
- Select **Investment** in the column selection list and click **Y** in the dialog.
- Select **Group** in the column selection list and click **Add** in the dialog.
- Select **Participant** in the column selection list and click **Add**.
- Now select **Participant** in the effects list and select **Group** in the column selection list. Then click **Nest** in the dialog. This action gives the Participant[Group] nested effect in the effect list.
- Select the **Participant[Group]** effect in the effect list and choose **Random Effect** from the **Attributes** menu in the dialog, as shown at the bottom in Figure 12.25.
- Finally, enter the interaction term. Select both **Group** and **Time** in the column selection list and click **Cross** in the dialog.

Figure 12.25 shows the Fit Model dialog for multivariate analysis at the top and the univariate repeated-measures analysis at the bottom.

> **Note:** The nested effect Participant[Group] is designated as a Random effect so that it will be used as the error term for F tests of all effects showing above it in the effects list. In this repeated-measures analysis, it is the appropriate error term for the Group effect. The F tests for effects below the random effect use the model error (residual) as the denominator.

Figure 12.25 Comparison of Model Specifications

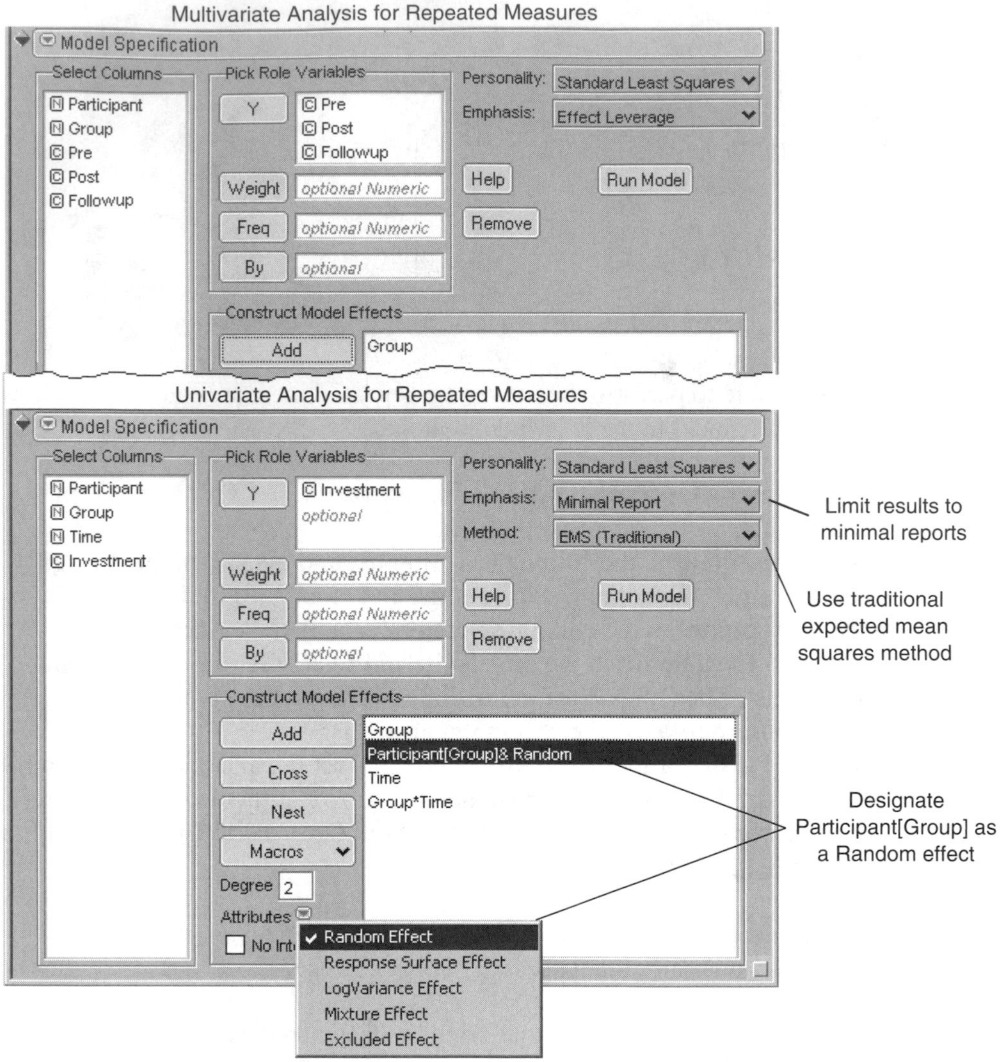

Click **Run Model** to see the univariate repeated-measures analysis. Figure 12.26 shows a comparison of the results for the two types of analyses.

Figure 12.26 Univariate *F* Tests for Repeated-Measures Analysis

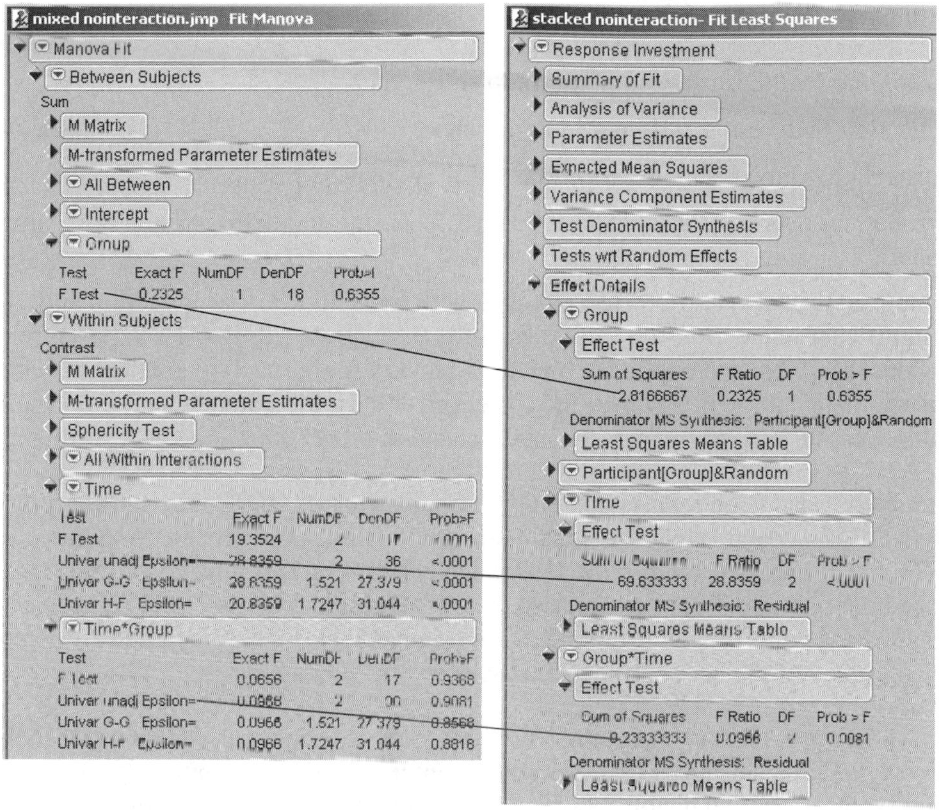

Summary

This chapter introduced a design that has both repeated-measures factors and between-subjects factors, sometimes called a mixed design. This two-way mixed design extends the one-way repeated-measures design presented in the previous chapter by adding one or more groups. Example data included one additional group, used as a control group. Adding a control group lets you test the plausibility of alternative explanations that could account for study results.

When there is more than one between-subjects factor (such as Group and Time in the example), the statistical analysis can perform significance tests to check for significant main effects, interaction, and simple effects (slices).

Auxiliary sections showed examples of the following:

- using the JMP Distribution platform to screen data
- tailoring JMP results with platform options
- rearranging tables with the **Summary** and **Transpose** commands
- producing overlay plots that show main effects and interactions using the rearranged data.

Example analyses were performed for data with both significant interaction and nonsignificant interaction. The analyses used multivariate fitting (MANOVA) from the Fit Model platform and explained how the MANOVA approach to analysis of repeated-measures data automatically uses the correct error term for statistical tests. A detailed description told how to perform the analysis and interpret the MANOVA results.

When there was a significant main effect (Group) with no interaction, the analysis included testing each level of this main effect with a one-way repeated-measures ANOVA. Dividing the alpha level by 2 when performing this kind of multiple testing reduced the risk of a Type I error. Also, it was suggested that the F test values be computed using the error term found in the original analysis that included all model effects instead of the error terms given by these one-way analyses.

A univariate approach to analyzing a two-way mixed design was shown as an alternative analysis method. This analysis required rearranging the data and building a different model in the Fit Model dialog that explicitly identifies the error term for the random effect, Group.

Assumptions for Factorial ANOVA with Repeated-Measures and Between-Subjects Factors

All of the statistical assumptions associated with a one-way anova, repeated-measures design are also required for a factorial anova with repeated-measures factors and between-subjects factors. In addition, the latter design also requires a homogeneity of covariances assumption for the multivariate. This section reviews the assumptions for the one-way repeated measures anova and introduces the new assumptions for the factorial anova, mixed design.

Assumptions for the Multivariate Test

Level of measurement

The response variable is a continuous numeric measurement. The predictor variables should be nominal variables (categorical variables). One predictor codes the within-subjects variable and the second codes the between-subjects variable.

Independent observations

A given participant's score in any one condition should not be affected by any other participant's score in any of the study's conditions. However, it is acceptable for a given participant's score to be dependent upon his or her own score under different conditions (under the within-subjects predictor variable).

Random sampling

Scores on the response variable should represent a random sample drawn from the populations of interest.

Multivariate normality

The measurements obtained from participants should follow a multivariate normal distribution. Under conditions normally encountered in social science research, violations of this assumption have only a very small effect on the Type I error rate (the probability of incorrectly rejecting a true null hypothesis).

Homogeneity of covariance matrices

In the population, the response-variable covariance matrix for a given group (under the between-subjects predictor variable) should be equal to the covariance matrix for each of the remaining groups. This assumption was discussed in greater detail in Chapter 10, "Multivariate Analysis of Variance (MANOVA) with One Between-Subjects Factor."

Assumptions for the Univariate Test

The univariate test requires all of the preceding assumptions as well as the following assumptions of sphericity and symmetry.

Sphericity

Sphericity is a characteristic of a difference-variable covariance matrix obtained when performing a repeated-measures ANOVA. The concept of sphericity was discussed in greater detail in Chapter 11, "One-Way ANOVA with One Repeated-Measures Factor." Briefly, two conditions must be satisfied for the covariance matrix to demonstrate sphericity. First, each variance on the diagonal of the matrix should be equal to every other variance on the diagonal. Second, each covariance off the diagonal should equal zero. This is analogous to saying that the correlations between the difference variables should be zero. Remember that, in a study with a between-subjects factor, there will be a separate difference-variable covariance matrix for each group under the between-subjects variable.

Symmetry conditions

The first symmetry condition is the sphericity condition just described. The second condition is that the difference-variable covariance matrices obtained for the various groups (under the between-subjects factor) should be equal to one another. For example, assume that a researcher has conducted a study that includes a repeated-measures factor with three conditions and a between-subjects factor with two conditions. Participants assigned to Condition 1 under the between-subjects factor are designated as "Group 1" and those assigned to Condition 2 are designated as "Group 2." With this research design, one difference-variable covariance matrix will be obtained for Group 1 and a second for Group 2. The nature of these difference-variable covariance matrices was discussed in Chapter 11. The symmetry conditions are met if both matrices demonstrate sphericity and each element in the matrix for Group 1 is equal to its corresponding element in the matrix for Group 2.

References

Keppel, G. 1982. *Design and Analysis: A Researcher's Handbook.* 2d ed. Englewood Cliffs, NJ: Prentice Hall.

Looney, S., and W. Stanley. 1989. "Exploratory Repeated Measures Analysis for Two or More Groups." *American Statistician* 43:220–225.

Rusbult, C. E. 1980. "Commitment and Satisfaction in Romantic Associations: A Test of the Investment Model." *Journal of Experimental Social Psychology* 16:172–186.

SAS Institute Inc. 2003. *JMP Statistics and Graphics Guide.* Cary, NC: SAS Institute Inc.

Winer, B. J. 1971. *Statistical Principles in Experimental Design.* 2d ed. New York: McGraw-Hill.

Multiple Regression

Overview. This chapter shows how to perform multiple regression analysis to investigate the relationship between a continuous response variable and multiple continuous predictor variables. It describes the different components of the multiple regression equation, and discusses the meaning of R^2 and other results from a multiple regression analysis. It shows how bivariate correlations, multiple regression coefficients, and uniqueness indices can be reviewed to assess the relative importance of predictor variables. Fictitious data are examined using the Multivariate platform and the Fit Model platform in JMP to show how the analysis can be conducted and to illustrate how the results can be summarized in tables and in text.

Introduction to Multiple Regression

Multiple regression is a highly flexible procedure that enables researchers to address many different types of research questions. One of the most common multiple regression analyses involves a single numeric continuous response variable and multiple numeric continuous predictors.

For example, suppose you want to find the relative importance of variables believed to predict income. To conduct your research, you obtain information for 1,000 adults. The response variable in your study is annual income for these participants. The predictor variables are Current Age, Years of Education, and Years in Workforce. In this study, the response variable and the predictor variables are all numeric continuous scales. Therefore, multiple regression is the appropriate data analysis procedure.

Analysis with multiple regression addresses a number of research questions. For example, it allows you to determine

- whether there is a significant relationship between the response variable and the multiple predictor variables examined as a group.

- whether the multiple regression coefficient for a given predictor variable is statistically significant. This coefficient represents the amount of weight given to a specific predictor, while holding constant the other predictors.

- whether a given predictor accounts for a significant amount of variance in the response, beyond the variance accounted for by the other predictors.

A multiple regression analysis tells you about the relative importance of the predictor variables included in the multiple regression equation. Researchers conducting nonexperimental research in the social sciences are often interested in learning about the relative importance of naturally occurring predictor variables such as age and income. This chapter shows how to perform such analyses.

There are many other types of regression analyses that are beyond the scope of this book. In the study dealing with annual income, all predictor variables are numeric continuous variables and there is a linear relationship between the predictor variables and the response. This chapter provides an introduction to multiple regression but does not cover circumstances that use nominal predictor variables or test nonlinear relationships. Once you learn the basics of multiple regression, you can learn more about advanced regression topics in Cohen, Cohen, West, and Aiken (2003) or Pedhazur (1982).

Multiple Regression and ANOVA

Chapters 8 through 12 presented analysis of variance (ANOVA) procedures, commonly used to analyze data from *experimental research*, in which one or more categorical independent predictor variables (experimental conditions) are manipulated to determine how they affect the study's dependent variable.

For example, imagine you are interested in studying *prosocial behavior*—actions intended to help others. Examples of prosocial acts might include donating money to the poor, giving blood, doing volunteer work at a hospital, and so forth. Suppose you develop an experimental treatment (such as promise of reward) you believe increases the likelihood that people will engage in prosocial acts. To investigate this treatment, you conduct an experiment in which you manipulate the independent variable (half of your participants are given the experimental treatment and half are given a placebo treatment). You then assess the dependent variable, which is the number of prosocial acts that the participants later perform. It would be appropriate to analyze data from this study using one-way ANOVA because there is a single numeric response (number of prosocial acts) and a single nominal predictor variable (experimental group—reward or no reward).

Multiple regression is similar to ANOVA in at least one important respect—both analyses require a numeric continuous response variable. Recall from Chapter 1 that a continuous variable can assume a large number of values. The number of prosocial acts performed over a six-month period can be considered a continuous variable because subjects demonstrate a wide variety of scores, such as 0, 4, 10, 11, 20, 25, 30, and so forth.

However, multiple regression also differs from ANOVA. When data are analyzed with ANOVA, the predictor variable is a categorical variable whose values indicate group membership. In contrast, the predictor variables in multiple regression are usually numeric continuous variables.

As an illustration, assume that you conduct a study in which you administer a questionnaire to a group of participants to assess the number of prosocial acts each has performed. You then obtain scores for the same participants on each of the following predictor variables:

- age
- income
- a questionnaire-type scale that assesses level of moral development.

Suppose you hypothesize that the number of prosocial acts performed is causally determined by these three predictor variables, as illustrated by the model in Figure 13.1.

Figure 13.1 A Model of Determinants of Prosocial Behavior

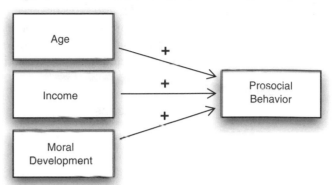

The predictor variables in this study (Age, Income, and Moral Development) are continuous. Continuous predictor variables and a continuous response variable mean you can analyze your data using multiple regression.

That is the most important distinction between the two statistical procedures. With ANOVA, the predictor variables are always categorical, but with multiple regression they are usually continuous. For more information, see Cohen, Cohen, West, and Aiken (2003) or Pedhazur (1982).

Multiple Regression and Naturally Occurring Variables

Multiple regression is particularly well suited for studying the relationship between *naturally occurring* predictor and response variables. The researcher does not manipulate naturally occurring variables—they are simply measured as they naturally occur. The preceding prosocial behavior study provides a good example of the naturally occurring predictor variables age, income, and level of moral development.

Research in the social sciences often focuses on naturally occurring variables, which makes multiple regression an important tool. Those kinds of variables cannot be experimentally manipulated. As another example, suppose you hypothesize that domestic violence (an act of aggression against a domestic partner) is caused by

- childhood trauma experienced by the abuser
- substance abuse
- low self-esteem.

Figure 13.2 illustrates the model for this hypothesis. In this example you cannot experimentally manipulate the predictor variables of the model and later observe the participants as adults to see if the manipulation affected their propensity for domestic violence. However, it is possible to measure these variables as they naturally occur and determine whether they are related to one another in the predicted fashion. Multiple regression allows you to do this.

Figure 13.2 A Model of the Determinants of Domestic Violence

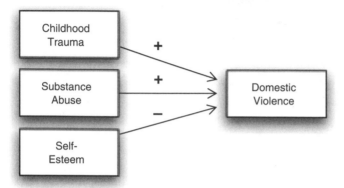

This does not mean that ANOVA is only for the analysis of manipulated predictor variables and multiple regression is only for the analysis of naturally occurring variables. Naturally occurring variables can be predictor variables in an ANOVA provided they are categorical. For example, ANOVA can be used to determine whether participant gender (a naturally occurring predictor variable) is related to relationship commitment (a response variable). In addition, JMP can analyze a categorical manipulated variable, such as an experimental condition, as a predictor variable in multiple regression, but the examples in this chapter cover only numeric continuous predictors. The main distinction to remember is this:

- ANOVA predictor variables must be categorical variables.

- Multiple regression predictor variables can be categorical but are usually continuous.

> **Note:** An analysis that has both categorical and numeric continuous predictor variables is often referred to as analysis of covariance (ANCOVA). In this kind of model, the continuous predictor variable is called a covariate.

How Large Must My Sample Be?

Multiple regression is a large-sample procedure. Unreliable results might be obtained if the sample does not include at least 100 observations, preferably 200. The greater the number of predictor variables included in the multiple regression equation, the greater the number of participants that will be necessary to obtain reliable results. Most experts recommend at least 15 to 30 participants per predictor variable. See Cohen (1992).

Cause-and-Effect Relationships

Multiple regression can determine whether a given set of variables is useful for predicting a response variable. Among other things, this means that multiple regression can be used to determine

- whether the relationship between the response variable and predictors variables (taken as a group) is statistically significant
- how much variance in the response is accounted for by the predictors
- which predictor variables are relatively important predictors of the response.

Although the preceding section often refers to causal models, it is important to remember that the procedures discussed in this chapter do not provide strong evidence concerning cause-and-effect relationships between predictor variables and the response. For example, consider the causal model for predicting domestic violence presented in Figure 13.2. Assume that naturally occurring data for these four variables are gathered and analyzed using multiple regression. Assume further that the results are significant—that multiple regression coefficients for all three of the predictor variables are significant and in the predicted direction. Even though these findings are consistent with your theoretical model, they do not *prove* that these predictor variables have a causal effect on domestic violence. Because the predictor variables are probably correlated, there can be more than one way to interpret the observed relationships between them. The most that you can say is that your findings are *consistent* with the causal model portrayed in Figure 13.2. It is incorrect to say that the results *prove* that the model is correct.

There are a number of reasons to analyze correlational data with multiple regression. Often, researchers are not really interested in testing a causal model. The purpose of the study could simply be to understand relationships between a set of variables.

However, even when the research is based on a causal model, multiple regression can still give valuable information. For example, suppose you obtain correlational data relevant to the domestic violence model of Figure 13.2, and none of the regression coefficients are significant. This is useful because it shows that the model failed to survive an analysis that investigated the predictive relationships between the variables.

However, if significant results are obtained, you can prepare a research report stating that the results were consistent with the hypothesized model. In other words, the model survived an attempt at disconfirmation. If the predictor variables can be ethically manipulated, you could conduct a follow-up experiment to determine whether the predictors appear to have a causal effect on the response variable under controlled circumstances.

In summary, it is important to remember that the multiple regression procedures discussed in this chapter do not provide evidence of cause-and-effect relationships. However, they are extremely useful for determining whether one set of variables can predict variation in a response.

> **Note:** The term *prediction* in multiple regression should not be confused with *causation*.

Predicting a Response from Multiple Predictors

The response variable (also called predicted or dependent variable) in multiple regression is represented with the symbol Y, and is therefore often referred to as the *Y variable*. The predictor variables (also known as independent variables or factors) are represented as X_1, X_2, X_3…X_n, and are referred to as the *X variables*. The purpose of multiple regression is to understand the relationship between the Y variable and the X variables when taken as a group.

A Simple Predictive Equation

Consider the model of prosocial behavior shown previously in Figure 13.1. The model hypothesizes that the number of prosocial acts performed by a person in a given period of time can be predicted by the person's age, income, and level of moral development.

Notice that each arrow (assumed predictive path) in the figure is identified with either a plus (+) sign or a minus (–) sign. A plus sign indicates that you expect the relevant predictor variable to demonstrate a positive relationship with the response, whereas a minus sign indicates that you expect the predictor to demonstrate a negative relationship.

The nature of these signs in Figure 13.1 shows that you expect

- a positive relationship between age and prosocial behavior, meaning that older participants tend to perform more prosocial acts
- a positive relationship between income and prosocial behavior, meaning that more affluent people tend to perform more prosocial acts
- a positive relationship between moral development and prosocial behavior, meaning that participants who score higher on the paper-and-pencil measure of moral development tend to perform more prosocial acts.

Assume that you administer a questionnaire to a sample of 100 participants to assess their level of moral development. Scores on this scale can range from 10 to 100, with higher scores reflecting higher levels of development. You also obtain the participants' ages and information about their income. You now want to use this information to predict the number of prosocial acts the participants will perform in the next six months. More specifically, you want to create a new variable, $\hat{Y}$, that represents your best guess of how many prosocial acts the participants will perform. Keep in mind that $\hat{Y}$ represents the *prediction* of the participants' standing on the response variable. Y is the participants' *actual* standing on the response variable.

Assume the predictor variables are positively related to prosocial behavior. One way to predict how many prosocial behaviors participants will engage in could be to simply add together the participants' scores on the three X variables. This sum, as expressed by the following equation, then constitutes your best guess as to how many prosocial acts they will perform.

$$\hat{Y} = X_1 + X_2 + X_3$$

where

$\hat{Y}$ = the participants' predicted scores on prosocial behavior

X_1 = the participants' actual scores on age

X_2 = the participants' actual scores on income (in thousands)

X_3 = the participants' actual scores on the moral development scale.

To make this more concrete, consider the fictitious data presented in Table 13.1. This table presents actual scores for four of the study's participants on the predictor variables.

Table 13.1 Fictitious Data, Prosocial Behavior Study

Participant	Age, X_1	Income (in thousands), X_2	Moral Development, X_3
Lars	19	15	12
Sally	24	32	28
Jim	33	45	50
.	.	.	.
.	.	.	.
.	.	.	.
Sheila	55	60	95

To arrive at an estimate of the number of prosocial behaviors predicted for the first participant (Lars), insert his scores on the three X variables into the preceding equation:

$$\hat{Y} = X_1 + X_2 + X_3$$

$$\hat{Y} = 19 + 15 + 12$$

$$\hat{Y} = 46$$

So your best guess is that the first participant, Lars, will perform 46 prosocial acts in the next six months. You can repeat this process for all participants and compute their $\hat{Y}$ scores in the same way. Table 13.2 presents the predicted scores on the prosocial behavior variable for some of the study's participants.

Table 13.2 Predicted Scores for Fictitious Data, Prosocial Behavior Study

Participant	Predicted Scores on Prosocial Behavior, $\hat{Y}$	Age, X_1	Income (in thousands), X_2	Moral Development, X_3
Lars	46	19	15	12
Sally	84	24	32	28
Jim	128	33	45	50
.	.	.	.	.
.	.	.	.	.
.	.	.	.	.
Sheila	210	55	60	95

Notice the general relationships between $\hat{Y}$ and the X variables in Table 13.2. If participants have low scores on age, income, and moral development, your equation predicts that they will engage in relatively few prosocial behaviors. However, if participants have high scores on these X variables, your equation predicts that they will engage in a relatively large number of prosocial behaviors. For example, Lars had relatively low scores on these X variables, and as a consequence, the equation predicts that he will perform only 46 prosocial acts over the next six months. In contrast, Sheila displayed higher scores than Lars on age, income, and moral development, so your equation predicts that she will engage in 210 prosocial acts.

You have created a new variable, $\hat{Y}$. Imagine that you now gather data regarding the actual number of prosocial acts these participants engage in over the following six months. This variable is represented with the symbol Y because it represents the participants' *actual* scores on the response and not their predicted scores. You can list these actual scores in a table with their predicted scores on prosocial behavior, as shown in Table 13.3.

Notice that, in some cases, the participants' predicted scores on Y are not close to the actual scores. For example, your equation predicted that Lars would engage in 46 prosocial activities, but in reality he engaged in only 10. Similarly, it predicted that Sheila would engage in 210 prosocial behaviors, while she actually engaged in only 130.

Despite these discrepancies, you should not lose sight of the fact that the new variable $\hat{Y}$ does appear to be correlated with the actual scores on Y. Notice that participants with low $\hat{Y}$ scores (such as Lars) also tend to have low scores on Y, and participants with high $\hat{Y}$ scores (such as Sheila) also tend to have high scores on Y. There is probably a moderately high product-moment correlation (r) between Y and $\hat{Y}$. This correlation supports your model because it suggests that there really is a relationship between Y and the three X variables when taken as a group.

Table 13.3 Actual and Predicted Scores for Fictitious Data, Prosocial Behavior Study

Participant	Actual Scores on Prosocial Behavior, Y	Predicted Scores on Prosocial Behavior, $\hat{Y}$
Lars	10	46
Sally	40	84
Jim	70	128
.	.	.
.	.	.
.	.	.
Sheila	130	210

The procedures (and the predictive equation) described in this section have been for illustration only. They do not describe the way that multiple regression is actually performed. However, they do illustrate some important basic concepts in multiple regression analysis. In multiple regression analysis,

- you create an artificial variable, $\hat{Y}$, to represent your best guess of the participant's standings on the response variable
- the relationship between this variable and the participant's actual standing on the response (Y) is assessed to indicate the strength of the relationship between Y and the X variables when taken as a group.

Multiple regression has many important advantages over the practice of simply adding together the X variables, as illustrated in this section. With true multiple regression, the various X variables are multiplied by optimal weights before they are added together to create $\hat{Y}$. This usually results in a more accurate estimate of the participants' standing on the response variable. In addition, you can use the results of a true multiple regression procedure to determine which of the X variables are more important and which are less important predictors of Y.

An Equation with Weighted Predictors

In the preceding section, each predictor variable has an equal weight (1) when computing scores on $\hat{Y}$. You did not, for example, give twice as much weight to income as you gave to age when computing $\hat{Y}$ scores. Assigning equal weights to the various predictors makes sense in some situations, especially when all of the predictors are equally predictive of the response.

However, what if your measure of moral development displayed a strong correlation with prosocial behavior ($r = 0.70$, for example), income demonstrated a moderate correlation with prosocial behavior ($r = 0.40$), and age demonstrated only a weak correlation ($r = 0.20$)? In this situation, it would make sense to assign different weights to the different predictors. For example, you might assign a weight of 1 to age, a weight of 2 to income, and a weight of 3 to moral development.

The predictive equation that reflects this weighting scheme is

$$\hat{Y} = (1)X_1 + (2)X_2 + (3)X_3$$

where X_1 = age, X_2 = income, and X_3 = moral development. To calculate a given participant's score for $\hat{Y}$, multiply the participant's score on each X variable by the appropriate weight and sum the resulting products. For example, Table 13.2 showed that Lars had a score of 19 on X_1, a score of 15 on X_2, and a score of 12 on X_3.

Using weights, his predicted prosocial behavior score, $\hat{Y}$, is calculated as

$$\begin{aligned}
\hat{Y} &= (1)X_1 + (2)X_2 + (3)X_3 \\
&= (1)19 + (2)15 + (3)12 \\
&= 85
\end{aligned}$$

This weighted equation predicts that Lars will engage in 85 prosocial acts over the next six months. You could use the same weights to compute the remaining participants' scores for $\hat{Y}$. Although this example has again been somewhat crude in nature, it illustrates the concept of optimal weighting, which is at the heart of multiple regression analysis.

Regression Coefficients and Intercepts

In linear multiple regression as performed by the Fit Model platform in JMP, optimal weights are automatically calculated by analysis.

The following symbols are used to represent the various components of an actual multiple regression equation,

$$\hat{Y} = b_1 X_1 + b_2 X_2 + b_3 X_3 \ldots b_n X_n + a$$

where

$\hat{Y}$ = the participants' predicted scores on the response variable

b_k = the nonstandardized multiple regression coefficient for the kth predictor variable

X_k = kth predictor variable

a = the intercept constant.

A *multiple regression coefficient* for a given X variable represents the average change in Y associated with a one-unit change in that X variable, while holding constant the remaining X variables.

This somewhat technical definition for a regression coefficient is explained in more detail in a later section. For the moment, it is useful to think of a regression coefficient as revealing the amount of *weight* that the X variable is given when computing $\hat{Y}$. In this text, nonstandardized multiple regression coefficient are referred to as *b weights*.

The symbol *a* represents the *intercept constant* of the equation. The intercept is a fixed value added to or subtracted from the weighted sum of X scores when computing $\hat{Y}$.

To develop a true multiple regression equation using the Fit Model platform in JMP, it is necessary to gather data on both the Y variable and the X variables. Assume that you do this in a sample of 100 participants. You analyze the data, and the results of the analyses show that the relationship between prosocial behavior and the predictor variables can be described by the following equation:

$$\hat{Y} = b_1X_1 + b_2X_2 + b_3X_3$$
$$= (0.10)X_1 + (0.25)X_2 + (1.10)X_3 + (-3.25)$$

The preceding equation indicates that your best guess of a given participant's score on prosocial behavior can be computed by multiplying the age score by 0.10, multiplying the income score 0.25, and multiplying the moral development by 1.10. Then, sum these products and subtract the intercept of 3.25 from this sum.

This process is illustrated by inserting Lars's scores on the X variables in the equation:

$$\hat{Y} = (0.10)19 + (0.25)15 + (1.10)12 + (-3.25)$$
$$= 1.9 + 3.75 + 13.2 (-3.25)$$
$$= 18.85 + (-3.25)$$
$$= 15.60$$

Your best guess is that Lars will perform 15.60 prosocial acts over the next six months. You can calculate the $\hat{Y}$ scores for the remaining participants in Table 13.2 by inserting their X scores in this same equation.

The Principle of Least Squares

At this point it is reasonable to ask, "How did the analysis program determine that the optimal b weight for X_1 was 0.10? How did it determine that the optimal weight for X_2 was 0.25? How did it determine that the 'optimal' intercept (a term) was -3.25?"

The answer is that these values are *optimal* in the sense that they minimize a function of the errors of prediction. An *error of prediction* refers to the difference between a participant's actual score on the response (Y) and that person's predicted score on the response ($\hat{Y}$). This difference can be written

$$Y - \hat{Y}$$

Remember that you must gather actual scores on Y in order to perform the multiple regression analysis that computes the error of prediction (the difference between Y and $\hat{Y}$) for each participant in the sample. For example, Table 13.4 reports several participants' actual scores on Y, their predicted scores on Y, based on the optimally weighted regression equation above, and their errors of prediction.

Table 13.4 Prediction Errors Using Optimally Weighted Multiple Regression Equation

Participant	Actual Scores on Prosocial Behavior, Y	Predicted Scores on Prosocial Behavior $\hat{Y}$	Errors of Prediction, $Y - \hat{Y}$
Lars	10	15.6	−5.60
Sally	40	37.95	2.05
Jim	70	66.30	3.70
.	.	.	.
.	.	.	.
.	.	.	.
Sheila	130	121.75	8.25

For Lars, the actual number of prosocial acts performed was 10, while the multiple regression equation predicted that he would perform 15.60 acts. The error of prediction for Lars is 10 − 15.60 = −5.60. The errors of prediction for the remaining participants are calculated in the same fashion.

Earlier it was stated that the *b* weights and intercept calculated by regression analysis are optimal in the sense that they minimize errors of prediction. More specifically, these weights and intercept are computed according to the principle of least squares. The *principle of least squares* says that $\hat{Y}$ values are calculated such that the sum of the squared errors of prediction is a minimal.

> **Note:** This principle of least squares means that no other set of *b* weights and intercept constant gives a smaller value for the sum of the squared errors of prediction.

The sum of the squared errors of prediction is written

$$\Sigma(Y - \hat{Y})^2$$

To compute the sum of the squared errors of prediction according to this formula, it is necessary to do the following:

1. Compute the error of prediction $(Y - \hat{Y})$ for a given participant.
2. Square this error.
3. Repeat this process for all remaining participants.
4. Sum the resulting squares. The purpose of squaring the errors before summing them is to eliminate the negative error values that cause the sum of errors to be zero.

When analyzing a JMP data table using multiple regression, the Fit Model platform applies a set of formulas that calculates the optimal *b* weights and the optimal intercept for those data. These calculations minimize the squared errors of prediction. That is, regression calculates optimal weights and intercepts. They are optimal in the sense that no other set of *b* weights or intercepts could do a better job of minimizing squared errors of prediction for the current set of data.

With these points established, it is finally possible to summarize what multiple regression actually lets you to do

> Multiple regression lets you examine the relationship between a single response variable and an optimally weighted linear combination of predictor variables

In the preceding statement,

- $\hat{Y}$ is the optimally weighted linear combination of predictor variables
- *linear combination* means the various X variables are combined or added together to arrive at $\hat{Y}$.
- *optimally weighted* means that X variables have weights that satisfy the principle of least squares.

Although we think of multiple regression as a procedure that examines the relationship between single response and multiple predictor variables, it is also a procedure that examines the relationship between just two variables, Y and $\hat{Y}$.

The Results of a Multiple Regression Analysis

The next sections describe important information given by the multiple regression analysis. Details include how to compute and interpret

- the multiple correlation coefficient
- the amount of variation accounted for by predictor variables
- the effect of correlated predictor variables.

The Multiple Correlation Coefficient

The *multiple correlation coefficient*, symbolized as R, represents the strength of the relationship between a response variable and an optimally weighted linear combination of predictor variables. Its values range from 0 through 1. It is interpreted like a Pearson product-moment correlation coefficient (*r*) except that R can assume only positive values. Values approaching zero indicate little relationship between the response and the predictors, whereas values near 1 indicate strong relationships. An R value of 1.00 indicates perfect or complete prediction of response values.

Conceptually, R is the product-moment correlation between Y and $\hat{Y}$. This can be symbolized as

$$R = r_{yy'}$$

In other words, if you obtain data from a sample of participants that include their scores on Y and a number of X variables, compute $\hat{Y}$ scores for each participant and then find the correlation of predicted response scores ($\hat{Y}$) with actual scores (Y). That bivariate correlation coefficient is R, the multiple correlation coefficient.

With bivariate regression, the square of the correlation coefficient between Y and X estimates the proportion of variance in Y accounted for by X. The result is sometimes called the *coefficient of determination*. For example, if $r = 0.50$ for a given pair of variables, then

$$\text{coefficient of determination} = r^2$$
$$= (0.50)^2$$
$$= 0.25$$

You can say that the X variable accounts for 25% of the variance in the Y variable.

An analogous coefficient of determination can be computed in multiple regression by squaring the observed multiple correlation coefficient, R. This R^2 value (often referred to as R squared) represents the percentage of variance in Y accounted for by the linear combination of predictor variables given by the multiple regression equation. The next section gives details about the concept of *variance accounted for*.

Variance Accounted for by Predictor Variables

In multiple regression analyses, researchers often speak of *variance accounted for*, which is the percent of variance in the response variable accounted for by the linear combination of predictor variables.

A Single Predictor Variable

This concept is easy to understand with a simple bivariate example. Assume that you compute the correlation between prosocial behavior and moral development and find that $r = 0.50$. You can determine the percentage of variance in prosocial behavior accounted for by moral development by squaring this correlation coefficient:

$$R^2 = (0.50)^2 = 0.25$$

Thus, variation in moral development accounts for 25% of the variance in the prosocial behavior response data. This can be graphically illustrated with a Venn diagram that uses circles to represent total variance in a variable. Figure 13.3 is a Venn diagram that represents the correlation between prosocial behavior and moral development.

The circle for moral development overlaps the circle for prosocial behavior, which indicates that the two variables are correlated. More specifically, the circle for moral development overlaps about 25% of the circle for prosocial behavior. This illustrates that moral development accounts for 25% of the variance in prosocial behavior.

Figure 13.3 Variance in Prosocial Behavior
Accounted for by Moral Development

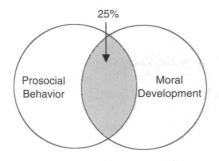

Multiple Predictor Variables with Correlations of Zero

The idea of *variance accounted for* can be expanded to the situation in which there are multiple X variables. Assume that you obtain data on prosocial behavior, age, income, and moral development from a sample of 100 participants and compute correlations between the four variables, as summarized in Table 13.5.

Table 13.5 Correlation Matrix: Zero Correlations between X Variables

Variable	Y	X_1	X_2	X_3
Y, Prosocial Behavior	1.00			
X_1, Age	0.30	1.00		
X_2, Income	0.40	0.00	1.00	
X_3, Moral Development	0.50	0.00	0.00	1.00

The correlations for a set of variables are often presented in the form of a *correlation matrix* such as the one presented in Table 13.5. The intersection of the row for one variable and the column of another identifies the correlation between the two variables. For example, where the row for X_1 (age) intersects with the column for Y (prosocial behavior), you see the correlation (*r*) between age and procial behavior is 0.30.

You can use these correlations to determine how much variance in Y is accounted for by the three X variables. For example, the correlation, *r*, between age and prosocial behavior is 0.30, and therefore R^2 is 0.09 ($0.30 \times 0.30 = 0.09$). This means that age accounts for 9% of the variance in prosocial behavior. Following the same procedure, you learn that income accounts for 16% of the variance in Y ($0.40 \times 0.40 = 0.16$), and moral development continues to account for 25% of the variance in Y.

Notice in Table 13.5 that each of the X variables demonstrates a correlation of zero with all of the other X variables. For example, where the row for X_2 (income) intersects with the column for X_1 (age), you see a correlation coefficient of zero. Table 13.5 also shows correlations of zero between X_1 and X_3, and between X_2 and X_3. Correlations of zero rarely occur with real data.

The correlations between the variables in Table 13.5 can be illustrated with the Venn diagram in Figure 13.4. Notice that each of the X variables intersects Y at the proportion given by its R square value with Y. Because the correlation between each pair of X variables is zero, none of their circles intersect with each other.

The Venn diagram in Figure 13.4 shows two important points:

- Each X variable accounts for some variance in Y
- No X variable accounts for any variance in any other X variable.

Because the X variables are uncorrelated, no X variable overlaps with any other X variable in the Venn diagram.

To determine the total variance accounted for in Figure 13.4, sum the percent of variance contributed by each of the three predictors:

   total variance accounted for $= 0.09 + 0.16 + 0.25 = 0.50$

The linear combination of X_1, X_2, and X_3 accounts for 50% of observed variance in Y, prosocial behavior. In most areas of research in the social sciences, 50% is considered a fairly large percentage of variance.

Figure 13.4 Variance in Prosocial Behavior Accounted for by Three Uncorrelated Predictors

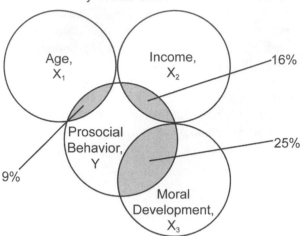

In the preceding example, the total variance in Y explained (accounted for) by the X variables is found by summing the squared bivariate correlations between Y and each X variable. It is important to remember that you can use this procedure to determine the total explained variance in Y only when the X variables are completely uncorrelated with one another (which is rare). When there is correlation between the X variables, this approach can give misleading results. The next section discusses the reasons why care must be taken when there is correlation between the X variables.

Variance Accounted for by Correlated Predictor Variables

The preceding example shows that it is easy to determine the percent of variance in a response accounted for by a set of predictor variables with zero correlation. In that situation, the total variance accounted for is the sum of the squared bivariate correlations between the X variables and the Y variable.

The situation becomes more complex when the predictor variables are correlated with one another. In this situation, you must look further to make statements about how much variance is accounted for by a set of predictors. This is partly because multiple regression equations with correlated predictors behave one way when the predictors include a suppressor variable and behave a different way when the predictors do not contain a suppressor variable.

A later section explains what a suppressor variable is, and describes the complexities introduced by this somewhat rare phenomenon. First, consider the simpler situation when the predictors in a multiple regression equation are correlated but do not contain a suppressor variable.

When Correlated Predictors Do Not Include a Suppressor Variable

In nonexperimental research you almost never observe a set of totally uncorrelated predictor variables. Remember that nonexperimental research involves measuring naturally occurring (nonmanipulated) variables that almost always display some degree of correlation between some pairs of variables.

For example, consider the nature of the variables studied here. It is likely that participant age is positively correlated with income because people tend to earn higher salaries as they grow older. Similarly, moral development is sometimes a function of maturation and is probably correlated with age. You might expect to see a correlation as high as 0.50 between age and income as well as a correlation of 0.50 between age and moral development. Table 13.6 shows these new (hypothetical) correlations. Notice that all X variables display the same correlations with Y displayed earlier. However, there is no correlation between age (X_1) and moral development (X_3).

Table 13.6 Correlation Matrix: Nonzero Correlations between X Variables

Variable	Y	X_1	X_2	X_3
Y, Prosocial Behavior	1.00			
X_1, Age	0.30	1.00		
X_2, Income	0.40	0.50	1.00	
X_3, Moral Development	0.50	0.00	0.50	1.00

The X variables of Table 13.6 display the same correlations with Y as previously displayed in Table 13.5. However, in most cases this does not mean that the linear combination of the three X variables still accounts for the same total percentage of variance in Y. Remember that in Table 13.5 the X variables are not correlated with one another, whereas in Table 13.6 there is now a correlation between age and income, and between income and moral development. These correlations can decrease the total variance in Y accounted for by the X variables. The Venn diagram in Figure 13.5 illustrates these correlations.

In Figure 13.5, plain gray areas represent variance in income that is accounted for by age only and moral development only. The white area with crosshatching represents variance in prosocial behavior accounted for by income only.

Notice that each X variable individually still accounts for the same percentage of variance in prosocial behavior—age still accounts for about 9%, income accounts for 16%, and moral development accounts for 25%.

Figure 13.5 Venn Diagram of Variance Accounted for by Correlated Predictors

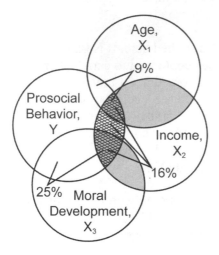

Figure 13.5 shows that some of the X variables are now correlated with one another. The area of overlap between age and income (gray and gray with crosshatching) shows that age and income now share about 25% of their variance (the correlation between these variables was 0.50, and $0.50^2 = 0.25$). In the same way, the area of overlap between income and moral development (also gray and gray with crosshatching) represents the fact that these variables share slightly less than 25% of their variance in common.

When there is a correlation between some of the X variables, a portion of the variance in Y accounted for by one X is now also accounted for by another X. For example, consider X_2, the income variable. By itself, income accounted for 16% of the variance in prosocial behavior. But notice how the circle for age overlaps part of the variance in Y accounted for by income (area with gray crosshatching). This means that some of the variance in Y that is accounted for by income is also accounted for by age. That is, the amount of variance in Y *uniquely* accounted for by income has decreased because of the correlation between age and income.

The same is true when you consider the correlation between income and moral development. The circle for moral development overlaps part of the variance in Y accounted for by income (gray with crosshatching). Some of the variance in Y that was accounted for by income is now also accounted for by moral development.

Because age and income are correlated, there is redundancy between them in the prediction of Y. There is also redundancy between income and moral development in the prediction of Y. The result of correlation between the predictor variables is a decrease in the total amount of variance in Y accounted for by the linear combination of X variables.

Compare the Venn diagram for the situation in which the X variables were uncorrelated (Figure 13.4) to the Venn diagram for the situation in which there was some correlation between the X variables (Figure 13.5). When the X variables were not correlated among themselves, they accounted for 50% of the variance in Y. Notice that the area in Y that overlaps with the X variables in Figure 13.5 is smaller, showing that the X variables account for less of the total variance in Y when they are correlated. The greater the correlation between the X variables, the smaller the amount of unique variance in Y accounted for by each individual X variable, and hence the smaller the total variance in Y that is accounted for by the combination of X variables.

The meaning of unique variance can be understood by looking at the crosshatched area in the Venn diagram of Figure 13.5. The two sections of the gray crosshatched area identify the variance in Y that is accounted for by income and age together, and by income and moral development together. The white crosshatched area shows the variance in Y that is uniquely accounted for by income. This area is quite small, which indicates that income accounts for very little variance in prosocial behavior that is not already accounted for by age and moral development.

One practical implication arising from this scenario is that the amount of variance accounted for in a response variable is larger to the extent that the following two conditions hold:

- The predictor variables shows strong correlations with the response variable.
- The predictor variables demonstrate weak correlations with each other.

These conditions hold true in general but they do not apply to the special case of a suppressor variable, discussed next.

A second implication is that there is generally a point of diminishing returns when adding new X variables to a multiple regression equation. Because so many predictor variables in social science research are correlated, only the first few predictors in a predictive equation are likely to account for meaningful amounts of unique variance in a response. Variables that are subsequently added tend to account for smaller and smaller percentages of unique variance. At some point, predictors added to the equation account for only negligible amounts of unique variance. For this reason, most multiple regression equations in social science research contain a relatively small number of variables, usually 2 to 10.

When Correlated Predictors Include a Suppressor Variable

The preceding section describes the results that you can usually expect to observe when regressing a response variable on multiple correlated predictor variables. However, suppressor variables are a special case in which the preceding generalizations do not hold. Although genuine suppressor variables are somewhat rare in the social sciences, it is important to understand the concept, so that you can recognize a suppressor variable.

A *suppressor variable* is a predictor variable that improves the predictive power of a multiple regression equation by controlling for unwanted variation that it shares with other predictors. Suppressor variables typically display these characteristics:

- zero or near-zero correlations with the response

- moderate to strong correlations with at least one other predictor variable.

Suppressor variables are interesting because, even though they can display a bivariate correlation with the response variable of zero, adding them to a multiple regression equation can result in a meaningful increase in R^2 for the model, which violates the generalizations given in the preceding section.

To understand how suppressor variables work, consider this fictitious example. Imagine that you want to identify variables that can be used to predict the success of firefighters. To do this, you conduct a study with a group of 100 firefighters. For each participant, you obtain a *Firefighter Success Rating*, which indicates how successful this person has been as a firefighter. These ratings are on a scale of 1 to 100, with higher ratings indicating greater success.

To identify variables that might be useful in predicting these success ratings, each firefighter completes a number of paper-and-pencil tests. One of these is a Firefighter Knowledge Test. High scores on this test indicate that the participant possesses the knowledge needed to operate a fire hydrant, enter a burning building safely, and perform other tasks related to firefighting. A second test is a Verbal Ability Test. This test has nothing to do with firefighting—high scores indicate that the participant has a good vocabulary and other verbal skills.

The three variables in this study could be represented with the following symbols:

Y = Firefighter Success Rating (the response variable)

X_p = Firefighter Knowledge Test (the predictor variable of interest)

X_s = Verbal Ability Test (the suppressor variable).

Imagine that you perform some analyses to understand the nature of the relationship among these three variables. First, you compute the Pearson correlation coefficient between the Firefighter Knowledge Test and the Verbal Ability Test, and find that $r = 0.40$. This is a moderately strong correlation and makes sense because both firefighter knowledge and verbal ability are assessed by a paper-and-pencil testing method. To some extent, getting a high score on either of these tests requires that the participant be able to read instructions, read questions, read possible responses, and perform other verbal tasks. The results of the two tests are correlated because scores on both tests are influenced by the participant's verbal ability.

Next, you perform a series of regressions, in which Firefighter Success Ratings ($\hat{Y}$) is the response, to determine how much variance in this response is accounted for by various regression equations. This is what you learn:

- When the regression equation contains only the Verbal Ability Test, it accounts for 0% of the variance in Y.

- When the regression equation contains only the Firefighter Knowledge Test, it accounts for 20% of the variance in Y.

- When the regression equation contains both the Firefighter Knowledge Test and the Verbal Ability Test, it accounts for 25% of the variance in Y.

Finding that the Verbal Ability Test accounts for none of the variance in the Firefighter Success Rating makes sense, because firefighting (presumably) does not require a good vocabulary or other verbal skills.

The second finding, that the Firefighter Knowledge Test accounts for a respectable 20% of the variance in the Firefighter Success Rating, also makes sense, because it is reasonable to expect more knowledgeable firefighters to be rated as better firefighters.

However, you run into difficulty when trying to make sense of finding that the equation with both the Verbal Ability Test and the Firefighter Knowledge Test accounts for 25% of the variance in $\hat{Y}$. How is it possible that the combination of these two variables accounts for 25% of the variance in Y when one accounted for only 20% and the other accounted for 0%?

The answer is that, in this situation, the Verbal Ability Test is serving as a suppressor variable. It is suppressing irrelevant variance in scores on the Firefighter Knowledge Test, thus *purifying* the relationship between the Firefighter Knowledge Test and Y. Here is how it works. At least two factors influence scores on the Firefighter Knowledge Test—their actual knowledge about firefighting and their verbal ability (ability to read instructions). Obviously, the first of these two factors is relevant for predicting Y, whereas the second factor is not. Because scores on the Firefighter Knowledge Test are to some extent *contaminated* by the effects of the participants' verbal ability, the actual correlation between the Firefighter Knowledge Test and the Firefighter Success Rating is somewhat lower than it would be if you could somehow purify Firefighter Knowledge Test scores of this unwanted verbal factor. That is exactly what a suppressor variable does.

In most cases, a suppressor variable is given a negative regression weight in a multiple regression equation. These weights are discussed in more detail later in this chapter. Partly because of the negative weight, including the suppressor variable in the equation adjusts each participant's predicted score on Y so that it comes closer to that participant's actual score on Y. In the present case, this means that a participant who scores above the mean on the Verbal Ability Test will have a predicted score on Y adjusted downward to penalize for scoring high on this irrelevant predictor. Alternatively, a participant who scores below the mean on the Verbal Ability Test has a predicted score on Y that is adjusted upward. Another way of thinking about this is to say that a person applying to be a firefighter who has a high score on the Firefighter Knowledge Test but a low score on the Verbal Ability Test would be preferred over an applicant with a high score on the knowledge test and a high score on the verbal test because the second candidate's score on the knowledge test was probably inflated due to having good verbal skills.

The net effect of these corrections is improved accuracy in predicting Y. This is why you earlier found that R^2 is 0.25 for the equation that contains the suppressor variable, but only 0.20 for the equation that does not contain it.

The possible existence of suppressor variables has implications for multiple regression analyses. For example, when attempting to identify variables that would make good predictors in a multiple regression equation, it is clear that you should not base the selection exclusively on the bivariate (Pearson) correlations between the variables. For example, even if two predictor variables are moderately or strongly correlated, it does not necessarily mean that they are always providing redundant information. If one of them is a suppressor variable, then the two variables are not entirely redundant. That noted, predictor variables that are highly correlated with other variables can cause problems with the analysis because of this *collinearity*. See the description of collinearity in the final section of this chapter, "Assumptions Underlying Multiple Regression."

In the same way, a predictor variable should not be eliminated from consideration as a possible predictor just because it displays a low bivariate correlation with the response. This is because a suppressor variable may display a zero bivariate correlation with Y even though it could substantially increase R^2 if added to a multiple regression equation. When starting with a set of possible predictor variables, it is generally safer to begin the analysis with a multiple regression equation that contains all predictors, on the chance that one of them serves as a suppressor variable. Choosing an *optimal* subset of predictor variables from a larger set is complex and beyond the scope of this text.

To provide a sense of perspective, true suppressor variables are somewhat rare in social science research. In most cases, you can expect your data to behave according to the generalizations made in the preceding section when there are no suppressors. In most cases you will find that R^2 is larger to the extent that the X variables are more strongly correlated with Y and less strongly correlated with one another. To learn more about suppressor variables, see Pedhazur (1982).

The Uniqueness Index

A *uniqueness index* represents the percentage of variance in a response that is uniquely accounted for by a given predictor variable, above and beyond the variance accounted for by the other predictor variables in the equation. A uniqueness index is one measure of an X variable's importance as a predictor—the greater the amount of unique variance accounted for by a predictor, the greater its usefulness.

The Venn diagram in Figure 13.6 illustrates the uniqueness index for the predictor variable income. It is identical to Figure 13.5 with respect to the correlations between income and prosocial behavior and the correlations between the three X variables. However, in Figure 13.6 only the area that represents the uniqueness index for income is shaded. You can see that this area is consistent with the previous definitions, which stated that the uniqueness index for a given variable represents the percentage of variance in the response (prosocial behavior) that is accounted for by the predictor variable (income) over and above the variance accounted for by the other predictors in the equation (age and moral development).

Figure 13.6 Venn Diagram: Uniqueness Index for Income

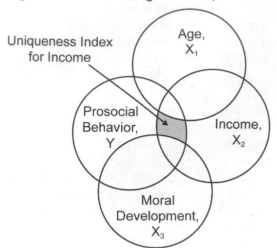

It is often useful to know the uniqueness index for each X variable in the multiple regression equation. These indices, along with other information, help explain the nature of the relationship between the response and the predictor variables. The Fit Model platform in JMP computes the uniqueness index and its significance level for each effect in the model.

Multiple Regression Coefficients

The *b* weights discussed earlier in this chapter referred to *nonstandardized* multiple regression coefficients. A *multiple regression coefficient* for a given X variable represents the average change in Y that is associated with a one-unit change in one X variable while holding constant the remaining X variables. *Holding constant* means that the multiple regression coefficient for a given predictor variable is an estimate of the average change in Y that would be associated with a one-unit change in that X variable if all participants had identical scores on the remaining X variables.

When conducting multiple regression analyses, researchers often want to determine which of the X variables are important predictors of the response. It is tempting to review the multiple regression coefficients estimated in the analysis and use these as indicators of importance. According to this logic, a regression coefficient represents the amount of *weight* that is given to a given X variable in the prediction of Y.

Nonstandardized Compared to Standardized Coefficients

However, caution must be exercised when using multiple regression coefficients in this way. Two types of multiple regression coefficients are produced in the course of an analysis—nonstandardized coefficients and standardized coefficients.

- *Nonstandardized multiple regression coefficients* (*b* weights) are the coefficients that are produced when the data analyzed are in raw score form. *Raw score* variables have not been standardized in any way. This means that different variables can have very different means and standard deviations. For example, the standard deviation for X_1 might be 1.35, while the standard deviation for X_2 might be 584.20.

  In general, it is not appropriate to use the relative size of nonstandardized regression coefficients to assess the relative importance of predictor variables, because the relative size of a nonstandardized coefficient for a given predictor variable is influenced by the size of that predictor's standard deviation. Variables with larger standard deviations tend to have smaller nonstandardized regression coefficients, while variables with smaller standard deviations tend to have larger regression coefficients. Therefore, the size of nonstandardized coefficients usually tells little about the importance of the predictor variables.

- Nonstandardized coefficients are often used to calculate participants' predicted scores on Y. For example, an earlier section presented a multiple regression equation for the prediction of prosocial behavior in which X_1 = age, X_2 = income, and X_3 = moral development. That equation is reproduced here.

$$\hat{Y} = (0.10)X_1 + (0.25)X_2 + (1.10)X_3 + (-3.25)$$

In this equation, the nonstandardized multiple regression coefficient for X_1 is 0.10, the coefficient for X_2 is 0.25, and so forth. If you have a participant's raw scores on the three predictor variables, these values can be inserted in the preceding formula to compute that participant's estimated score on Y. The resulting $\hat{Y}$ value is also in raw score form. It is an estimate of the number of prosocial acts you expect the participant to perform over a six-month period.

In summary, you should not refer to the nonstandardized regression coefficients to assess the relative importance of predictor variables. A better alternative (though still not perfect) is to refer to the standardized coefficients. *Standardized multiple regression coefficients* (called Beta weights in this book) are produced when the data are in a standard score form. Standard score form (or *z* score form) means that the variables have been standardized so that each has a mean of zero and a standard deviation of 1. This is important because all variables (Y variables and X variables alike) now have the same standard deviation (a standard deviation of 1). This means they are now measured on the same scale of magnitude. Variables no longer display large regression coefficients simply because they have small standard deviations. To some extent, the size of standardized regression coefficients does reflect the relative importance of the various predictor variables. These coefficients should be consulted to interpret the results of a multiple regression analysis.

For example, assume that the analysis of the prosocial behavior study produced the following multiple regression equation with standardized coefficients:

$$\hat{Y} = (0.70)X_1 + (0.20)X_2 + (0.20)X_3$$

In this equation, X_1 shows the largest standardized coefficient, which can be interpreted as evidence that it is a relatively important predictor variable, compared to X_2 and X_3.

You can see that the preceding equation, like all regression equations with standardized coefficients, does not contain an intercept constant. This is because the intercept is always equal to zero in a standardized equation. If a researcher presents a multiple regression equation in a research article but does not indicate whether it is a standardized or a nonstandardized equation, look for the intercept constant. If there is an intercept in the equation, it is almost certainly a nonstandardized equation. If there is no intercept, it is probably a standardized equation. Also, remember that the lowercase letter *b* is often used to represent nonstandardized regression coefficients, while standardized coefficients are referred to as Beta weights (sometimes represented by the capital letter *B* or the Greek letter β).

Interpretability of Multiple Regression Coefficients

The preceding section noted that standardized coefficients reflect the importance of predictors only to some extent.

Regression coefficients can be difficult to interpret because when multiple regression using the same variables is performed on data from more than one sample, different estimates of the multiple regression coefficients are often obtained in the different samples. This is the case for standardized as well as nonstandardized coefficients.

For example, assume that you recruit a sample of 50 participants, measure the variables discussed in the preceding section (prosocial behavior, age, income, moral development), and compute a multiple regression equation in which prosocial behavior is the response and the remaining variables are predictors (assume that age, income, and moral development are X_1, X_2, and X_3, respectively). With the analysis completed, it is possible that your output would reveal the following standardized regression coefficients for the three predictors:

$$\hat{Y} = (0.70)X_1 + (0.20)X_2 + (0.20)X_3$$

The relative size of the coefficients in the preceding equation suggests that X_1 (with a Beta weight of 0.70) is the most important predictor of Y, while X_2 and X_3 (each with Beta weights of 0.20) are much less important.

However, if you replicate your study with a different group of 50 participants, you would compute different Beta weights for the X variables. For example, you might obtain

$$\hat{Y} = (0.30)X_1 + (0.50)X_2 + (0.10)X_3$$

In the second equation, X_2 has emerged as the most important predictor of Y, followed by X_1 and X_3.

When the same study is performed on different samples, researchers usually obtain coefficients of different sizes. This means that the interpretation of these coefficients must always be made with caution. Specifically, multiple regression coefficients become increasingly unstable as the analysis is based on smaller samples and as the X variables become more correlated with one another. Unfortunately, much of the research carried out in the social sciences involves the use of small samples and correlated X variables. For this reason, the standardized regression coefficients (Beta weights) are only a few of the pieces of information to review when assessing the relative importance of predictor variables. The use of these coefficients should be supplemented with a review of the simple bivariate correlations between the X variables and Y and the uniqueness indices for the standardized X variables. The following sections show how to do this.

Example: A Test of the Investment Model

Fictitious data from a correlational study based on the investment model (Rusbult, 1980a, 1980b) illustrate the statistical procedures in this chapter. Recall that the investment model data were used to illustrate a number of other statistical procedures in previous chapters.

The investment model identifies a number of variables believed to predict a person's level of commitment to a romantic relationship (as well as to other types of relationships). *Commitment* refers to the individual's intention to maintain the relationship and remain with a current partner. One version of the investment model asserts that commitment is affected by the following four variables:

- Rewards: the number of *good things* that the participant associates with the relationship—the positive aspects of the relationship

- Costs: the number of *bad things* or hardships associated with the relationship

- Investment Size: the amount of time and personal resources that the participant has put into the relationship

- Alternative Value: the attractiveness of the participant's alternatives to the relationship—the attractiveness of alternative romantic partners.

The illustration in Figure 13.7 shows the hypothesized relationship between commitment and these four predictor variables.

Figure 13.7 One Version of the Investment Model

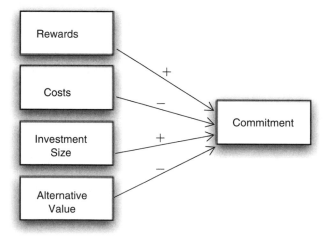

If the four predictor variables in Figure 13.7 do have a causal effect on commitment, it is reasonable to assume that they should also be useful for predicting commitment in a simple correlational study. Therefore, the present study is correlational in nature and uses multiple regression procedures to assess the nature of the predictive relationship between these variables. If the model survives this test, you might then wish to follow it up by performing a new study to test the proposed causal relationships between the variables—perhaps a true experiment. However, the present chapter focuses only on multiple regression procedures.

Overview of the Analysis

To better understand the nature of the relationship between the five variables presented in Figure 13.7, the data are analyzed using several JMP platforms.

First, the Multivariate platform computes Pearson correlations between variables. Simple bivariate statistics are useful for understanding the big picture.

Next, the Fit Model platform performs a multiple regression in which commitment is simultaneously regressed on the predictor variables. This provides a number of important pieces of information.

- First, you determine whether there is a significant relationship between commitment and the linear combination of predictors. You want to know if there is a significant relationship between commitment and the four variables taken as a group.

- You can also review the multiple regression coefficients for each of the predictors to determine which are statistically significant and which standardized coefficients are relatively large.

- Examine the uniqueness of each predictor to see which variables account for a significant amount of variation in the response when included in the regression equation.

- The Stepwise platform gives a multiple regression analysis for every possible combination of predictor variables. From these results you can determine the amount of variance in commitment accounted for any combination of predictor variables. These results are the difference in R^2 values between the full model and the model that is reduced by one or more predictors. Stepwise gives statistical tests that determine which of the uniqueness indices are significantly different from zero.

Gathering Data

Assume that you conduct a study of 48 college students who are currently involved in a romantic relationship. Each participant completes a 24-item questionnaire designed to assess the five investment model constructs—commitment, rewards, costs, investment size, and alternative value.

Each investment model construct is assessed with four questions. For example, the following four items assess the *commitment* construct:

---

21. How committed are you to your current relationship?

Not at All   1 2 3 4 5 6 7 8 9   Extremely Committed
Committed

22. How long do you intend to remain in your current relationship?

A Very Short Time   1 2 3 4 5 6 7 8 9   A Very Long Time

23. How often do you think about breaking up with your current partner?

Frequently   1 2 3 4 5 6 7 8 9   Never

24. How likely is it that you will maintain your current relationship for a long time?

Extremely Unlikely   1 2 3 4 5 6 7 8 9   Extremely Likely

Notice that a higher response number (such as 8 or 9) indicates a higher level of commitment, whereas a lower response number (such as 1 or 2) indicates a lower level of commitment. The sum of these four items for a participant gives a single score that reflects that participant's overall level of commitment. This cumulative score serves as the commitment variable in your analyses. Scores on this variable can range from 4 to 36, with higher values indicating higher levels of commitment.

Each of the remaining investment model constructs is assessed the same way. The sum of four survey items creates a single overall measure of the construct. Scores can range from 4 to 36, and higher scores indicate higher levels of the construct being assessed.

The JMP Data Table

The summarized (fictitious) data are in the JMP data table called **commitment regression.jmp**, shown in Figure 13.8. There are scores for variables commitment, reward, cost, investment, and alternative, for 48 participants. Assume the scores in the commitment regression table were obtained as follows:

- Forty-eight participants completed a 24-item questionnaire designed to evaluate commitment to an existing relationship. In particular, five sets of four items assessed five commitment constructs. The score for a construct is the sum of its four items.

- The sum of the items for five of the constructs, commitment, reward, cost, investment, and alternative, are the scores in the commitment regression data table for each of the 48 participants.

Figure 13.8 Partial Listing of the Commitment Data Table

	commitment	reward	cost	investment	alternative
1	34	25	13	25	12
2	32	27	14	32	13
3	34	24	21	30	14
4	38	23	18	27	10
5	4	25	10	11	34
6	31	30	22	31	13
7	22	31	18	27	14
8	32	27	9	31	8
9	33	31	11	28	13
10	36	26	14	19	12
11	19	22	23	23	4
12	36	32	14	34	4
13	30	30	20	30	23
14	35	22	20	24	19
15	36	33	7	31	4

Columns (5/0)
- commitment
- reward
- cost
- investment
- alternative

Rows
- All Rows — 48
- Selected — 0
- Excluded — 0
- Hidden — 0
- Labelled — 0

Computing Simple Statistics and Correlations

It is often appropriate to begin a multiple regression analysis by computing simple univariate statistics on the study's variables and all possible bivariate correlations between them. A review of these statistics and correlations can help you understand the big picture concerning relationships between the response variable and the predictor variables and among the predictor variables themselves.

Univariate Statistics

Use the Distribution platform to look at univariate statistics on the study variables. Chapter 4, "Exploring Data with the Distribution Platform," describes the Distribution platform in detail. To see simple statistics:

- Choose **Analyze → Distribution**.
- When the launch dialog appears, select all the variables in the variable selection list and click **Y, Columns** in the dialog to see them in the analysis list.
- Click **OK** in the launch dialog to see the Distribution results.

The platform results in Figure 13.9 show only the Quantiles and Moments tables. The Histograms and Outlier Box plots are not shown here, but offer a visual approach to identifying possible outliers in the data.

The distribution results are discussed in the section "Reviewing Simple Statistics and Correlation Results."

Figure 13.9 Univariate Statistics on Study Variables

Quantiles		commitment	reward	cost	investment	alternative
100.0%	maximum	36.000	33.000	26.000	34.000	34.000
99.5%		36.000	33.000	26.000	34.000	34.000
97.5%		36.000	33.000	25.775	34.000	33.775
90.0%		36.000	32.000	24.000	32.000	28.000
75.0%	quartile	35.000	30.000	21.000	30.000	20.750
50.0%	median	32.000	27.500	17.500	27.000	16.000
25.0%	quartile	22.500	24.000	14.000	20.500	12.000
10.0%		6.900	19.900	7.000	16.800	7.700
2.5%		4.000	10.575	5.000	11.000	4.000
0.5%		4.000	9.000	5.000	11.000	4.000
0.0%	minimum	4.000	9.000	5.000	11.000	4.000

Moments	commitment	reward	cost	investment	alternative
Mean	27.708333	26.645833	16.583333	25.333333	16.604167
Std Dev	10.195866	5.0507618	5.4959689	6.0821795	7.5052877
Std Err Mean	1.4716464	0.7290147	0.7932748	0.877887	1.083295
upper 95% Mean	30.668904	28.112422	18.179196	27.099414	18.783475
lower 95% Mean	24.747763	25.179245	14.98747	23.567252	14.424858
N	48	48	48	48	48

Bivariate Correlations

Chapter 5, "Measures of Bivariate Association," discussed using the Multivariate platform to compute correlations. This analysis proceeds the same way. With the commitment regression table active (the front-most table):

- Choose **Analyze → Multivariate Methods → Multivariate**.
- When the launch dialog appears, select all the variables in the variable selection list and click **Y, Columns** in the dialog to see them in the analysis list.

- Click **OK** in the launch dialog to see the initial Multivariate results.
- In the initial results window, close the Scatterplot matrix, and choose **Pairwise Correlations** from the menu on the Multivariate title bar to see the results shown in Figure 13.10.

Figure 13.10 Bivariate Correlations and Simple Statistics for All Study Variables

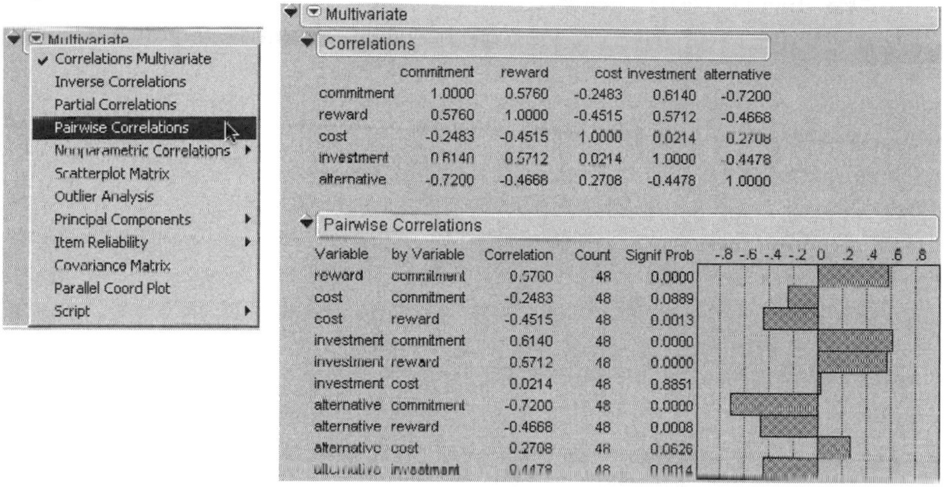

	commitment	reward	cost	investment	alternative
commitment	1.0000	0.5760	-0.2483	0.6140	-0.7200
reward	0.5760	1.0000	-0.4515	0.5712	-0.4668
cost	-0.2483	-0.4515	1.0000	0.0214	0.2708
investment	0.6140	0.5712	0.0214	1.0000	-0.4478
alternative	-0.7200	-0.4668	0.2708	-0.4478	1.0000

Variable	by Variable	Correlation	Count	Signif Prob
reward	commitment	0.5760	48	0.0000
cost	commitment	-0.2483	48	0.0889
cost	reward	-0.4515	48	0.0013
investment	commitment	0.6140	48	0.0000
investment	reward	0.5712	48	0.0000
investment	cost	0.0214	48	0.8851
alternative	commitment	-0.7200	48	0.0000
alternative	reward	-0.4668	48	0.0008
alternative	cost	0.2708	48	0.0626
alternative	investment	-0.4478	48	0.0014

Reviewing Simple Statistics and Correlation Results

1. **Make sure everything looks reasonable.** It is always important to review descriptive statistics to help verify that there are no obvious errors in a data table. The Distribution platform results in Figure 13.9 give tables of basic univariate statistics.

   The Moments table gives means, standard deviations, and other descriptive statistics for the five study variables, and the Quantiles table lists values for common percentiles. It is easy to verify that all of the figures are reasonable. In particular, check the Minimum and Maximum values for evidence of problems. The lowest score that a participant could possibly receive on any variable was 4. If any variable displays a Minimum score below 4, an error must have been made when entering the data. Similarly, no valid score in the Maximum column should exceed 36.

2. **Determine the size of the sample producing the correlations.** The N in the Moments table shows that all computations were completed for 48 observations. If any variable had a different number of observations associated with it (had missing values), then all simple statistics involving that variable are based on its reduced N value. Observations with missing values are not used for computations of any correlations in the Correlations table (Figure 13.10). These Pearson product-moment

correlations are based only on observations that have values for all variables. If there are missing values, the Pairwise Correlations table shows correlations based on all nonmissing values for each pair of variables, and indicates the N for each pairwise correlation. Because there were no missing values in this example, the values in the Correlations table and in the Pairwise Correlations table are the same.

3. **Review the correlation coefficients.** The correlations among the five variables appear in the 5 by 5 matrix at the top of the Multivariate results in Figure 13.10. In this matrix, where the row for one variable intersects with the column of a second variable, you find the cell that gives the correlation coefficient for these variables. The Pairwise Correlations table shows the probabilities associated with each correlation.

   For example, the correlation between commitment (the response) and reward is 0.576 The p value associated with this correlation, seen in the Pairwise Correlations table, is less than 0.0001, which means there is less than 1 chance in 10,000 of obtaining a coefficient this large by chance alone in another random sample of the same size. In other words, this correlation of 0.576 is statistically significant at the 0.0001 level.

Interpreting the Correlations

The correlations in the commitment column of the Correlations table tell you about the pattern of bivariate correlations between commitment and the four predictors. Notice that commitment has a positive correlation with reward and investment. As you expect, participants who report higher levels of rewards and investment size also report higher levels of commitment. However, commitment has negative relationships with cost and alternative. This is also reasonable because participants who report higher levels of costs and alternative value report lower levels of commitment. Each of these correlations is in the direction predicted by the investment model (see Figure 13.7).

Notice that some predictors are more strongly related to commitment than others. Specifically, alternative displays the strongest correlation at –0.72, followed by investment (0.61) and reward (0.58). These correlations are statistically significant at the 0.0001 level.

The correlation between commitment and cost is lower at approximately –0.25 and is not statistically significant ($p = 0.0889$). Based on these results, you could not reject the null hypothesis that commitment and cost are uncorrelated in the population.

The correlations in Figure 13.10 also give important information about the correlations among the four predictor variables. When using multiple regression, an ideal predictive situation is one in which each predictor variable displays a relatively strong correlation with the response while the predictor variables display relatively weak correlations among themselves, as discussed earlier.

With this in mind, the correlations presented in Figure 13.10 indicate that the association among variables in the current data table is less than ideal. It is true that three of the predictors display relatively strong correlations with commitment, but it is also true that most of the predictors display relatively strong correlations with each other. Notice that $r = -0.45$ for reward and cost, $r = 0.57$ between reward and investment, and $r = -0.47$ for reward and alternative. These correlations are moderately large, as is the correlation between investment and alternative. Moderate correlations sometimes result in nonsignificant multiple regression coefficients for at least some predictor variables, as seen in the analyses reported in the following section.

Estimating the Full Multiple Regression Equation

The Fit Model platform in JMP can perform multiple regression analysis. This procedure estimates multiple regression coefficients for the various predictors, calculates R^2, tests for significance, and shows additional information relevant to the analysis.

Using the Fit Model Platform

Chapters 8, 9, and 10 showed how to use the Fit Model platform for analysis of variance, analysis of factorial designs, and multivariate analysis of variance. The process is the same for doing multiple regression.

- Select **Analyze → Fit Model**.
- When the Fit Model dialog appears, select **commitment** as Y and the four predictors as effects, as shown in Figure 13.11.
- Click **Run Model** in the dialog to see the multiple regression analysis.

Note: It is important to remember that JMP uses the modeling type of the analysis variables to determine the correct analysis. When the response (Y) variable and all the predictors (X variables or effects) are continuous numeric variables, the Fit Model platform automatically selects a standard least squares multiple regression analysis unless you specify otherwise.

Figure 13.11 Fit Model Dialog for Multiple Regression

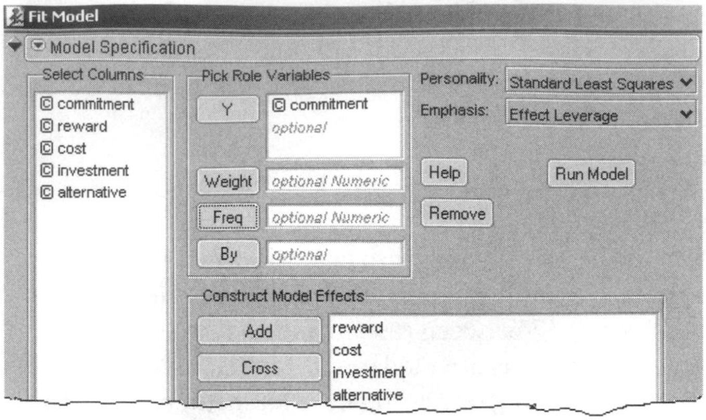

Reviewing the Multiple Regression Results

When you click **Run Model** in the Fit Model dialog, you see the results shown in Figure 13.12.

Figure 13.12 Multiple Regression Results

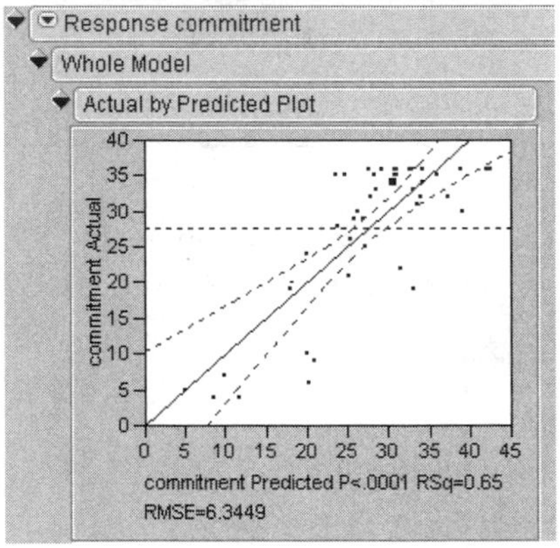

Summary of Fit

RSquare	0.645701
RSquare Adj	0.612743
Root Mean Square Error	6.344889
Mean of Response	27.70833
Observations (or Sum Wgts)	48

Analysis of Variance

Source	DF	Sum of Squares	Mean Square	F Ratio
Model	4	3154.8393	788.710	19.5916
Error	43	1731.0774	40.258	Prob > F
C. Total	47	4885.9167		<.0001

Use the following guidelines to review these results:

1. **Make sure that everything looks reasonable.** The title bar of the results shows that the response variable is commitment. The value for Observations in the Summary of Fit table shows that there were 48 participants. In the Analysis of Variance table, the heading DF (degrees of freedom) shows that the C. total (corrected total) is 47, which is equal to N – 1, where N is the total number of participants who provided usable data

for the analysis. There are four predictor variables so there should be 4 degrees of freedom for the model, as shown by the DF for Model in the Analysis of Variance table.

2. **Review the Whole Model results.** The initial results of the multiple regression are shown in Figure 13.12. The Actual by Predicted Plot and Analysis of Variance table tell you if the linear combination of predictor variables accounts for a significant amount of variation in the response, commitment.

Note that the Actual by Predicted leverage plot shows you at a glance that the multiple regression model is statistically significant. The horizontal reference line on the plot represents the condition that all effects are zero. The fitted line and its confidence lines cross the reference line (they do not encompass the reference line), which indicates a significant regression model.

To test the null hypothesis that all effects are zero, look at the line labeled Model in the Analysis of Variance table. You see an F value (F Ratio) of 19.59, and its associated probability (Prob > F) is less than 0.0001. Remember that the p value is the probability that you would obtain an F value this large or larger if the null hypothesis were true. In this case, the p value is very small (< 0.0001), so you reject the null hypothesis and conclude that the predictors, taken as a group, do account for a significant amount of variation in the response.

3. **Review the obtained value of R^2.** You see the R^2 value in two places— the Actual by Predicted plot in Figure 13.12 lists Rsq – 0.65, and the Summary of Fit table gives Rsquare as 0.645701. Recall that this R^2 value indicates the percentage of variance in the response variable accounted for by the linear combination of predictor variables. The R^2 of 0.6457 indicates that the linear combination of reward, cost, investment, and alternative accounts for (explains) about 65% of the variance in commitment. The significance test for Model, discussed above, also tests the hypothesis $R^2 = 0$.

The regression analysis determines whether the linear combination of predictor variables accounts for a significant amount of variance in the response. You also want to determine whether the predictors account for a large enough amount of variance to be considered meaningful. How large must an R^2 value be to be considered meaningful? That depends, in part, on what has been found in prior research concerning the response variable being investigated. If, for example, predictor variables in earlier investigations have routinely accounted for 50% of the variance in the response but the predictors in your study have accounted for only 10%, these new findings might not be viewed as being very important. On the other hand, if the predictors of earlier studies have routinely accounted for only 5% of the variance but the variables of your study have accounted for 10%, this might be considered a meaningful amount of variance.

This issue of *statistical significance* versus *percentage of variance accounted for* is important because it is possible to obtain an R^2 value that is very small (say, 0.03), but is still statistically significant. This occurs when analyzing data from very large samples.

Note: When assessing the importance of your findings, always review both the statistical significance of the model equation, and the total amount of variance accounted for by the predictors.

4. **Review the adjusted value of R^2.** The Summary of Fit table also lists the adjusted R^2, labeled Rsquare Adj. The adjusted R^2 is a ratio of mean squares instead of a ratio of sums of squares, and therefore uses degrees of freedom (the number of model parameters) in its calculations. Adjusting the R^2 of a model so that it is in units of the number of its parameters gives a meaningful way to compare the analyses of models with different numbers of parameters. That is, the adjusted R^2 is comparable across models. The unadjusted R^2 does not give this useful comparison.

5. **Review the intercept and nonstandardized regression coefficients.** The parameter estimates, in the Parameter Estimates table shown here, are the terms that constitute the multiple regression equation.

    - The first column of information, called Term, lists the names of the terms in the regression equation—the intercept and the four predictor variables (reward, cost, investment, and alternative).

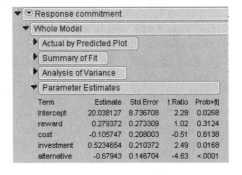

Term	Estimate	Std Error	t Ratio	Prob>|t|
Intercept	20.038127	8.736708	2.29	0.0268
reward	0.279372	0.273309	1.02	0.3124
cost	-0.105747	0.208003	-0.51	0.6138
investment	0.5234654	0.210372	2.49	0.0168
alternative	-0.67943	0.146704	-4.63	<.0001

- The column called Estimate lists the intercept estimate and the nonstandardized multiple regression coefficients for the predictors. The intercept is 20.038, the nonstandardized regression coefficient for reward is 0.279 and for cost is –0.106, and so forth. You can write the multiple regression equation in this way:

$$\hat{Y} = 0.279(\text{reward}) - 0.106(\text{cost}) + 0.523(\text{investment}) - 0.679(\text{alternative}) + 20.038$$

Remember that the multiple regression coefficient for a given predictor indicates the amount of change in Y associated with a one-unit change in that predictor while holding the remaining predictors constant. Nonstandardized coefficients represent the change observed when the variables are in raw score form (the different variables have different means and standard deviations). The nonstandardized regression equation is used to predict participants' scores on commitment so that the resulting scores will be on the same scale of magnitude as observed with the raw data. However, the coefficients in this equation cannot be used to assess the *relative* importance of the predictor variables.

6. **Review the leverage plots for the regression coefficients.** Researchers usually want to determine whether regression coefficients for the various predictor variables are significantly different from zero. A zero coefficient for a predictor variable indicates the variable has no effect on the response. A statistically significant coefficient suggests that the corresponding predictor variable is an important predictor of the response.

The Fit Model results include effect leverage plots, which tell you at a glance whether effects are significant. Figure 13.13 shows leverage plots for the four effects in the multiple regression analysis.

Leverage plots were discussed in Chapter 9, "Factorial ANOVA with Two Between-Subjects Factors." The leverage plot shows the line of fit, its 95% confidence curves, and the horizontal reference line that represents the null hypothesis. If the 95% curves cross the horizontal mean reference line, then the model fits better than the simple mean fit. If the curves do not cross the mean line—if they include or encompass the mean line—the model line fits the data no better than the mean line itself. Note that neither reward nor cost appears significant in the leverage plots, but investment and alternative are statistically significant.

Figure 13.13 Leverage Plots for Four Effects in the Multiple Regression Analysis

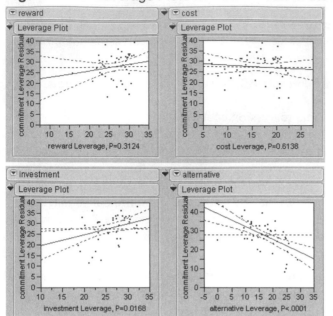

7. **Examine the significance tests for the coefficients.** The Parameter Estimates table gives details about the regression coefficients. The column called t Ratio lists a *t* test for the intercept and each predictor variable. These test the null hypothesis that the regression coefficient is equal to zero. The *p* value corresponding to each *t* statistic is in the column Prob>|t|.

For example, the nonstandardized regression coefficient estimate for reward is 0.279, its *t* test value is 1.02, and the corresponding *p* value is 0.3124. Because this *p* value is greater than 0.05, you cannot reject the null hypothesis and conclude that the regression coefficient for reward is not significantly different from zero. However, investment has a nonstandardized regression coefficient of 0.523. The *t* value for this coefficient is 2.49 with a corresponding *p* value of 0.0168. Because this value is less than 0.05, you reject the null hypothesis and tentatively conclude that the coefficient for Investment is significantly different from zero.

Parameter Estimates				
Term	Estimate	Std Error	t Ratio	Prob>|t|
Intercept	20.038127	8.736708	2.29	0.0268
reward	0.279372	0.273309	1.02	0.3124
cost	-0.105747	0.208003	-0.51	0.6138
investment	0.5234654	0.210372	2.49	0.0168
alternative	-0.67943	0.146704	-4.63	<.0001

Recall that the statistical significance of a regression coefficient suggests that variable's importance as a predictor. This statement includes a qualification to emphasize that caution must be used when interpreting statistical significance as evidence of a predictor's importance. There are at least two reasons for this:

- First, multiple regression coefficients can be unstable and difficult to interpret when the sample size is small. Small sample sizes often produce inflated standard errors and correlation between predictor variables.

- Second, a multiple regression coefficient can be statistically significant even when the standardized coefficient is relatively small in absolute magnitude and has little predictive value. This is more likely to happen when sample sizes are very large

For these reasons, the statistical significance of regression coefficients should be viewed as only one indicator of a variable's importance and should always be combined with additional information such as the size of the standardized regression coefficients and uniqueness indices.

8. **Review the standardized regression coefficients (Beta weights).** The values of nonstandardized regression coefficients often indicate little about the relative importance of predictor variables because the different predictors often have different scales of measurement. These differences affect the size of the nonstandardized coefficients. To avoid this difficulty review the standardized multiple regression coefficients (Beta weights). Beta weights are the regression coefficients that result when all the variables are standardized to have mean of zero and standard deviation of one. It is appropriate to review the Beta weights when you want to compare the relative importance of predictor variables.

To include standardized regression coefficients (Beta weights) in the results, right-click anywhere in the Parameter Estimates table to see the menu shown in Figure 13.14. When you choose **Std Beta** from this menu, the standardized parameter estimates then appear in the Parameter Estimates table. Note that the intercept for this equation is zero—this is always the case with a standardized regression equation:

- The Beta weight for reward is 0.138.
- The Beta weight for cost is –0.057.
- The Beta weight for investment is 0.312.
- The Beta weight for alternative is –0.500.

Based on these findings, you can tentatively rank the predictors from most to least influential Y predictors as alternative, investment, reward, and cost.

The Parameter Estimates table (Figure 13.14) does not report significance tests for the standardized regression coefficients because when the *t* test for the nonstandardized coefficient is significant, then the corresponding standardized coefficient for that variable is also significantly different from zero.

A more cautious approach to understanding the relative importance of predictor variables involves combining information from a variety of sources, including bivariate correlations, standardized regression coefficients, and the uniqueness indices.

Figure 13.14 Parameter Estimates Table with Standardized Beta

Term	Estimate	Std Error	t Ratio	Prob>|t|	Std Beta
Intercept	20.038127	8.736708	2.29	0.0268	0
reward	0.279372	0.273309	1.02	0.3124	0.138393
cost	-0.105747	0.208003	-0.51	0.6138	-0.057
investment	0.5234654	0.210372	2.49	0.0168	0.312265
alternative	-0.67943	0.146704	-4.63	<.0001	-0.50014

Right-click on Parameter Estimates table to see context menu

Uniqueness Indices for the Predictors

This section covers how to determine which predictors are most important in a multiple regression analysis.

The multiple regression analysis shown in the previous section also gives statistics that tell which effects are significant in the multiple regression equation. The Effect Tests table, shown in Figure 13.15, gives the *F* statistic for testing that each individual effect is zero. The *F* ratios, shown in the Effect Tests table, test the uniqueness index for each predictor.

The *F* value is the ratio of the mean square for the effect and the mean square for error in the ANOVA. The effect's mean square is its sum of squares divided by its degrees of freedom.

You can verify this by looking at the quantities in the tables shown in Figure 13.15. For example, the test for the alternative effect shows it is a significant predictor, with $F = 21.4489$ and $p < 0.0001$.

That is,

$$F = \frac{SS_{\text{alternative}} \div df_{\text{alternative}}}{\text{Error Mean Square}} = \frac{863.48323 \div 1}{40.25761} = 21.4489$$

Note: To adjust the number of decimal places in any column of numbers showing in a report, double-click on the column of numbers and complete the Column Numeric Format dialog that appears.

This F ratio, showing in the Effect Tests table, tests the uniqueness index for alternative.

Figure 13.15 Effect Tests Table for Multiple Regression Analysis

Response commitment

Summary of Fit

Analysis of Variance

Source	DF	Sum of Squares	Mean Square	F Ratio
Model	4	3154.8393	788.7098	19.5916
Error	43	1731.0774	40.25761	Prob > F
C. Total	47	4885.9167		<.0001

Parameter Estimates

Effect Tests

Source	Nparm	DF	Sum of Squares	F Ratio	Prob > F
reward	1	1	42.00359	1.0440	0.3124
cost	1	1	10.40510	0.2585	0.6138
investment	1	1	249.25704	6.1916	0.0168
alternative	1	1	863.48323	21.4489	<.0001

Summarizing the Results

There are a number of different ways to summarize the results of a multiple regression analysis in a paper. The following summarized results use table formats that are fairly representative of those appearing in journals of the American Psychological Association (APA, 2003).

Correlation Results from the Multivariate Platform

It is usually desirable to present simple descriptive statistics, such as means and standard deviations, along with the correlation matrix. If reliability estimates (such as coefficient alpha estimates) are available for the predictors, they should be included on the diagonal of the correlation matrix, in parentheses. The results for the present fictitious study are summarized in Table 13.8.

Table 13.8 Means, Standard Deviations, Correlations, and Alpha Reliability Estimates

Variable	Mean	Std Dev	1	2	3	4	5
			Correlations				
1 Commitment	27.71	10.20	(0.84)				
2 Rewards	26.65	5.05	0.58**	(0.75)			
3 Costs	16.58	5.50	−0.25	−0.45*	(0.72)		
4 Investment Size	25.33	6.08	0.61**	0.57**	0.02	(0.83)	
5 Alternative Value	16.60	7.50	−0.72**	−0.47**	0.27	−0.45*	(0.91)

Note: N = 48. *$p < 0.01$. **$p < 0.001$. Reliability estimates appear on the diagonal above correlation coefficients.

Regression Results from the Fit Model Platform

Depending on the nature of the research problem, it is sometimes feasible to report standardized regression coefficients (Beta weights, labeled Std Beta in the Parameter Estimates table) and the uniqueness indices in a single table, as in Table 13.9.

Table 13.9 Nonstandardized Estimates (*b* Weights), Beta Weights, and Uniqueness Indices from Multiple Regression Analyses

Predictor	*b* Weight	SE *b* Weight	Std Beta	*t*	Uniqueness Index
1 Rewards	0.28	0.27	0.14	1.02	0.009
2 Costs	−0.11	0.21	−0.06	−0.51	0.002
3 Investment size	0.52	0.21	0.31	2.49*	0.051*
4 Alternative value	−0.68	0.15	−0.50	−4.64**	0.177**

Note: $p < 0.05$ **. $p < 0.001$. $R^2 = 0.65$, $F(4, 43) = 19.59$, $p < 0.01$ for predictor variables.

Getting the Big Picture

It is instructive to reflect on the big picture of your findings before you proceed to summarize them in text form. First, notice the bivariate correlations from Table 13.8. The correlations between commitment and rewards, between commitment and investment size, and between commitment and alternative value are all significant and in the predicted direction. Only the correlation between commitment and costs is nonsignificant. These findings provide partial support for the investment model.

A somewhat similar pattern of results can be seen in Table 13.9, which shows that the Beta weights and uniqueness indices for investment size and alternative value are both significant and in the predicted direction. However, unlike correlations, Table 13.9 shows that neither the Beta weight nor the uniqueness index for rewards is statistically significant. This might come as a surprise because the correlation between commitment and rewards was moderately strong at 0.58. With such a strong correlation, how could the multiple regression coefficient and uniqueness index for rewards be nonsignificant?

A possible answer may be found in the correlations of Table 13.8. Notice that the correlation between commitment and rewards is somewhat weaker than the correlation between commitment and either investment size or alternative value. In addition, it can be seen that the correlations between rewards and both investment size and alternative value are fairly substantial at $r = 0.57$ and $r = -0.47$, respectively. In short, amount of reward shares a great deal of variance in common with investment size and alternative value and is a poorer predictor of commitment. In this situation, it is unlikely that a multiple regression equation that already contains investment size and alternative value would need a variable like rewards to improve the accuracy of prediction. In other words, any variance in commitment that is accounted for by rewards has probably already been accounted for by Investment size and alternative value. As a result, the rewards variable is redundant and consequently displays a nonsignificant Beta weight and uniqueness index.

Was this a real test of the investment model? It must be emphasized again that the results of the investment model presented here are entirely fictitious and should not be viewed as legitimate tests of that conceptual framework. Most published studies of the investment model have been very supportive of its predictions. For representative examples of this research, interested readers are referred to Rusbult (1980a, 1980b), Rusbult and Farrell (1983), and Rusbult, Johnson, and Morrow (1986).

Formal Description of Results for a Paper

There are a number of ways to summarize the results of these analyses within the text of the paper. The amount of detail in the description should be dictated by the statistical sophistication of your audience. If the audience is likely to be familiar with the use of multiple regression, then less detail is needed.

The following format is fairly typical:

Results were analyzed using both bivariate correlation and multiple regression. Means, standard deviations, and Pearson correlations appear in Table 13.8. The bivariate correlations reveal three predictor variables that were significantly related to commitment: rewards ($r = 0.58$); investment size ($r = 0.61$); and alternative value ($r = -0.72$). All correlations are significant at $p < 0.001$, and all are in the expected direction. The correlation between costs and commitment is nonsignificant, at $\alpha = 0.05$, with $r = -0.25$.

Using multiple regression, commitment scores were then regressed on the linear combination of rewards, costs, investment size, and alternative value. The equation containing these four variables accounted for 65% of observed variance in commitment, $F(4, 43) = 19.60$, $p < 0.001$, adjusted $R^2 = 0.61$.

Beta weights (standardized multiple regression coefficients) and uniqueness indices were subsequently reviewed to assess the relative importance of the four variables in the prediction of commitment. The uniqueness index for a given predictor is the percentage of variance in the response accounted for by that predictor, beyond the variance accounted for by the other predictor variables. Beta weights and uniqueness indices are presented in Table 13.9.

Table 13.9 shows that only investment size and alternative value display significant Beta weights. Alternative value demonstrates a somewhat larger Beta weight at -0.50 ($p < 0.001$), while the Beta weight for investment size is 0.31 ($p < 0.05$). Both coefficients are in the predicted direction.

Findings regarding uniqueness indices correspond to those for Beta weights in that only investment size and alternative value display significant indices. Alternative value accounted for approximately 18% of the variance in commitment, beyond the variance accounted for by the other three predictors, $F(1, 43) = 21.51$, $p < 0.001$. In contrast, investment size accounts for only 5% of the unique variance in commitment, $F(1, 43) = 6.20$, $p < 0.05$.

Summary

This chapter provides an elementary introduction to multiple regression, one of the most flexible and powerful research tools in the social sciences. It discusses only the situation in which a response variable is being predicted from continuous predictor variables, all of which display a linear relationship with the response. At this time, you are ready to move on to a more comprehensive treatment of the topic that deals with curvilinear relationships, interactions between predictors, dummy coding, effect coding, and regression equations that contain both continuous and categorical predictor variables. Cohen, Cohen, West, and Aiken (2003) and Pedhazur (1982) provide authoritative treatments of these and other advanced topics in multiple regression. Also see Freund, Littell, and Creighton (2003) and Sall, Creighton, and Lehman (2004).

Assumptions Underlying Multiple Regression

Level of measurement

For multiple regression, both the predictor variables and the response variable should be numeric continuous variables.

Random sampling

Each participant in the sample contributes one score on each predictor variable and one score on the response variable. These sets of scores represent a random sample drawn from the population of interest.

Normal distribution of the response variable

For any combination of values on the predictor variables, the response variable should be normally distributed.

Homogeneity of variance

For any combination of values of the predictor variables, the response variable should demonstrate a constant variance.

Independent observations

A given observation should not be affected by (or related to) any other observation in the sample. For example, this assumption would be violated if the various observations represented repeated measurements taken from a single participant. It would also be violated if the study included multiple participants, some of whom contributed more that one observation to the data set (some participants contributed more than one set of scores on the response variable and predictor variables).

Linearity

The relationship between the response variable and each predictor variable should be linear. This means that the mean response scores at each value of a given predictor should fall on a straight line.

Errors of prediction

The errors of prediction should be normally distributed and the distribution of errors should be centered at zero. Error of prediction associated with a given observation should not be correlated with those associated with the other observations. Errors of prediction should demonstrate a consistent variance. Errors of prediction should not be correlated with the predictor variables.

Absence of measurement error

The predictor variables should be measured without error. Pronounced violations of this assumption lead to underestimation of the regression coefficient for the corresponding predictor.

Absence of specification errors

The term *specification error* generally refers to situations in which the model represented by the regression equation is not theoretically tenable. In multiple regression, specification errors most frequently result from omitting relevant predictor variables from the equation or including irrelevant predictor variables in the equation. Specification errors also result when researchers posit a linear relationship between variables that are actually involved in a nonlinear relationship.

It is infrequent that all of these assumptions will be fully satisfied in applied research. Fortunately, regression analysis is generally robust against minor violations of most of these assumptions. However, it is less robust against violations of the assumptions involving independent observations, measurement error, or specification errors (Pedhazur, 1982).

In addition to considering the preceding assumptions, researchers are also advised to inspect their data for possible problems involving outliers or *collinearity*.

- An outlier is an unusual observation that does not fit the regression model well. Outliers are often the result of mistakes made when entering data, and can profoundly bias parameter estimates (such as regression coefficients). The Outlier analysis in the Multivariate platform could be used to detect possible outliers.

- As previously noted, *collinearity* exists when two or more predictor variables demonstrate a high degree of correlation, such as $r > 0.90$, with one another. Collinearity can cause regression coefficient estimates to fail to demonstrate statistical significance, be biased, or even demonstrate the incorrect sign. Cohen, Cohen, West, and Aiken (2003) discuss these problems, and show how to detect outliers, collinearity, and other problems sometimes encountered in regression analysis.

References

American Psychological Association. 2001. *Publication Manual of the American Psychological Association.* 5th ed. Washington: American Psychological Association.

Cohen, J. 1992. "A Power Primer." *Psychological Bulletin* 112:155–159.

Cohen, J., P. Cohen, S. G. West, and L. S. Aiken. 2003. *Applied Multiple Regression/ Correlation Analysis for the Behavioral Sciences.* 3d ed. Hillsdale, NJ: Lawrence Erlbaum Associates.

Freund R., R. Littell, and L. Creighton. 2003. *Regression Using JMP.* Cary, NC: SAS Institute Inc.

Pedhazur, E. J. 1982. *Multiple Regression in Behavioral Research.* 2d ed. New York: Holt, Rinehart, and Winston.

Rusbult, C. E. 1980a. "Commitment and Satisfaction in Romantic Associations: A Test of the Investment Model." *Journal of Experimental Social Psychology* 16:172–186.

Rusbult, C. E. 1980b. "Satisfaction and Commitment in Friendships." *Representative Research in Social Psychology* 11:96 -105.

Rusbult, C. E., and D. Farrell. 1983. "A Longitudinal Test of the Investment Model: The Impact on Job Satisfaction, Job Commitment, and Turnover of Variations in Rewards, Costs, Alternatives, and Investments." *Journal of Applied Psychology* 68:429 -438.

Rusbult, C. E., D. J. Johnson, and G. D. Morrow. 1986. "Predicting Satisfaction and Commitment in Adult Romantic Involvements: An Assessment of the Generalizability of the Investment Model." *Social Psychology Quarterly* 49:81–89.

Sall, J., L. Creighton, and A. Lehman. 2004. *JMP Start Statistics.* Cary, NC: SAS Institute Inc.

Principal Component Analysis

> **Overview.** This chapter provides an introduction to principal component analysis, a variable-reduction procedure similar to factor analysis. It provides guidelines regarding the necessary sample size and number of items per component. It shows how to determine the number of components to retain, interpret the rotated solution, create factor scores, and summarize the results. Fictitious data are analyzed to illustrate these procedures. The present chapter deals only with the creation of orthogonal (uncorrelated) components.

Introduction to Principal Component Analysis

Principal component analysis is an appropriate procedure when you have measures on a number of observed variables and want to develop a smaller number of variables (called principal components) that account for most of the variance in the observed variables. The principal components can then be used as predictors or response variables in subsequent analyses.

A Variable Reduction Procedure

Principal component analysis is a variable reduction procedure. It is useful when you obtain data for a number of variables (possibly a large number of variables) and believe that there is redundancy among those variables. In this case, redundancy means there is correlation among subsets of variables. Because of this redundancy, you believe it should be possible to reduce the observed variables into a smaller number of principal components that will account for most of the variance in the observed variables.

Because principal component analysis is a variable reduction procedure, it is similar to exploratory factor analysis. In fact, the steps followed in a principal component analysis are identical to those followed when conducting an exploratory factor analysis. However, there are significant conceptual differences between the two procedures. It is important not to claim you are performing factor analysis when you are actually performing principal component analysis. The section in this chapter called "Principal Component Analysis Is *Not* Factor Analysis" describes the differences between these two procedures.

An Illustration of Variable Redundancy

The following fictitious research example illustrates the concept of variable redundancy. Imagine that you develop a seven-item questionnaire like the one shown in Table 14.1, designed to measure job satisfaction. You administer this questionnaire to 200 employees and use their responses to the seven items as seven separate variables in subsequent analyses.

There are a number of problems with conducting the study in this manner. One of the more important problems involves the concept of redundancy, mentioned previously. Examine the content of the seven items in the questionnaire closely. Notice that items 1 to 4 deal with the employees' satisfaction with their supervisors. In this way, items 1 to 4 are somewhat redundant. Similarly, notice that items 5 to 7 appear to deal with the employees' satisfaction with their pay.

Table 14.1 Questionnaire to Measure Job Satisfaction

```
Please respond to each of the following statements by placing a rating in
the space to the left of the statement. In making your ratings, use any
number from 1 to 7, where 1 = "strongly disagree" and 7 = "strongly
agree."

_____ 1. My supervisor(s) treats me with consideration.

_____ 2. My supervisor(s) consults with me concerning important
            decisions that affect my work.

_____ 3. My supervisor(s) gives me recognition when I do a good job.

_____ 4. My supervisor(s) gives me the support I need to do my job well.

_____ 5. My pay is fair.

_____ 6. My pay is appropriate, given the amount of responsibility that
            comes with my job.

_____ 7. My pay is comparable to the pay earned by other employees whose
            jobs are similar to mine.
```

Empirical findings might further support the notion that there is redundancy among items. Assume that you administer the questionnaire to 200 employees and compute all possible correlations between responses to the seven items. Table 14.2 shows correlations from this fictitious data.

When correlations among several variables are computed, they are typically summarized in the form of a *correlation matrix* such as the one presented in Table 14.2. The rows and columns of Table 14.2 correspond to the seven variables included in the analysis. Row 1 and column 1 represent variable 1, row 2 and column 2 represents variable 2, and so forth. The correlation for a pair of variables appears where a given row and column intersect. For example, where the row for variable 2 intersects with the column for variable 1, you find a correlation of 0.75. This means that the correlation between variable 1 and variable 2 is 0.75.

Table 14.2 Correlations among Seven Job Satisfaction Items for 200 Responses

	Correlations						
Variable	1	2	3	4	5	6	7
1	*1.00*						
2	0.75	*1.00*					
3	0.83	0.82	*1.00*				
4	0.68	0.92	0.88	*1.00*			
5	0.03	0.01	0.04	0.01	*1.00*		
6	0.05	0.02	0.05	0.07	0.89	*1.00*	
7	0.02	0.06	0.00	0.03	0.91	0.76	*1.00*

The correlations of Table 14.2 show that the seven items seem to form two distinct groups. First, notice that items 1 to 4 show relatively strong correlations with one another. This could be because items 1 to 4 measure the same construct. Similarly, items 5 to 7 correlate strongly with one another, a possible indication that they also measure the same construct. Also notice that items 1 to 4 show very weak correlations with items 5 to 7. This is what you expect to see if items 1 to 4 and items 5 to 7 are measuring two different constructs.

Given this apparent redundancy, it is likely that the seven items on the questionnaire do not measure seven different constructs. More likely, items 1 to 4 measure a single construct that could reasonably be labeled "satisfaction with supervision," whereas items 5 to 7 measure a different construct that could be labeled "satisfaction with pay."

If responses to the seven items actually displayed the redundancy suggested by the pattern of correlations in Table 14.2, it would be advantageous to reduce the number of variables so that items 1 to 4 collapse into a single new variable and items 5 to 7 collapse into a second new variable. You could then use these two new variables (instead of the seven original variables) as predictor variables in a multiple regression or any other type of analysis.

Number of Components Extracted

The number of components extracted (created) in a principal component analysis is equal to the number of observed variables being analyzed. This means that an analysis of the seven-item questionnaire can result in seven components.

However, in most analyses only the first few components account for meaningful amounts of variance so only these first few components are interpreted and used in a subsequent analyses such as a multiple regression. For example, in the analysis of the seven-item job satisfaction questionnaire, it is likely that only the first two components will reflect the redundancy in the data and account for a meaningful amount of variance. Therefore only these would be retained for use in subsequent analysis. You assume that the remaining five components account for only trivial amounts of variance. These latter components are not retained, interpreted, or further analyzed.

Principal component analysis is a mathematical way of collapsing variables. It allows you to reduce a set of observed variables into a set of variables called principal components. A subset of those principal components can then be used in subsequent analyses.

What Is a Principal Component?

Technically, a *principal component* can be defined as a linear combination of optimally weighted observed variables. In order to understand the meaning of this definition, it is necessary to first describe how participant scores on a principal component are computed.

How Principal Components Are Computed

In the course of performing a principal component analysis, it is possible to calculate a score for each participant on a given principal component. Participants' actual scores on the seven questionnaire items are optimally weighted and then summed to compute their scores on a given component.

Following is the general form for the formula to compute scores on the first component extracted in a principal component analysis:

$$C_1 = b_{11}(X_1) + b_{12}(X_2) + \ldots b_{1p}(X_p)$$

where

C_1 = the participant's score on principal component 1 (the first component extracted)

b_{1p} = the regression coefficient (or weight) for observed variable p, as used in creating principal component 1

X_p = the participant's score on observed variable p.

For example, you can determine each participant's score on the first principal component by using the following fictitious formula:

$$C_1 = 0.44(X_1) + 0.40(X_2) + 0.47(X_3) + 0.32(X_4)$$
$$+ 0.02(X_5) + 0.01(X_6) + 0.03(X_7)$$

In this case, the observed variables (the X variables) are responses to the seven job satisfaction questions: X_1 represents question 1, X_2 represents question 2, and so forth. Notice that different regression coefficients are assigned to the different questions to scores on C_1 (principal component 1). Questions 1 to 4 have larger regression weights that range from 0.32 to 0.44, whereas questions 5 to 7 have smaller weights ranging from 0.01 to 0.03.

A different equation with different weights computes scores on component 2. Here is a fictitious illustration of this formula:

$$C_2 = 0.01(X_1) + 0.04(X_2) + .02(X_3) + 0.02(X_4)$$
$$+ 0.48(X_5) + 0.31(X_6) + 0.39(X_7)$$

This equation shows that, in creating scores on the second component, items 5 to 7 have higher weights and items 1 to 4 have lower weights.

Notice how the weighting for component 1 and component 2 reflect the questionnaire design and variable redundancy shown in the correlation table (Table 14.2):

- Component 1 accounts for the most of the variability in the "satisfaction with supervision" items. Satisfaction with supervision is assessed by questions 1 to 4. You expect that items 1 to 4 have more weight in computing participant scores on this component and items 5 to 7 have less weight. Component 1 will be strongly correlated with items 1 to 4.

- Component 2 accounts for much of the variability in the "satisfaction with pay" items. You expect that items 5 to 7 have more weight in computing participant scores on this component and items 1 to 4 have less weight. Component 2 will be strongly correlated with items 5 to 7.

Optimal Weights for Principal Components

At this point, it is reasonable to wonder how the regression weights from the preceding equations are determined. The JMP Multivariate platform generates these weights by computing eigenvalues. An eigenvalue is the amount of variance captured by a given component or factor. The eigenvalues are optimal weights in that, for a given set of data, no other weights produce a set of components that are more effective in accounting for variance among observed variables. The weights satisfy a principle of least squares similar (but not identical) to the principle of least squares used in multiple regression (see Chapter 13, "Multiple Regression"). Later, this chapter shows how the Multivariate platform in JMP can be used to extract (create) principal components

It is now possible to understand the definition provided at the beginning of this section more fully. A principal component is defined as a linear combination of optimally weighted observed variables. The term *linear combination* means that the component score is created by adding together weighted scores on the observed variables being analyzed. *Optimally weighted* means that the observed variables are weighted in such a way that the resulting components account for a maximum amount of observed variance in the data. No other set of weights can account for more variation.

Characteristics of Principal Components

The first principal component extracted in a principal component analysis accounts for a maximal amount of total variance in the observed variables. Under typical conditions, this means that the first component is correlated with at least some of the observed variables. In fact, it is often correlated with many of the variables.

The second principal component extracted has two important characteristics.

- The second component accounts for a maximal amount of variance in the data not accounted for by the first component. Under typical conditions, this means that the second component is correlated with some of the observed variables that did not display strong correlations with the first component.

- The second characteristic of the second component is that it is *uncorrelated* with the first component. If you compute the correlation between component 1 and component 2, that correlation is zero.

The remaining components extracted in the analysis display these same two characteristics—each component accounts for a maximal amount of variance in the observed variables that was not accounted for by the preceding components and is uncorrelated with all of the preceding components. A principal component analysis proceeds in this manner with each new component accounting for progressively smaller amounts of variance. This is why only the first few components are retained and interpreted. When the analysis is complete, the resulting components display varying degrees of correlation with the observed variables, but are completely uncorrelated with one another.

What Is the Total Variance in the Data?

To understand the meaning of *total variance* as it is used in a principal component analysis, remember that the observed variables are standardized in the course of the analysis. This means that each variable is transformed so that it has a mean of zero and a standard deviation of one (and hence a variance of one). The total variance in the data is the sum of the variances of these observed variables. Because they have been standardized to have a standard deviation of one, each observed variable contributes one unit of variance to the total variance in the data. Therefore, the total variance in a principal component analysis always equals the number of observed variables analyzed. For example, if seven variables are analyzed, the total variance is seven.

> **Note:** The total number of components extracted in the principal component analysis partitions the total variance. That is, the sum of the variance accounted for by all principal components is the total variation in the data.

Principal Component Analysis Is *Not* Factor Analysis

Principal component analysis is often confused with factor analysis. There are many important similarities between the two procedures. Both methods attempt to identify groups of observed variables. Both procedures can be performed with the Multivariate platform and often provide similar results.

The purpose of both common factor analysis and principal component analysis is to reduce the original variables into fewer composite variables, called factors or principal components. However, these two analyses have different underlying assumptions about the variance in the original variables, and the obtained composite variables serve different purposes.

In principal component analysis, the objective is to account for the maximum portion of the variance present in the original set of variables with a minimum number of composite variables called principal components. The assumption is that the error (unique) variance represents a small portion of the total variance in the original set of the variables. Principal components analysis does not make a distinction between common (shared) and unique parts of the variation in a variable.

In factor analysis, a small number of factors are extracted to account for the correlations among the observed variables—to identify the latent dimensions that explain why the variables are correlated with each other. The observed variables are only indicators of the latent constructs to be measured. The assumption is that the error (unique) variance represents a significant portion of the total variance and that the variation in the observed variables is due to the presence of one or more latent variables (factors) that exert causal influence on these observed variables. An example of such a causal structure is presented in Figure 14.1.

The ovals in Figure 14.1 represent the latent (unmeasured) factors of "satisfaction with supervision" and "satisfaction with pay." These factors are latent in the sense that they are assumed to actually exist in the employees' belief systems but are difficult to measure directly. However, they do exert an influence on the employees' responses to the seven items in the job satisfaction questionnaire in Table 14.1. These seven items are represented as the squares labeled V1 to V7 in the figure. It can be seen that the "supervisor" factor exerts influence on items V1 to V4 and the "pay" factor exerts influence on items V5 to V7.

Figure 14.1 Example of the Underlying Causal Structure
Assumed in Factor Analysis

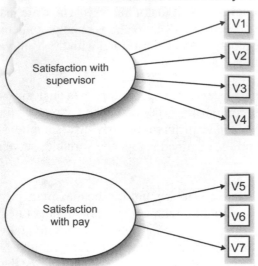

Researchers use factor analysis when they believe latent factors exist that exert causal influence on the observed variables they are studying. Exploratory factor analysis helps the researcher identify the number and nature of these latent factors.

In contrast, principal component analysis makes no assumption about an underlying causal structure. Principal component analysis is a variable reduction procedure that can result in a relatively small number of components that account for most of the variance in a set of observed variables.

In summary:

- Factor analysis groups variables under the assumption that the groups represent latent constructs. Factor analysis attempts to identify these constructs but is not expected to extract all variability from the data. A factor analysis only uses the variability in an item that it has in common with the other items (its communality).

- Principal component analysis produces a set of orthogonal (uncorrelated) linear combinations of the variables. The first principal component accounts for the most possible variance in the data. The second component accounts for the most variance not accounted for by the first component, and so on until all variability is accounted for. The first few components account for most of the total variation in the data, and can be used for subsequent analysis.

Both factor analysis and principal component analysis have important roles to play in social science research, but their conceptual foundations are quite distinct. In most cases, these two methods yield similar results. However, principal components analysis is often preferred as a method of data reduction, while principal factors analysis is often preferred when the goal of the analysis is to detect structure.

The Prosocial Orientation Inventory

Assume that you have developed an instrument called the Prosocial Orientation Inventory (POI) that assesses the extent to which a person has engaged in types of "helping behavior" over the preceding six-month period. The instrument contains six items and is presented in Table 14.3.

When this instrument was first developed, you intended to administer it to a sample of participants and use their responses to the six items as separate predictor variables in a multiple regression equation. As previously stated, you learned that this is a questionable practice and decide instead to perform a principal component analysis on responses to the six items to see if a smaller number of components can successfully account for most of the variance in the data. If this is the case, you will use the resulting components as the predictor variables in your regression analyses.

Table 14.3 The Prosocial Orientation Inventory (POI)

```
Instructions: Below are a number of activities in which people sometimes
engage. For each item, please indicate how frequently you have engaged in
this activity during the past six months. Make your rating by circling the
appropriate number to the left of the item using the following response
format:
    7 = Very Frequently
    6 = Frequently
    5 = Somewhat Frequently
    4 = Occasionally
    3 = Seldom
    2 = Almost Never
    1 = Never

1 2 3 4 5 6 7      1. Went out of my way to do a favor for a coworker.
1 2 3 4 5 6 7      2. Went out of my way to do a favor for a relative.
1 2 3 4 5 6 7      3. Went out of my way to do a favor for a friend.
1 2 3 4 5 6 7      4. Gave money to a religious charity.
1 2 3 4 5 6 7      5. Gave money to a charity not associated with a religion.
1 2 3 4 5 6 7      6. Gave money to a panhandler.
```

At this point, it may be instructive to review the content of the six items that constitute the prosocial orientation inventory (POI) to make an informed guess as to what you are likely to observe from the principal component analysis. Imagine that, when you first constructed the instrument, you assumed that the six items were assessing six different types of prosocial behavior. However, inspection of items 1 to 3 shows that these three items appear to have something in common—they deal with "going out of one's way to do a favor for someone else." It would not be surprising to learn that these three items are highly correlated and group together in the principal component analysis to be performed. In the same way, a review of items 4 to 6 shows that all of these items involve "giving money to those in need." These three items might also be correlated and form the basis of a second principal component.

In summary, the nature of the items suggests that it may be possible to account for the variance in the POI with just two components that will be a "helping others" component and a "financial giving" component. At this point, this is only speculation. Only a formal analysis can determine the number of components measured by the POI.

> **Note:** Keep in mind that the principal component analysis does not assume the underlying constructs described above ("helping others" and "financial giving"). These speculative constructs are used here only as a possible explanation for the correlations between variables 1 to 3 and between variables 4 to 6 that ultimately result in useful principal components.

The preceding instrument is fictitious and used for purposes of illustration only. It should not be regarded as an example of a good measure of prosocial orientation. Among other problems, this questionnaire obviously deals with very few forms of helping behavior.

Preparing a Multiple-Item Instrument

The preceding section illustrates an important point about how *not* to prepare a multiple-item measure of a construct. Generally speaking, it is poor practice to throw together a questionnaire, administer it to a sample, and then perform a principal component analysis (or factor analysis) to determine what the questionnaire measures.

Better results are much more likely when you make *a priori* decisions about what you want the questionnaire to measure and then take steps to ensure that it does. For example, you would be more likely to obtain desirable results if you

- begin with a thorough review of theory and research on prosocial behavior

- use that review to determine how many types of prosocial behavior probably exist

- write multiple questionnaire items to assess each type of prosocial behavior.

Using this approach, you could have made statements such as "there are three types of prosocial behavior—acquaintance helping, stranger helping, and financial giving." You could have then prepared a number of items to assess each of these three types and administered the questionnaire to a large sample. A principal component analysis might then produce three major components and a subsequent factor analysis could be used to verify that three factors did, in fact, emerge.

Number of Items per Component

The weight given a variable such as a questionnaire item in a principal component analysis (or a factor in a factor analysis) is its *load* on that component or factor. For example, if the item "Went out of my way to do a favor for a coworker" is given a large weight on the first principal component, we say that the item loads on that component.

It is desirable to have at least three (and preferably more) variables loading on each retained component when the principal component analysis is complete (see Clark and Watson, 1995). Because some of the items could be dropped during the course of the analysis, it is good practice to write at least five items for each construct you want to measure. This increases the chance that at least three items per component will survive the analysis. Note that we have violated this recommendation by writing only three items for each of the two *a priori* components constituting the POI.

The recommendation of three items per scale offered here should be viewed as an absolute minimum and not as an optimal number of items per scale. In practice, test and attitude scale developers usually desire that their scales contain many more than just three items to measure a given construct. It is not unusual to see individual scales that include 10, 20, or more items to assess a single construct (O'Rourke and Cappeliez, 2002). The recommendation of three items per scale should therefore be viewed as a lower bound, appropriate only if practical concerns (such as total questionnaire length) prevent you from including more items. For more information on scale construction, see Spector (1992).

Minimally Adequate Sample Size

Principal component analysis is a large-sample procedure. To obtain useful results, the minimum number of participants for the analysis should be the larger of 100 participants or five times the number of variables being analyzed (Streiner, 1994).

To illustrate, assume that you want to perform an analysis on responses to a 50-item questionnaire. Remember that the number of variables to be analyzed is equal to the number of items on the questionnaire. Five times the number of items on the questionnaire equals 250. Therefore, on a 50-item questionnaire the final sample should provide usable (complete) data from at least 250 participants. Keep in mind that any participant who fails to answer even a single item does not provide usable data for the principal component analysis. To ensure that the final sample includes at least 250 usable responses, you would be wise to administer the questionnaire to perhaps 300 to 350 participants.

These rules regarding the number of participants per variable again constitute a lower bound. Some researchers argue that these rules should apply only when many variables are expected to load on each component and when variable communalities are high. Under less optimal conditions, even larger samples might be required. In factor analysis, the proportion of variance of an item that is due to common factors (shared with other items) is called communality. The principal component analysis in JMP also shows item communalities. A communality refers to the percent of variance in an observed variable that is accounted for by the retained components (or factors). A given variable displays a large communality if it loads heavily on at least one of the study's retained components. The concept of variable communality is more relevant in a factor analysis than in principal component analysis, but communalities are estimated in both procedures.

Using the Multivariate Platform

You can perform a principal component analysis using either the Multivariate or the Spinning Plot platform. This chapter shows how to perform the analysis using the Multivariate platform, but also shows advantages to using the Spinning Plot platform to visualize the extracted components.

The Multivariate platform uses the raw data in the data table and can perform the principal component analysis on those raw data, a correlation matrix derived from the data or a covariance matrix derived from the data. The Multivariate platform has options for you to choose the basis for a principal component analysis. The following example performs a principal component analysis on the correlations of standardized data given in the data table called **prosocial behavior.jmp**.

Assume that you administered the POI to 50 participants and entered their responses into a JMP table. Figure 14.2 shows the first 10 observations in the job satisfaction table.

Figure 14.2 The Prosocial Behavior Data Table and Multivariate Menu

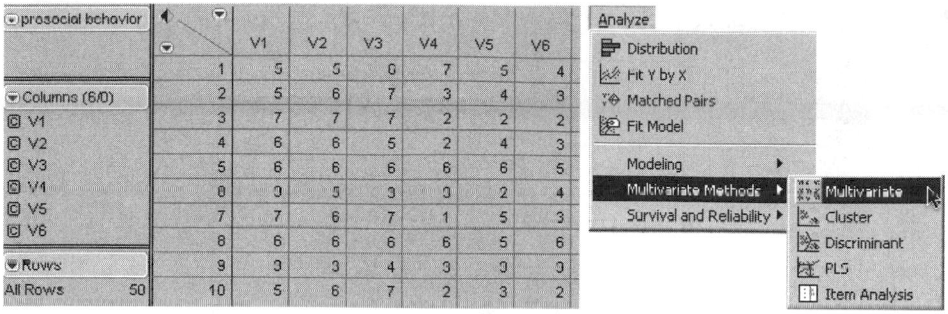

	V1	V2	V3	V4	V5	V6
1	5	5	6	7	5	4
2	5	6	7	3	4	3
3	7	7	7	2	2	2
4	6	6	5	2	4	3
5	6	6	6	6	6	5
6	3	5	3	3	2	4
7	7	6	7	1	5	3
8	6	6	6	6	5	6
9	3	3	4	3	3	3
10	5	6	7	2	3	2

The following steps show how to produce the principal component analysis in JMP.

- With the prosocial behavior table active, choose **Multivariate Methods → Multivariate** from the **Analyze** menu.
- When the Multivariate launch dialog appears, select **V1** through **V6** in the variable list and add them to the analysis list. Click **OK** in the launch dialog to see the initial results in Figure 14.3.

The Multivariate platform begins with a correlation matrix followed by a Scatterplot Matrix (Figure 14.3) that gives a visual representation of the correlations. Recall that the first three questionnaire items (V1–V3) address the "satisfaction with supervisor" component, so you expect them to exhibit a higher correlation with each other than with the last three items, Likewise, items V4–V6 address the "satisfaction with pay" component and show higher correlations with each other than with the first three items. The Scatterplot Matrix clearly shows these correlations.

Each cell in the Scatterplot Matrix represents the correlation between the two variables listed on the horizontal and vertical axis of that cell. Each cell has a 95% density ellipse overlaid on its points. If the variables are bivariate normally distributed, this ellipse encloses about 95% of the points. Variables that have higher correlations show an ellipse that is flattened and elongated on the diagonal. Variables with little correlation (correlation close to zero) show a rounded ellipse that is not diagonally oriented.

You can see in Figure 14.3 that items V1–V3 show higher correlations among themselves and items V4–V6 are also correlated, as expected by the design of the questionnaire.

Figure 14.3 Correlations and Scatterplot Matrix for Job Satisfaction Data

Multivariate

Correlations

	V1	V2	V3	V4	V5	V6
V1	1.0000	0.4944	0.7134	-0.1041	0.1141	0.0762
V2	0.4944	1.0000	0.3882	0.0535	-0.0597	0.1423
V3	0.7134	0.3882	1.0000	-0.0247	0.2038	0.0583
V4	-0.1041	0.0535	-0.0247	1.0000	0.6201	0.6353
V5	0.1141	-0.0597	0.2038	0.6201	1.0000	0.4551
V6	0.0762	0.1423	0.0583	0.6353	0.4551	1.0000

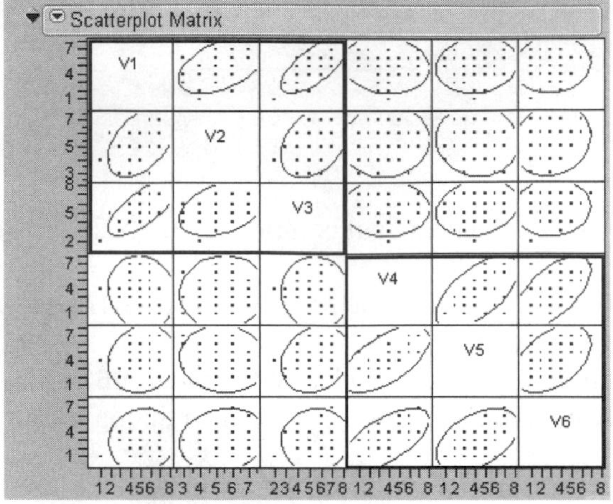

Conducting the Principal Component Analysis

Principal component analysis is normally conducted in a sequence of steps, with somewhat subjective decisions being made at various points. Because this is an introductory treatment of the topic, it does not provide a comprehensive discussion of all of the options available to you at each step. Instead, specific recommendations are made, consistent with practices commonly followed in applied research. For a more detailed treatment of principal component analysis and factor analysis, see Kim and Mueller (1978a, 1978b), Rummel (1970), or Stevens (1986).

Step 1: Extract the Principal Components

After launching the Multivariate platform as described in the previous section, proceed with the principal component analysis. In principal component analysis, the number of components extracted is equal to the number of variables being analyzed. Because six variables are analyzed in the present study, six components are extracted. The first component can be expected to account for a large amount of the total variance. Each succeeding component then accounts for progressively smaller amounts of variance. Although a large number of components can be extracted in this way, only the first few components will be sufficiently important to be retained for interpretation.

To continue with the principal components analysis, select **Principal Components → on Correlations** from the menu on the Multivariate title bar, as shown to the right.

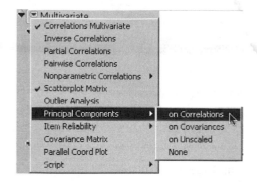

The principal components analysis then appears appended to the Multivariate platform, as shown in Figure 14.4.

The analysis begins with the eigenvalue table. An *eigenvalue* represents the amount of variance captured by a given component. The row labeled Eigenvalue shows each component's eigenvalue. The first column gives information about the first principal component, the second column gives information about the second component, and so forth.

The eigenvalue for component 1 is 2.2664 and the eigenvalue for component 2 is 1.9746. Note that the eigenvalue of each successive component is less than the previous one. This pattern is consistent with the statement made earlier that the first components tend to account for relatively large amounts of variance whereas the later components account for relatively smaller amounts.

A previous section called "How Principal Components Are Computed" defined a principal component as a linear combination of optimally weighted observed variables. These optimal weights are such that no other set of weights can produce principal components that successively account for more variation.

Figure 14.4 Results of Principal Component Analysis

▼ ▽ Multivariate						
▶ Correlations						
▶ ▽ Scatterplot Matrix						
▼ ▽ Principal Components / Factor Analysis						
▼ Principal Components: on Correlations						
Eigenvalue	2.2664	1.9746	0.7973	0.4392	0.2913	0.2312
Percent	37.7739	32.9102	13.2883	7.3197	4.8544	3.8534
Cum Percent	37.7739	70.6842	83.9725	91.2922	96.1466	100.0000
Eigenvectors						
V1	0.38301	0.49862	-0.15631	-0.22728	-0.45080	0.57042
V2	0.31940	0.37476	0.69419	0.45464	-0.09417	-0.24510
V3	0.39938	0.44011	-0.37485	-0.08435	0.65087	-0.27483
V4	0.42410	-0.45840	0.13677	0.27373	0.41245	0.58844
V5	0.45432	-0.31678	-0.47850	0.39738	-0.43381	-0.34378
V6	0.45333	-0.32673	0.32479	-0.70831	-0.07729	-0.27309

The Eigenvectors table in the Principal Components report gives these optimal weights. The principal components are computed on standardized values. Suppose S_n represents the standardized values of V_n. That is,

$$S_n = (V_n - V_n) \div \text{stdev}(V_n)$$

Then, the formulas for the first two principal components (using rounded coefficients) are

$$C_1 = 0.383(S_1) + 0.319(S_2) + 0.399(S_3) + 0.424(S_4) + 0.454(S_5) + 0.453(S_6)$$
$$C_2 = 0.499(S_1) + 0.375(V_2) + 0.440(S_3) - 0.458(S_4) - 0.317(S_5) - 0.327(S_6)$$

Using the eigenvalues, six principal component scores can be computed for each participant in the study.

Step 2: Determine the Number of Meaningful Components

The number of components extracted is equal to the number of variables analyzed. In general, you expect that only the first few components account for meaningful amounts of variance and that the later components tend to account for only trivial variance. The next step of the analysis is to determine how many meaningful components should be retained for interpretation.

This section describes the following criteria that can be used in making this decision:

- the eigenvalue-one criterion
- the scree test
- the proportion of variance accounted for.

Retain Components Based on the Eigenvalue-One Criterion

In principal component analysis, one of the most commonly used criteria for solving the number-of-components problem is the eigenvalue-one criterion, also known as the Kaiser-Guttman criterion (Kaiser, 1960). With this approach, you retain and interpret any component with an eigenvalue greater than 1.00.

The rationale for this criterion is straightforward. Each observed variable contributes one unit of variance to the total variance in the data. Any component that displays an eigenvalue greater than 1.00 accounts for a greater amount of variance than had been contributed by one variable. Such a component therefore accounts for a meaningful amount of variance and is worthy of further consideration.

On the other hand, a component with an eigenvalue less than 1.00 accounts for less variance than contributed by one variable. The purpose of principal component analysis is to reduce a number of observed variables into a relatively smaller number of components. This cannot be effectively achieved if you retain components that account for less variance than had been contributed by individual variables. For this reason, components with eigenvalues less than 1.00 are viewed as trivial and are not retained.

The eigenvalue-one criterion has a number of positive features that contribute to its popularity. Perhaps the most important reason for its widespread use is its simplicity. You do not make subjective decisions but merely retain components with eigenvalues greater than one.

Further, it has been shown that this criterion often results in retaining the correct number of components, particularly when a small to moderate number of variables are analyzed and the variable communalities are high. Stevens (1986) reviews studies that have investigated the accuracy of the eigenvalue-one criterion and recommends its use when fewer than 30 variables are being analyzed and communalities are greater than 0.70, or when the analysis is based on more than 250 observations and the mean communality is greater than 0.59.

However, there are a number of problems associated with the eigenvalue-one criterion. This approach sometimes leads to retaining the wrong number of components under circumstances that are often encountered in research, such as when many variables are analyzed or when communalities are small. Also, the mindless application of this criterion can lead to retaining a certain number of components when the actual difference in the eigenvalues of successive components is trivial. For example, if Component 2 displays an eigenvalue of 1.01 and Component 3 displays an eigenvalue of 0.99, then Component 2 is retained but Component 3 is not. This can mislead you to believe that the third component was meaningless when, in fact, it accounted for almost exactly the same amount of variance as the second component. In short, the eigenvalue-one criterion can be helpful when used judiciously, but thoughtless application can lead to errors of interpretation.

Figure 14.4 shows the eigenvalues for Components 1, 2, and 3 are 2.27, 1.97, and 0.80, respectively. Only Components 1 and 2 demonstrated eigenvalues greater than 1.00, so the eigenvalue-one criterion leads you to retain and interpret only these two components.

The application of the criterion is fairly unambiguous in this case. The second (and last) component retained displays an eigenvalue of 1.97, which is substantially greater than 1.00. The next component (3) displays an eigenvalue of 0.80, which is much lower than 1.00. In this analysis you are not faced with the difficult decision of whether to retain a component that demonstrates an eigenvalue approaching 1.00. In situations such as this, the eigenvalue-one criterion can be used with confidence.

Retaining Components Based on the Scree Test

The scree test (Cattell, 1966) plots the eigenvalues associated with each component and looks for a definitive break in the plot between the components with relatively large eigenvalues and those with small eigenvalues. The components that appear *before* the break are assumed to be meaningful and are retained for rotation, whereas those appearing *after* the break are assumed to be unimportant and are not retained. Sometimes a scree plot displays several large breaks. When this is the case, you should look for the last break before the eigenvalues begin to level off. Only the components that appear before this last large break should be retained.

The scree test provides reasonably accurate results, provided that the sample is large (over 200) and most of the variable communalities are large (Stevens, 1986). However, this criterion has its own weaknesses, most notably the ambiguity that is sometimes displayed by scree plots under typical research conditions. Often, it is difficult to determine exactly where in the scree plot a break exists, or even if a break exists at all. In contrast to the eigenvalue-one criterion, the scree test is usually more subjective.

Figure 14.5 presents a fictitious scree plot from a principal component analysis of 17 variables. Notice that there is no obvious break in the plot that separates the meaningful components from the trivial components. Most researchers would agree that Components 1 and 2 are likely to be meaningful whereas Components 13 to 17 are probably trivial. Scree plots such as this one are common in social science research. This example underscores the qualitative nature of judgments based solely on the scree test. The scree test must be supplemented with additional criteria such as the variance accounted for criterion, described later.

Figure 14.5 A Scree Plot with No Obvious Break

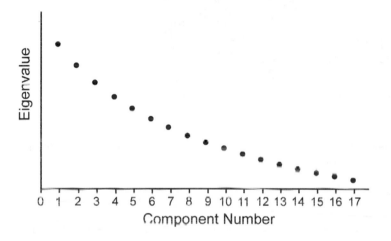

A scree plot for the fictitious data used in this example shows how this kind of plot can be useful. To produce a scree plot in JMP, do the following:

- Create a JMP data table from the Multivariate platform principal components analysis. To do this, right-click (control-click on the Macintosh) and select **Make into Data Table** from the menu that appears.

Figure 14.6 shows the Multivariate principal components results, the context menu, and the JMP data table that contains the principal component results. You can create a JMP table from any platform results in this way.

Figure 14.6 Creating a JMP Data Table from Platform Results

Principal Components: on Correlations						
Eigenvalue	2.2664	1.9746	0.7973	0.4392	0.2913	0.2312
Percent	37.7739	32.9102	13.2883	7.3197	4.8544	3.8534
Cum Percent	37.7739	70.6842	83.9725	91.2922	96.1466	100.0000
Eigenvectors						
V1	0.38301	0.49862	-0.15631	-0.22728	-0.45080	0.57042
V2	0.31940	0.37476	0.69419	0.45464	-0.09417	-0.24510
V3	0.39938	0.44011	-0.37485	-0.08435	0.65087	-0.27483
V4	0.42410	-0.45840	0.13677	0.27373	0.41245	0.58844
V5	0.45432	-0.31678	-0.47850	0.39738	-0.43381	-0.34378
V6	0.45333	-0.32673	0.32479	-0.70831	-0.07729	-0.27309

Table Style ▸
Columns ▸
Sort by Column...
Make into Data Table
Make Into Matrix

		Column 1	Column 2	Column 3	Column 4	Column 5	Column 6	Column 7
	1	Eigenvalue	2.2664	1.9746	0.7973	0.4392	0.2913	0.2312
	2	Percent	37.7739	32.9102	13.2883	7.3197	4.8544	3.8534
	3	Cum Percent	37.7739	70.6842	83.9725	91.2922	96.1466	100.0000
	4	V1	0.3830	0.4986	-0.1563	-0.2273	-0.4508	0.5704
	5	V2	0.3194	0.3748	0.6942	0.4546	-0.0942	-0.2451
	6	V3	0.3994	0.4401	-0.3749	-0.0844	0.6509	-0.2748
	7	V4	0.4241	-0.4584	0.1368	0.2737	0.4125	0.5884
	8	V5	0.4543	-0.3168	-0.4785	0.3974	-0.4338	-0.3438
	9	V6	0.4533	-0.3267	0.3248	-0.7083	-0.0773	-0.2731

Next, use the **Transpose** command from the **Tables** menu to rearrange the data table so it can be used with the Chart platform.

- ◦ Choose **Tables → Transpose** and select numeric variables (Column 2 – Column 7) to be transposed in this new transposed data table.
- ◦ Select **Column 1** in the Columns list on the left of the dialog and click **Label** to specify that you want the values of Column 1 as column names in the transposed table (see Figure 14.7).

The Transpose dialog appears with all the numeric variables listed in the **Columns to be transposed** list. Transpose can only transpose variables with the same data type (all numeric or all character). You can add or remove variables from this list, transpose only highlighted rows, and choose a column whose values you want to show as column names in the new transposed table. In this example, all numeric columns in the source table are to be transposed.

Figure 14.7 shows the Transpose dialog and the new transposed principal components table. Note that the highlighted column in the new transposed table contains the eigenvalues you want to plot.

Figure 14.7 Transpose the Data Table of Principal Components

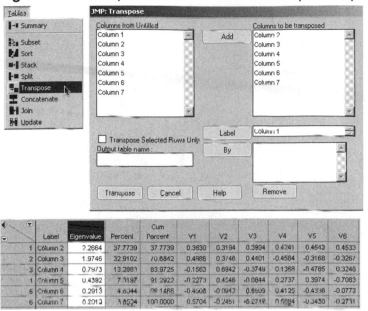

	Label	Eigenvalue	Percent	Cum Percent	V1	V2	V3	V4	V5	V6
1	Column 2	2.2664	37.7739	37.7739	0.3830	0.3194	0.3994	0.4241	0.4543	0.4533
2	Column 3	1.9746	32.9102	70.6842	0.4986	0.3746	0.4401	-0.4584	-0.3168	-0.3267
3	Column 4	0.7973	13.2883	83.9725	-0.1563	0.6942	-0.3749	0.1368	-0.4765	0.3246
4	Column 5	0.4392	7.3197	91.2922	-0.2273	0.4546	-0.0844	0.2737	0.3974	-0.7083
5	Column 6	0.2913	4.8544	96.1466	-0.4508	-0.0942	0.6509	0.4125	-0.4338	-0.0773
6	Column 7	0.2012	3.8534	100.0000	0.5704	-0.2451	-0.2749	0.5884	-0.3430	-0.2731

The transposed table is ready to chart.

🖰 Use the **Chart** command from the **Graph** menu to create the scree plot.

The Chart launch dialog, shown in Figure 14.8, prompts you for the statistics to chart. In this case you want to see the eigenvalues themselves, not a summary of these values.

🖰 To complete the Chart launch dialog, highlight **Eigenvalue** from the Select Columns list, and choose **Data** from the **Statistics** menu.

🖰 Select **Line Chart** as the chart type, as shown in Figure 14.8.

🖰 Click **OK** in the Chart launch dialog to see the scree plot in Figure 14.9.

Figure 14.8 Chart Command and Chart Launch Dialog

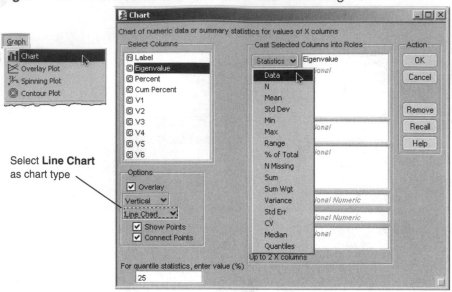

Select **Line Chart** as chart type

The scree plot in Figure 14.9 shows the component numbers (row numbers) on the horizontal axis and eigenvalues on the vertical axis. Notice there is a small break between Components 1 and 2, and a larger break following Component 2. The breaks between Components 3, 4, 5, and 6 are fairly small. Because the large break in this plot appears between Components 2 and 3, the scree test leads you to retain only Components 1 and 2.

Figure 14.9 Screen Plot for Example Data

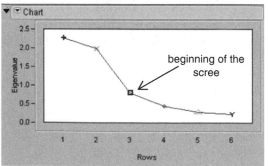

Why is it called a scree test? The word *scree* refers to the loose rubble that lies at the base of a cliff or glacier. You hope that the scree plot takes the form of a cliff. The eigenvalues at the top represent the few meaningful components, followed by a definitive break (the base of the cliff). The eigenvalues for the trivial components form the scree at the bottom of the cliff.

Retaining Components Based on Proportion of Variance Accounted For

A third criterion in choosing the number of principal components to consider involves retaining a component if it exceeds a specified proportion (or percentage) of variance in the data. For example, you might decide to retain any component that accounts for at least 10% of the total variance. This proportion can be calculated with a simple formula:

$$\text{Proportion} = \frac{\text{eigenvalue for component of interest}}{\text{total eigenvalues of the correlation matrix}}$$

In principal component analysis, the eigenvalues sum to the number of variables being analyzed because each variable contributes one unit of variance to the analysis. The proportion of variance captured by each component is printed in the eigenvalue table in the row called Percent (Figure 14.4). A portion of the table showing information for the *first four* principal components is shown here. You can see that the first component accounts for 37.7739% of the total variance,

Principal Components: on Correlations			
Eigenvalue	2.2664	1.9746	0.7973	0.4392
Percent	37.7739	32.9102	13.2883	7.3197
Cum Percent	37.7739	70.6842	83.9725	91.2922

the second component accounts for 32.9102%, the third component accounts for 13.2883%, and the fourth component accounts for 7.3197%. Assume that you decide to retain any component that accounts for at least 10% of the total variance in the data. For the present results this criterion causes you to retain Components 1, 2, and 3. Notice that this results in retaining more components than would be retained with either the eigenvalues-one criterion or the scree test.

An alternative criterion is to retain enough components so that the *cumulative* percent of variance accounted for is equal to some minimal value. For example, Components 1, 2, 3, and 4 account for approximately 38%, 33%, 13%, and 7% of the total variance, giving a total of 91%. This means that the *cumulative* percent of variance accounted for by Components 1, 2, 3, and 4 is 91%.

When researchers use the *cumulative percent of variance accounted for* as the criterion for solving the number-of-components problem, they usually retain enough components so that the cumulative percent of variance accounted for is at least 70% (and sometimes 80%). Retaining only Components 1 and 2 accounts for 70.6842% of the variance, as seen in the row labeled Cum Percent. Each value in that row indicates the percent of variance accounted for by the present component as well as all preceding components.

The entry for component 3 is 83.9725, which means that approximately 84% of the variance is accounted for by Components 1, 2, and 3.

The proportion of variance criterion has a number of positive features. For example, in most cases, you do not want to retain a group of components that account for only a fraction of variance in the data (say, 30%). Nonetheless, the critical values discussed earlier (10% for individual components and 70% to 80% for the combined components) are arbitrary. Because of these and related problems, this approach has sometimes been criticized for its subjectivity (Kim and Mueller, 1978b).

Recommendations

Given the preceding options, what procedure should you follow in solving the number-of-components problem? First, look at the eigenvalue-one criterion. Use caution if the break between the components with eigenvalues above 1.00 and those below 1.00 is not clear-cut.

Next, perform a scree test and look for obvious breaks in the eigenvalues. Because there is often more than one break in the scree plot, it might be necessary to examine two or more possible solutions.

Next, review the amount of variance accounted for by each individual component. You probably should not rigidly use some specific but arbitrary cutoff point such as 5% or 10%. Still, if you are retaining components that account for as little as 2% or 3% of the variance, it is wise to take a second look at the solution and verify that these latter components are truly of substantive importance. In the same way, it is best if the combined components account for at least 70% of the cumulative variance. If the components you choose to retain account for less than 70% of the cumulative variance, it might be prudent to consider alternative solutions that include a larger number of components.

Step 3: Rotate to a Final Solution

Ideally, you want to review the correlations between the variables and components, and use this information to interpret the components. That is, you want to determine what construct seems to be measured by Component 1, what construct seems to be measured by Component 2, and so forth. When more than one component is retained in an analysis, the interpretation of an unrotated pattern can be difficult. To facilitate interpretation, you can perform an operation called a rotation. A *rotation* is a linear transformation performed on the principal components for the purpose of making the solution easier to interpret.

The menu on the Principal Components title bar has the **Factor Rotation** command. When you choose this command, a dialog prompts you for the number of factors to rotate. Follow the guidelines described previously for this example and enter 2 (number of retained components) as the number of factors to rotate.

The rotation method used by JMP is called a *varimax* rotation and is commonly used in factor analysis (Kaiser, 1958). A varimax rotation is an orthogonal rotation, which means that the rotated components are uncorrelated. Compared to other types of rotations, a varimax rotation tends to maximize the variance of a column of the factor pattern matrix (as opposed to a row of the matrix). This rotation is a commonly used orthogonal rotation in the social sciences. Figure 14.10 shows the results of the varimax rotation.

Figure 14.10 Factor Rotation Report for Principal Component Analysis

Factor pattern matrix

Rotated Factor Pattern

V1	-0.004293	0.907414
V2	0.033282	0.712335
V3	0.067203	0.859927
V4	0.902738	-0.087397
V5	0.910548	0.094736
V6	0.818340	0.083034

Communalities

V1	0.82342
V2	0.50853
V3	0.74399
V4	0.82257
V5	0.66596
V6	0.67658

Rotation Matrix

0.76914	0.63908
0.63908	0.76914

Std Score Coefs

V1	-0.031093	0.435511
V2	0.007261	0.340709
V3	0.003879	0.410435
V4	0.425151	-0.070867
V5	0.376101	0.019475
V6	0.380202	0.013610

Variance	Percent	Cum Percent
2.1472	35.787	35.787
2.0938	34.897	70.684

Note: Using a principal component analysis, retaining components with eigenvalues greater than 1, and then performing a varimax rotation to produce uncorrelated factors is called a *little jiffy factor analysis* (Kaiser, 1970). Thus, the following steps and discussion apply equally well to factor analysis and principal component analysis.

Step 4: Interpret the Rotated Solution

A rotated solution helps you determine what is measured by each of the retained components (two in this example). You want to identify the variables that exhibit high loadings for a given component and determine what these variables have in common.

At this stage, the first decision to be made is how large a factor loading must be to be considered "large." Stevens (1986) discusses some of the issues relevant to this decision and even provides guidelines for testing the statistical significance of factor loadings. Given that this is an introductory treatment of principal component analysis, simply consider a loading to be "large" if its absolute value exceeds 0.40.

1. **Read the loading values for the first variable.** The Rotated Factor Pattern table lists each variable (V1–V6) and shows its loading for two factors. The loading values for V1 are –0.004293 and 0.907414. V1 has a small loading on the first factor and a large loading (greater than 0.40) on the second factor. If a given variable has a meaningful loading on more than one component, ignore it in your interpretation. In this example V1 loads on only one factor, so you retain this variable.

2. **Repeat this process for the remaining variables and eliminate any variable that loads on more than one component.** In this analysis, none of the variables have high loadings on more than one component, so none have to be deleted from the interpretation. In other words, there are no complex items.

3. **Review all of the variables with high loadings on Component 1 to determine the nature of this component.** From the rotated factor pattern, you can see that only items 4, 5, and 6 (V4–V6) load on the first component. It is now necessary to turn to the questionnaire itself and review the content in order to decide what a given component should be named. What do questions 4, 5, and 6 have in common? What common construct do they appear to be measuring? For illustration, the questions being analyzed in the present

Rotated Factor Pattern		
V1	-0.004293	0.907414
V2	0.033282	0.712335
V3	0.067203	0.859927
V4	0.902738	-0.087397
V5	0.810548	0.094736
V6	0.818340	0.083034

case are reproduced here. Read questions 4, 5, and 6 (V4–V6) to see what they have in common.

Questions 4, 5, and 6 all seem to deal with "giving money to those in need." It is therefore reasonable to label Component 1 the "financial giving" component.

4. **Repeat this process to name the remaining retained components.** In the present case, there is only one remaining component to name—Component 2. This component has high loadings for questions 1, 2, and 3 (V1–V3). Each of these items seems to deal with helping friends, relatives, or other acquaintances. It is therefore appropriate to name this the "helping others" component.

```
1 2 3 4 5 6 7    1. Went out of my way to do a favor for a coworker.
1 2 3 4 5 6 7    2. Went out of my way to do a favor for a relative.
1 2 3 4 5 6 7    3. Went out of my way to do a favor for a friend.
1 2 3 4 5 6 7    4. Gave money to a religious charity.
1 2 3 4 5 6 7    5. Gave money to a charity not associated with a
                    religion.
1 2 3 4 5 6 7    6. Gave money to a panhandler.
```

Step 5: Create Factor Scores or Factor-Based Scores

Once the analysis is complete, it is often desirable to assign scores to participants to indicate where they stand on the retained components. For example, the two components retained in the present study can be interpreted as "financial giving" and "helping others." You might want to now assign one score to each participant to indicate that participant's standing on the "financial giving" component and a different score to indicate that participant's standing on the "helping others" component. With this done, these component scores could be used either as predictor variables or as response variables in subsequent analyses.

Before discussing the options for assigning these scores, it is important to first draw a distinction between *factor scores* and *factor-based scores*.

- In principal component analysis, a *factor score* (or *component score*) is a linear composite of the optimally weighted observed variables.

- A *factor-based score* is a linear composite of the variables that demonstrates meaningful loadings for the component in question. In the preceding analysis, items 4, 5, and 6 demonstrated meaningful loadings for the "financial giving" component. Therefore, you could calculate the factor-based score on this component for a given participant by simply adding together a participant's responses to items 4, 5, and 6. The observed variables are not multiplied by optimal weights before they are summed.

Computing Factor Scores

You can save each participant's factor scores for the two components as new columns in the prosocial behavior data table using the **Save Rotated Components** command found on the menu on the title bar of the Principal Components/Factor Analysis report. This command multiplies participant

responses to the questionnaire items by the optimal weights given in the principal component analysis and sums the products. This sum is a given participant's score on the component of interest. Remember that a separate equation with different weights is developed for each retained component. The two sets of scores are saved in the data table and named Factor1 and Factor2.

To see how these factor scores relate to the study's original observed variables, look at the correlations between the factor scores and the original variables.

- ☝ Choose **Multivariate Methods → Multivariate** from the **Analyze** menu.
- ☝ Select **V1** through **V6** and the new variables **Factor1** and **Factor2** as analysis variables. Click **OK** in the Multivariate launch dialog to see the correlations shown in Figure 14.11.

The correlations between Factor1 and Factor2 and the original observed variables appear toward the bottom of the multivariate report. You can see that the correlations between Factor1 and V1 to V6 are identical to the factor loadings of V1 to V6 on Factor1 shown as the Rotated Factor Pattern table in Figure 14.10. This makes sense, because the elements of a factor pattern (in an orthogonal solution) are simply correlations between the observed variables and the components themselves. Similarly, you can see that the correlations between Factor2 and V1 to V6 are also identical to the corresponding factor loadings from the Rotated Factor Pattern table.

Of special interest is the correlation between Factor1 and Factor2. Notice the observed correlation between these two components is zero. This is as expected because the rotation method used in the principal component analysis is the *varimax* method, which produces orthogonal (uncorrelated) components.

Figure 14.11 Correlations between Rotated Factors and Original Variables

	V1	V2	V3	V4	V5	V6	Factor1	Factor2
V1	1.0000	0.4944	0.7134	-0.1041	0.1141	0.0762	-0.0043	0.9074
V2	0.4944	1.0000	0.3882	0.0535	-0.0597	0.1423	0.0333	0.7123
V3	0.7134	0.3882	1.0000	-0.0247	0.2038	0.0583	0.0672	0.8599
V4	-0.1041	0.0535	-0.0247	1.0000	0.6201	0.6353	0.9027	-0.0874
V5	0.1141	-0.0597	0.2038	0.6201	1.0000	0.4551	0.8105	0.0947
V6	0.0762	0.1423	0.0583	0.6353	0.4551	1.0000	0.8183	0.0830
Factor1	-0.0043	0.0333	0.0672	0.9027	0.8105	0.8183	1.0000	0.0000
Factor2	0.9074	0.7123	0.8599	-0.0874	0.0947	0.0830	0.0000	1.0000

Computing Factor-Based Scores

A second (and less sophisticated) approach to scoring involves the creation of new variables that contain factor-based scores instead of true principal component scores. A variable that contains factor-based scores is sometimes called a *factor-based scale*.

Although factor-based scores can be created in a number of ways, the following method has the advantage of being relatively straightforward and is commonly used.

1. To calculate factor-based scores for Component 1, first determine which questionnaire items had high loadings on that component.

2. For a given participant, add together that participant's responses to these items. The result is that participant's score on the factor-based scale for Component 1.

3. Repeat these steps to calculate each participant's score on the remaining retained components.

You can do this in JMP by creating two new columns in the current data table (the table used to create rotated components). You have looked at the factor loadings and found that survey items 4, 5, and 6 loaded on Component 1 ("financial giving" component) and items 1, 2, and 3 loaded on Component 2 ("helping others" component).

To create the factor-based scores, create two new variables in the **prosocial behavior.jmp** data table and name them financial aid and helping others.

- Choose **New Column** from the **Cols** menu. When the New Column dialog appears, enter "financial aid" as the column name.
- Click **New Property** in the New Column dialog and select **Formula** from its menu.
- Click **Edit Formula** in the New Column dialog to see the Formula Editor.
- Create the formula to compute the factor-based score for financial giving. Click **V4** in the list of variables, then click the plus operator. Click **V5**, then the plus operator, and finally click **V6**.

Figure 14.12 shows the New Column dialog, the Formula Editor, and its formula to compute the factor-based scores for the first component.

Follow the same steps to create the second new column called "helping others," but use the column V1 + V2 + V3 to compute the factor-based scores for the second component.

Here is a shortcut method to create a new column and a formula:

- To create a new column, double-click anywhere in the empty area of the data table to the right of the existing columns.
- Highlight the column name area and type a new name.
- To see a column editor for the new column, right-click in the column heading area and choose **Formula** from the menu that appears.
- Enter the formula as described previously, and close the Formula Editor.

Figure 14.12 New Column Dialog and Formula Editor

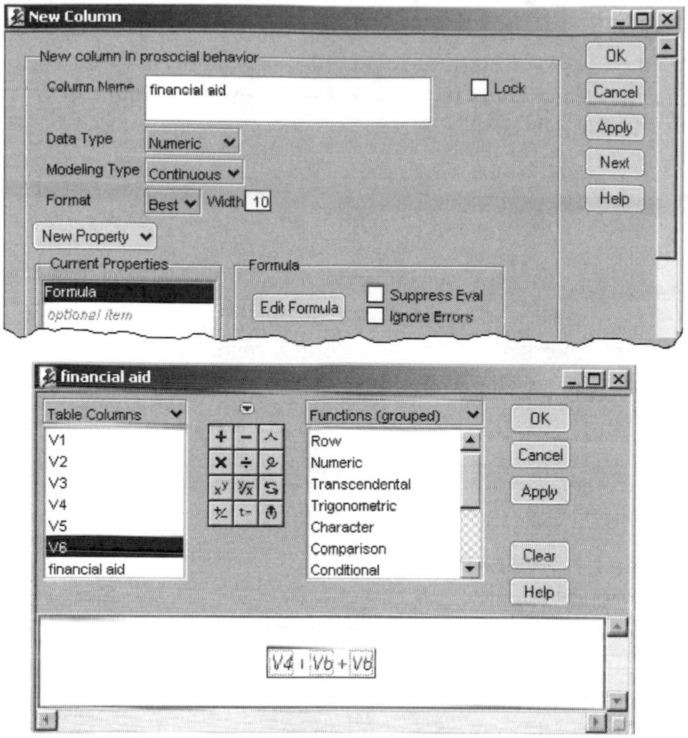

The first variable, called Financial Aid, lists each participant's factor-based score for financial giving. The second variable, called Helping Others, lists each participant's factor-based score for helping others. These new variables could be used as predictor variables in subsequent analyses.

However, the correlation between Financial Aid and Helping Others is not zero, unlike the factor scores, Factor 1 and Factor 2. This is because Factor 1 and Factor 2 are true principal components (created in an orthogonal solution) and are optimally weighted equations to be uncorrelated. In contrast, Financial Aid and Helping Others are not true principal components. They are variables based on factors identified by the factor loadings of a principal component analysis. These factor-based scores do not use optimal weights that ensure orthogonality.

Recoding Reversed Items Prior to Analysis

It is generally best to recode any reversed items before conducting any of the analyses described here. In particular, it is essential that reversed items be recoded prior to the program statements that produce factor-based scales. For example, the three questionnaire items that assess financial giving appear again here.

```
1 2 3 4 5 6 7    4. Gave money to a religious charity.
1 2 3 4 5 6 7    5. Gave money to a charity not associated
                    with a religion.
1 2 3 4 5 6 7    6. Gave money to a panhandler.
```

None of the previous items are reversed. With each item, a response of 7 indicates a high level of financial giving. However, in the following list, item 4 is a reversed item—a response of 7 indicates a low level of giving.

```
1 2 3 4 5 6 7    4. Refused to give money to a religious
                    charity.
1 2 3 4 5 6 7    5. Gave money to a charity not associated
                    with a religion.
1 2 3 4 5 6 7    6. Gave money to a panhandler.
```

If you perform a principal component analysis on responses to these items, the factor loading for item 4 would most likely have a sign that is the opposite of the sign of the loadings for items 5 and 6. That is, if items 5 and 6 had positive loadings, item 4 would have a negative loading. This complicates the creation of a factor-based scale. You would not want to sum these three items as they are presently coded. First, it is necessary to reverse item 4. You can do this in the JMP table by creating another formula based variable—call it **V4 New**—using the formula

$$8 - V4$$

Values of this new version of V4 are 8 minus the value of the old version of V4. A participant whose score on the old version of V4 is 1 has a value of 7 ($8 - 1 = 7$) on the new version, whereas someone who had a score of 7 now has a score of 1 ($8 - 7 = 1$). Note that the constant (8) used to create the new values is 1 greater than the number of possible responses to an item. If you use a 4-point response format, the constant is 5. If you use a 9-point scale, the constant is 10.

If you have prior knowledge about which items are will show reversed component loadings in your results, it is best to recode them before the analysis. This makes interpretation of the components more straightforward because it eliminates significant loadings with opposite signs from appearing on the same component.

Step 6: Summarize Results in a Table

For reports that summarize the results of your analysis, it is desirable to prepare a table that presents the rotated factor pattern. When analyzed variables contain responses to questionnaire items, it can be helpful to actually reproduce the questionnaire items within this table. This is done in Table 14.4.

Table 14.4 Rotated Factor Pattern and Final Communality Estimates from Principal Component Analysis of Prosocial Orientation Inventory

Components			
1	**2**	h^2	**Items**
0.00	0.91	0.82	1. Went out of my way to do a favor for a coworker.
0.03	0.71	0.51	2. Went out of my way to do a favor for a relative.
0.07	0.86	0.74	3. Went out of my way to do a favor for a friend.
0.90	−0.09	0.82	4. Gave money to a religious charity.
0.81	0.09	0.67	5. Gave money to a charity not associated with religion.
0.82	0.08	0.68	6. Gave money to a panhandler.

Note: $N = 50$. Communality estimates appear in column headed h^2.

The final communality estimates from the analysis are presented under the heading h^2 in the table. Often, the items that constitute the questionnaire are so lengthy, or the number of retained components is so large, that it is not possible to present the factor pattern, the communalities, and the items themselves in the same table. In such situations it might be preferable to present the factor pattern and communalities in one table and the items in a second (or in the text of the paper). Shared item numbers can then be used to associate each item with its corresponding factor loadings and communality.

Step 7: Prepare a Formal Description of Results for a Paper

The preceding analysis could be summarized in the following way:

```
Principal component analysis was applied to responses to the six-item
questionnaire. The principal axis method was used to extract the
components, and this was followed by a varimax (orthogonal) rotation.
Only the first two components exhibited eigenvalues greater than 1;
results of a scree test also suggested that only the first two were
meaningful. Therefore, only the first two components were retained for
rotation. Combined, Components 1 and 2 accounted for 71% of the total
variance.

Questionnaire items and corresponding factor loadings are presented in
Table 14.4. In interpreting the rotated factor pattern, an item was said
to load on a given component if the factor loading was .40 or greater for
that component and less than 0.40 for the other. Using these criteria,
three items were found to load on the first component, which was
subsequently labeled "financial giving." Three items also loaded on the
second component, labeled "helping others."
```

A Graphical Approach to Principal Components

Most JMP platforms offer ways to visualize statistical results. The Multivariate platform shows the Scatterplot Matrix to picture correlation results quickly. JMP has additional features that let you *see* the principal components and how they relate to the original variables in a three-dimensional spinning plot.

Using the Spinning Plot Platform for Principal Components

You can also use the Spinning Plot platform to perform and visualize principal components. Using the **prosocial behavior.jmp** table:

- Select **Spinning Plot** from the **Graph** menu.

- When the launch dialog appears, select V1 through V6 and click **Y, Columns**. Click **OK** in the launch dialog to see the initial three-dimensional plot shown in Figure 14.14.

Chapter 14: Principal Component Analysis **453**

🖰   Select **Axis Labels** from the menu accessed by the icon next to the spin control panel to show the variable names (V1–V3) on the axes instead of x, y and z.

🖰   To plot different variables, drag the X, Y, and Z icons from the Components list to different boxes in the list on the left of the spinning plot.

🖰   Use the spin icons on the spin control panel or the grabber tool found on the **Tools** menu to spin the plot.

Figure 14.13 Spinning Plot of Responses to Three Survey Items

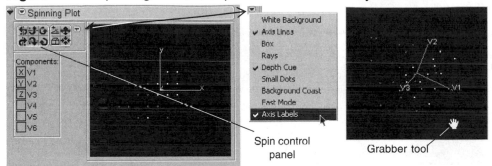

Spin control panel

Grabber tool

The Spinning Plot platform can produce the same principal component analysis as the Multivariate platform.

🖰   Select **Principal Components** from the menu on the Spinning Plot title bar, as shown here. This command produces the same report as shown in Figure 14.4, given by the Multivariate platform.

> **Note:** The **Principal Components** command produces principal components on standardized variables, just like selecting the **Principal Components → on Correlations** option on the Multivariate platform. The **Std Prin Components** command available on the Spinning Plot platform refers to a standardization method used on the axes in the spinning plot. See the *JMP Statistics and Graphics Guide* for details.

The Components list of variables on the left of the spinning plot now shows six additional variables that represent the six principal components. The spinning plot overlays rays that represent the principal components (*p-rays*). These rays approximate the principal components as functions of the variables on the axes. The length of a ray is proportional to the amount of variation accounted for by that principal component.

You can change the orientation of the plot using either the hand tool from the **Tools** menu, or the spin controls on the Spinning Plot panel. The plot on the left in Figure 14.14 shows V1–V3 and the p-rays in the default home position. The middle plot shows an altered orientation, and the plot on the right has the axes turned off. The **Axis Lines** command on the **Spinning Plot** menu toggles the axes on and off. You can see that the first and second principal components represent a large proportion of the variation in the data, as shown by their greater lengths in the spinning plot.

Figure 14.14 Spinning Plot of First Three Variables and All Principal Components

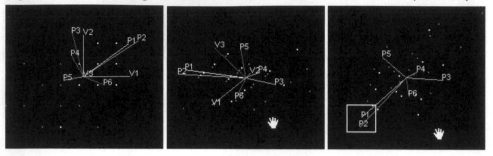

Looking at a Biplot

The Spinning Plot platform can help you visualize clusters of variables. Change the axes to be the first three principal components instead of the variables.

🖱 Drag the X, Y, and Z axis icons, shown in the Components list on the left of the spinning plot, to the first three principal components, as shown in Figure 14.15.

Arranging the plot axes in this way produces a *biplot*. A biplot displays the selected principal components as the (orthogonal) axes, and shows all the variables as rays. The biplot is attributed to Gabriel (1982). When you plot the variables in this way, the variables that were originally plotted as the axes now become closer to each other. The variables that are the most correlated cluster together in a group. These clusters of variables are what analysts hope to find when they do factor analysis.

You can see in Figure 14.15 that V1–V3 cluster together and V4–V6 form a second cluster. This biplot is what you expect because the questionnaire was designed such that V1–V3 address the "helpfulness to others" component and V4–V6 address the "financial giving" component.

Figure 14.15 Biplot of First Three Principal Components and All Variables

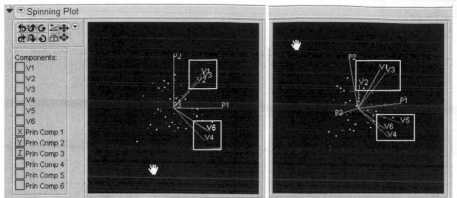

Rotating Components

The Spinning Plot platform can also produce and plot rotated components. To do this:

  Select **Rotated Components** from the menu on the Spinning Plot title bar.
  When the dialog appears to specify the number of components to rotate, specify 2 (the default) and click **OK**.

The rotated components now appear in the Components list in the left of the plot. The information for the factor rotation is appended to the Spinning Plot principal components report. It is the same information shown previously in Figure 14.10.

The goal of rotating components is to form orthogonal combinations of the original principal components that correspond to directions of variable clusters. Visually, this takes the axes of the number of principal components in a biplot you specify and rotates them so that each axis points as closely as possible to a variable ray. This is the same varimax rotation discussed previously.

Figure 14.16 shows the biplot with two rotated components on the plot. To see the plot to the right, use the **Axis Lines** command to toggle off the principal components that were showing as the axes. Then rotate the plot and see the rotated components aligning with the variables they represent.

Figure 14.16 Biplot with Two Rotated Principal Components

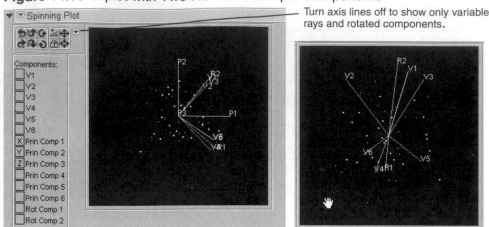

Turn axis lines off to show only variable rays and rotated components.

Summary

This chapter explained the problems with analyzing data when there are a number of predictor variables and some of them are correlated. Principal component analysis was shown as a way to reduce the number of observed variables to a smaller number of uncorrelated variables that account for most of the variance in a set of data.

A *principal component* was defined as a linear combination of optimally weighted observed variables. You saw how principal components are computed (extracted) from the data. The principal component procedure finds weights for the observed variables such that

- the first principal component extracted accounts for a maximal amount of total variance in the observed variables

- the second component accounts for a maximal amount of variance in the data not accounted for by the first component and is uncorrelated to the first component

- the third component is uncorrelated with the first two components and accounts for a maximal amount of the remaining variance, and so forth.

You learned how to use the Multivariate platform in JMP to do a principal component analysis that extracts a component for each of the original variables. Several methods were presented to determine the subset of meaningful components to retain and use for further analysis. The Multivariate platform gives the option to rotate the retained components. You learned that factor rotation can facilitate interpretation of the relationship between the components and possible underlying characteristics in the data represented by the retained components. You learned how to interpret the rotated solution, create factor scores, and summarize the results.

The Spinning Plot platform was also used to do principal component analysis, with the option of looking at a biplot that shows the first three components or rotated component on the same plot with three of the original variables. The biplot is a visualization of the relationship of the components to the predictor variables.

Assumptions Underlying Principal Component Analysis

Because a principal component analysis is performed on a matrix of Pearson correlation coefficients, the data should satisfy the assumptions for this statistic. These assumptions were described in detail in Chapter 6, "Measures of Bivariate Association," and are briefly reviewed here.

Continuous numeric measurement

All variables should be assessed on continuous numeric measurements.

Random sampling

Each participant contributes one score on each observed variable. These sets of scores should represent a random sample drawn from the population of interest.

Linearity

The relationship between all observed variables should be linear.

Bivariate normal distribution

Each pair of observed variables should display a bivariate normal distribution (they should form an elliptical point cloud when plotted).

References

Cattell, R. B. 1966. "The Scree Test for the Number of Factors." *Multivariate Behavioral Research* 1:245–276.

Clark, L. A., and D. Watson. 1995. "Constructing Validity: Basic Issues in Objective Scale Development." *Psychological Assessment* 7:309–319.

Gabriel, R. K. 1982. "Biplot." *Encyclopedia of Statistical Sciences*. Vol. 1, ed. N. L. Johnson and S. Kotz. New York: John Wiley & Sons.

Kaiser, H. F. 1958. "The Varimax Criterion for Analytic Rotation in Factor Analysis." *Psychometrika* 23:187–200.

Kaiser, H. F. 1960. "The Application of Electronic Computers to Factor Analysis." *Educational and Psychological Measurement* 20:141–151.

Kaiser, H. F. 1970. "A Second Generation Little Jiffy." *Psychometrika* 35:401–415.

Kim, J. O., and C. W. Mueller. 1978a. *Introduction to Factor Analysis: What It Is and How to Do It.* Beverly Hills, CA: Sage Publications.

Kim, J. O., and C. W. Mueller. 1978b. *Factor Analysis: Statistical Methods and Practical Issues.* Beverly Hills, CA: Sage Publications.

O'Rourke, N., and P. Cappeliez. 2002. "Development and Validation of a Couples Measure of Biased Responding: The Marital Aggrandizement Scale." *Journal of Personality Assessment* 78:301–320.

Rummel, R. J. 1970. *Applied Factor Analysis.* Evanston, IL: Northwestern University Press.

Rusbult, C. E. 1980. "Commitment and Satisfaction in Romantic Associations: A Test of the Investment Model." *Journal of Experimental Social Psychology* 16:172–186.

SAS Institute Inc. 2003. *JMP Statistics and Graphics Guide, Version 5.1.* Cary, NC: SAS Institute Inc.

Spector, P. E. 1992. *Summated Rating Scale Construction: An Introduction.* Newbury Park, CA: Sage Publications.

Stevens, J. 1986. *Applied Multivariate Statistics for the Social Sciences.* Hillsdale, NJ: Lawrence Erlbaum Associates.

Streiner, D. L. 1994. "Figuring Out Factors: The Use and Misuse of Factor Analysis." *Canadian Journal of Psychiatry* 39:135–140.

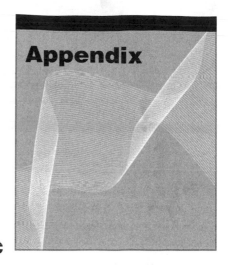

Appendix

Choosing the Correct Statistic

> **Overview.** This appendix presents a structured method of choosing the correct statistical approach to use when analyzing data. The choice of a specific statistic is based on the number and scale of the response (dependent) variables in the study considered, in conjunction with the number and scale of the predictor (independent) variables. Commonly used statistics are grouped into three tables based on the number of response and predictor variables in the analysis.

Introduction: Thinking about the Number and Scale of Your Variables

Researchers are often confused by the task of choosing the correct statistical procedure for analyzing a given data set. There are so many procedures from which to choose that it is easy to become frustrated, not even knowing where to begin. This appendix addresses this problem by providing a relatively simple system for classifying statistics. It provides a structured approach that should make it easier to find the appropriate statistical procedure for a wide variety of circumstances.

In a sense, most statistical procedures involve investigating the relationship between two variables (or two sets of variables). In a given study, the outcome variable that you are interested in is called either a *response variable* (in nonexperimental research) or a *dependent variable* (in experimental research). In nonexperimental research, you study the relationship between the response variable and some *predictor variable* whose values are used to predict scores on the response. In experimental research, a manipulated *independent variable* is the counterpart to this predictor variable. In general, nonexperimental research involves examining the relationship between a response variable and a predictor variable, whereas experimental research examines the relationship between a dependent variable and an independent variable(s). To simplify matters, this appendix blurs the distinction between nonexperimental and experimental research; it uses "response variable" to represent response variables as well as dependent variables, and it uses "predictor variable" to represent predictor variables as well as independent variables.

Thinking about Your Response Variables

Two of the primary factors that determine the selection of an appropriate statistical procedure are how many response variables you have and the data type (type of measurement or scale) of the response variables. Scale or data type refers to the level of measurement used in assessing these response variables (nominal, ordinal, or continuous).

For example, assume that you want to conduct a study to learn about variables that predict success in college. In your study, you may choose to use just one response variable as an index of college success such as grade point average (GPA). Alternatively, you may choose to use several response variables so that you will have multiple indices of success such as college GPA, college class rank, and whether or not participants are inducted into some college honorary society (yes or no). Here, you can see the *number* of response variables varies. In the first case there was only one response variable; in the second there were three.

Notice, however, that the scale used to assess college success also varies. The response variable college GPA has a continuous numeric scale, college class rank was assessed on an ordinal scale, and induction into an honorary society was assessed on a nominal scale. The number of response variables used in the analysis and the data type or scale used to measure those variables help determine which statistic you can use to analyze your data.

Thinking about Your Predictor Variables

However, you still do not have enough information to choose the appropriate statistic. Two additional factors that determine choice of the correct statistic are how many predictor variables you have and the data type (scale or measurement type) of the predictor variables. Again consider the college success study in which you want to learn about the variables that predict college success. Assume that you decide to use just one response variable as a measure of success—college grade point average (GPA). You might choose to design a study that also includes just one predictor variable—high school GPA. Alternatively, you could design a study that includes multiple predictor variables such as high school GPA, scores on the SAT, high school rank, and whether the student received a scholarship (yes or no).

Notice that in the previous paragraph, the number of predictor variables that can be included in a study varies. The first study included just one predictor, while the second included multiple predictors. Note that the scale of these predictors also varies. Predictors were assessed on a continuous numeric scale (high school GPA, SAT scores), an ordinal scale (high school rank), and a nominal scale (whether the student received a scholarship). The number of predictor variables included in your study and the data type (scale of measurement) also help determine the appropriate statistic.

Putting It Together

The preceding discussion provides context for the following recommendation. When choosing the appropriate statistic for an analysis, always consider both

- the number and data type (scale of measurement) of the response variables
- the number and data type (scale of measurement) of the predictor variables.

For example, suppose you use only one measure of college success (GPA) and one predictor variable (SAT scores) in your study. Because you have a single continuous numeric response variable and a single continuous numeric predictor variable, you know that the appropriate statistic is the Pearson correlation coefficient (assuming that a few additional assumptions are met). But what if you modified your study so that it still contained only one response variable but now contains two predictor variables, both continuous numeric variables such as SAT scores and high school GPA? In that case, it is more appropriate to analyze your data using multiple regression.

To select the right statistic, you must consider the number and nature of both your response and predictor variables. To facilitate this decision-making process, this appendix includes three tables:

- a table that lists statistics for studies that involve a single response variable and a single predictor
- a table for studies that involve a single response variable and multiple predictors
- a table for studies with multiple response variables.

A few words of caution are warranted before presenting the tables. First, these tables are not designed to present an exhaustive list of statistical procedures. They focus only on the tests that are considered to be the most commonly reported in the social sciences. A good number of statistical procedures that did not fit neatly into this format (such as principal component analysis) do not appear. Also, these tables do not necessarily provide you with all of the information you need to make the final selection of a statistical procedure. Many statistical procedures require that a number of assumptions be met concerning the data for the procedure to be appropriate, and these assumptions are often too numerous to include in a short appendix such as this. The purpose of this appendix is to help you locate the statistic that might be correct for your situation given the nature of the variables. It is then up to you to learn more about the assumptions for that statistic to determine whether your data satisfy those assumptions.

Guidelines for Choosing the Correct Statistic

Single Response Variable and a Single Predictor Variable

Table A.1 lists some of the simplest (and most common) studies conducted in the social sciences. These statistical procedures are covered in this text, but the list is not exhaustive.

Table A.1 has three columns:

- The first column describes the type of response or dependent variable in a study.
- The second describes the type of predictor or independent variable.
- The third describes the statistic that could be appropriate for that study.

For example, the first entry shows that if your predictor variable is nominal and your response is also nominal, it might be appropriate to evaluate the relationship between these variables using the chi-square test of independence. To understand this, assume that in your study you use one nominal-scale index of college success. You might have chosen "graduation" as this response where the code for the graduation variable is "Yes" if the student did graduate from college and "No" if not. Also, assume that you use scholarship status as the one nominal-scale predictor variable in your study, which has the code "Athletic" to represent students who received athletic scholarship, "Academic" for students who received academic scholarships, and "None" for students who received no scholarship. You analyze your data with a chi-square test of independence to determine whether there is a significant relationship between scholarship status and graduation. If the results give a significant value of chi square, inspection of the cells of the two-way classification table might show that students in the academic scholarship group are more likely to graduate than students in the athletic scholarship or no scholarship group.

The first row of Table A.1 deals with the chi-square test. The next row describes the appropriate conditions for a Kruskal-Wallis test. Notice that the entry in the Response/Dependent Variable column is "Nominal- or ordinal-scale grouping variable." This entry indicates that the Kruskal-Wallis test could be appropriate if you have a single response variable that has either an ordinal scale, an interval scale, or a continuous numeric scale.

Notice that in the Type of Analysis column the entry "Kruskal-Wallis test" is flagged with an asterisk (*). Tests that are flagged with an asterisk are not described in this text. However, flagged statistical procedures can be analyzed using JMP.

Table A.1 Studies with a Single Response Variable and a Single Predictor Variable

Response/Dependent Variable	Predictor/Independent Variable	Type of Analysis
Nominal- or ordinal-scale grouping variable	Nominal- or ordinal-scale grouping variable	Chi-square test
Continuous numeric ordinal-scale or non-normal continuous variable	Continuous numeric ordinal-scale variable (such as ranks)	*Kruskal-Wallis test
Continuous numeric variable	Nominal- or ordinal-scale grouping variable (two groups only)	*t* test
Continuous numeric variable	Nominal- or ordinal-scale grouping variable	One-way analysis of variance (ANOVA)
Numeric ordinal or non-normal continuous variable	Numeric ordinal-scale or non-normal continuous variable	Spearman correlation coefficient
Continuous numeric variable	Continuous numeric variable	Pearson correlation coefficient

* Not covered in this text.

Single Response Variable and Multiple Predictor Variables

Table A.2 lists some procedures that are appropriate when the analysis includes a single response variable and multiple predictors. For example, the last row of the table is for multiple regression. The entry in the predictor variable column is more than one numeric continuous variable. The chapter on multiple regression indicates that it is appropriate to use multiple regression to analyze data that included college GPA as the response variable, and SAT scores and high school GPA as predictors. Note how this is consistent with the guidelines of Table A.2. There is a single continuous numeric response variable (college GPA) and multiple numeric continuous predictor variables (SAT scores and high school GPA).

Table A.2 Studies with a *Single* Response Variable and *Multiple* Predictor Variables

Response/Dependent Variable	Multiple Predictor/ Independent Variables	Type of Analysis
One nominal- or ordinal-scale grouping variable	More than one continuous variable	*Logistic regression or Discriminant analysis
One continuous variable	More than one nominal- or ordinal-scale grouping variable	Factorial analysis of variance (ANOVA)
One continuous variable	At least one nominal- or ordinal-scale grouping variable and at least one continuous variable	Analysis of covariance (ANCOVA)
One continuous variable	More than one numeric continuous variable	Multiple regression

* Not covered in this text.

Multiple Response Variables

The response variable column of Table A.3 indicates that all of the procedures in this table are appropriate for studies that include multiple response variables. Note, however, that only the last three analytic procedures (factorial MANOVA, MANCOVA, and canonical correlation) involve multiple predictor variables. The first procedure (one-way MANOVA) requires only a single predictor variable on a nominal scale.

Table A.3 Studies with *Multiple* Response Variables

Response/Dependent Variable	Multiple Predictor/ Independent Variables	Type of Analysis
More than one continuous variable	One nominal- or ordinal-scale grouping variable	One-way multivariate analysis of variance (MANOVA)
More than one continuous variable	More than one nominal- or ordinal-scale grouping variable	*Factorial multivariate analysis of variance (MANOVA)
More than one continuous variable	More than one continuous variable	*Multivariate analysis of covariance (MANCOVA)
More than one continuous variable	More than one continuous variable	*Canonical correlation

* Not covered in this text.

Summary

This appendix is intended to serve as a starting point for choosing appropriate statistics. You can use the preceding tables to identify statistical procedures that might be appropriate for your research design. It is then up to you to learn more about the assumptions associated with the statistic (whether it requires data drawn from a normal population, whether it requires independent observations). These tables, when used in conjunction with the "assumptions" sections at the ends of chapters in this text, should help you find the right statistical procedure for analyzing the types of data most frequently encountered in social science research.

Index

Books Available from SAS Press

support.sas.com/pubs

Integrating Results through Meta-Analytic Review Using SAS® Software
by **Morgan C. Wang**
and **Brad J. Bushman** . . . . Order No. A55810

The Little SAS® Book: A Primer, Second Edition
by **Lora D. Delwiche**
and **Susan J. Slaughter** . . . . Order No. A56649
(updated to include Version 7 features)

The Little SAS® Book: A Primer, Third Edition
by **Lora D. Delwiche**
and **Susan J. Slaughter** . . . . Order No. A59216
(updated to include SAS 9.1 features)

Logistic Regression Using the SAS® System: Theory and Application
by **Paul D. Allison** . . . . Order No. A55770

Longitudinal Data and SAS®: A Programmer's Guide
by **Ron Cody** . . . . Order No. A58176

Maps Made Easy Using SAS®
by **Mike Zdeb**. . . . . Order No. A57495

Multiple Comparisons and Multiple Tests Using SAS® Text and Workbook Set
(*books in this set also sold separately*)
by **Peter H. Westfall, Randall D. Tobias, Dror Rom, Russell D. Wolfinger,**
and **Yosef Hochberg** . . . . Order No. A55770

Multiple-Plot Displays: Simplified with Macros
by **Perry Watts** . . . . Order No. A58314

Multivariate Data Reduction and Discrimination with SAS® Software
by **Ravindra Khattree,**
and **Dayanand N. Naik** . . . . Order No. A56902

Output Delivery System: The Basics
by **Lauren E. Haworth** . . . . Order No. A58087

Painless Windows: A Handbook for SAS® Users, Third Edition
by **Jodie Gilmore** . . . . Order No. A58783
(*updated to include Version 8 and SAS 9.1 features*)

PROC TABULATE by Example
by **Lauren E. Haworth** . . . . Order No. A56514

support.sas.com/pubs

Professional SAS Programmer's Pocket Reference, Fifth Edition
by **Rick Aster** . . . . Order No. A60075

Professional SAS® Programming Shortcuts
by **Rick Aster** . . . . Order No. A59353

Quick Results with SAS/GRAPH® Software
by **Arthur L. Carpenter**
and **Charles E. Shipp**. . . . . Order No. A55127

Quick Results with the Output Delivery System
by **Sunil Gupta** . . . . Order No. A58458

Reading External Data Files Using SAS®: Examples Handbook
by **Michele M. Burlew** . . . . Order No. A58369

Regression and ANOVA: An Integrated Approach Using SAS® Software
by **Keith E. Muller**
and **Bethel A. Fetterman** . . . . Order No. A57559

SAS® for Forecasting Time Series, Second Edition
by **John C. Brocklebank**
and **David A. Dickey** . . . . Order No. A57275

SAS® for Linear Models, Fourth Edition
by **Ramon C. Littell, Walter W. Stroup,**
and **Rudolf Freund** . . . . Order No. A56655

SAS® for Monte Carlo Studies: A Guide for Quantitative Researchers
by **Xitao Fan, Ákos Felsővályi, Stephen A. Sivo,**
and **Sean C. Keenan** . . . . Order No. A57323

SAS® Functions by Example
by **Ron Cody** . . . . Order No. A59343

SAS® Macro Programming Made Easy
by **Michele M. Burlew** . . . . Order No. A56516

SAS® Programming by Example
by **Ron Cody**
and **Ray Pass** . . . . Order No. A55126

SAS® Survival Analysis Techniques for Medical Research, Second Edition
by **Alan B. Cantor** . . . . Order No. A58416

SAS® System for Elementary Statistical Analysis,
Second Edition
by **Sandra D. Schlotzhauer**
and **Ramon C. Littell**. . . . . Order No. A55172

SAS® System for Mixed Models
by **Ramon C. Littell, George A. Milliken, Walter W.
Stroup,** and **Russell D. Wolfinger** . . Order No. A55235

SAS® System for Regression, Second Edition
by **Rudolf J. Freund**
and **Ramon C. Littell**. . . . . Order No. A56141

SAS® System for Statistical Graphics, First Edition
by **Michael Friendly** . . . . Order No. A56143

The SAS® Workbook and Solutions Set
(*books in this set also sold separately*)
by **Ron Cody** . . . . Order No. A55594

Selecting Statistical Techniques for Social Science
Data: A Guide for SAS® Users
by **Frank M. Andrews, Laura Klem, Patrick M. O'Malley,
Willard L. Rodgers, Kathleen B. Welch,**
and **Terrence N. Davidson** . . . . Order No. A55854

Statistical Quality Control Using the SAS® System
by **Dennis W. King**. . . . . Order No. A55232

A Step-by-Step Approach to Using the SAS® System
for Factor Analysis and Structural Equation Modeling
by **Larry Hatcher**. . . . . Order No. A55129

A Step-by-Step Approach to Using the SAS® System
for Univariate and Multivariate Statistics,
Second Edition
by **Larry Hatcher, Norm O'Rourke,**
and **Edward J. Stepanski** . . . . Order No. A58929

Step-by-Step Basic Statistics Using SAS®: Student
Guide and Exercises
(*books in this set also sold separately*)
by **Larry Hatcher**. . . . . .Order No. A57541

Survival Analysis Using the SAS® System:
A Practical Guide
by **Paul D. Allison** . . . . Order No. A55233

Tuning SAS® Applications in the OS/390 and z/OS
Environments, Second Edition
by **Michael A. Raithel** . . . . Order No. 8172

Univariate and Multivariate General Linear Mo
Theory and Applications Using SAS® Software
by **Neil H. Timm**
and **Tammy A. Mieczkowski** . . . . Order No. A5

Using SAS® in Financial Research
by **Ekkehart Boehmer, John Paul Broussard,**
and **Juha-Pekka Kallunki** . . . . Order No. A57601

Using the SAS® Windowing Environment:
A Quick Tutorial
by **Larry Hatcher**. . . . . Order No. A57201

Visualizing Categorical Data
by **Michael Friendly** . . . . Order No. A56571

Web Development with SAS® by Example
by **Frederick Pratter** . . . . Order No. A58694

Your Guide to Survey Research Using the
SAS® System
by **Archer Gravely**. . . . . Order No. A55688

JMP® Books

JMP® for Basic Univariate and Multivariate Statistics:
A Step-by-Step Guide
by **Ann Lehman, Norm O'Rourke, Larry Hatcher,**
and **Edward J. Stepanski** . . . . Order No. A59814

JMP® Start Statistics, Third Edition
by **John Sall, Ann Lehman,**
and **Lee Creighton** . . . . Order No. A58166

Regression Using JMP®
by **Rudolf J. Freund, Ramon C. Littell,**
and **Lee Creighton** . . . . Order No. A58789

support.sas.com/pubs